PRACTICAL METHODS OF REGIONAL SCIENCE AND EMPIRICAL APPLICATIONS

Also by Walter Isard and published by New York University Press

LOCATION ANALYSIS AND GENERAL THEORY: Economic, Political, Regional and Dynamic
Edited by Christine Smith

Practical Methods of Regional Science and Empirical Applications

Selected Papers of Walter Isard, Volume 2

Edited by
Christine Smith

NEW YORK UNIVERSITY PRESS
Washington Square, New York

First published in the U.S.A. in 1990 by
NEW YORK UNIVERSITY PRESS
Washington Square
New York, N.Y. 10003

Library of Congress Cataloging-in-Publication Data
Isard, Walter
[Selections. 1990]
Selected papers of Walter Isard / edited by Christine Smith.
p. cm.
Includes bibliographical references.
Contents: v. 1. Location analysis and general theory — v. 2. Practical methods of regional science and empirical applications
ISBN 0–8147–7899–2 (st);
1. Regional economics. 2. Regional planning. I. Smith, Christine, 1956– II. Title.
HT388.I8 1990
338.9—dc20 89–1397
CIP

Printed in Great Britain

Walter Isard — founder of regional science
(© *Koby Photographers, Cambridge, Mass.*)

Contents

List of Figures

List of Tables

Acknowledgements

In the development of ideas covered in these volumes, Walter Isard wishes to acknowledge the stimulation and knowledge gained from his mentor, Abbott Payon Usher, Alvin H. Hansen and Wassily W. Leontief. He wishes to thank once again his many co-authors and former students whose contributions were seminal to the development of his scholarship. His gratitude to his family for their infinite patience and encouragement while he pursued his academic pursuits with at times excessive intensity is immeasurable.

Permission to reprint materials has been given by editors of the *Quarterly Journal of Economics* for Walter Isard and Caroline Isard, 'Economic Implications of Aircraft', LIX, February (1945), pp. 145–69; Walter Isard, 'Some Economic Implications of Atomic Energy', vol. LXII, February (1948), pp. 202–28; the editors of the *Review of Economics and Statistics* for Walter Isard, 'Interregional and Regional Input-Output Analysis: A Model of a Space Economy', vol. XXXIII, November (1951), pp. 318–28; and Walter Isard and Robert E. Kuenne, 'The Impact of Steel upon the Greater New York–Philadelphia Industrial Region', vol. XXXV, November (1953), pp. 289–301; the editors of the *Journal of Political Economy* for Walter Isard, 'Some Locational Factors in the Iron and Steel Industry Since the Early Nineteenth Century', Vol. LVI, no. 3, June (1948), pp. 203–17; the editors of the *Journal of Regional Science* for Walter Isard and Eugene W. Schooler, 'Industrial Complex Analysis, Agglomeration Economies and Regional Development', vol. 1, Spring (1959), pp. 19–34; the editors of *Geoforum* for Walter Isard, Roger Van Zele and Phyllis Kaniss, 'Potentials and Problems of Economic-Ecologic Models for Management of Multi-region Systems', vol. 7, (1986), pp. 203–13; the Gower Publishing Company for Walter Isard and Christine Smith, 1984, 'Policy Analysis Using an Integrated Multiregional Model', in M. Chatterji, *et al.* (eds), *Spatial, Environmental and Resource Policy in the Developing Countries* (Gower, Aldershot UK), pp. 64–78; the officers of the Regional Science Association for Walter Isard, 'The Scope and Nature of Regional Science', *Papers and Proceedings of the Regional Science Association*, VI, (1960), pp. 9–34; Walter Isard and Thomas W. Langford, JR., 'Impact of Vietnam War Expenditures on the Philadelphia Economy: Some Initial Experiments with the Inverse of the Philadelphia Input-Output Table', *Papers of the Regional Science Association*, vol. 23, (1969), pp. 217–65; Walter Isard, 'Some Notes on the Linkage of the Ecologic and Economic Systems', *Papers of the Regional Science Association*, vol. 22, (1968), pp. 85–96; Walter Isard and Christine Smith, 'Linked Integrated Multiregion Models at the International Level', *Papers of the Regional Science Association*, vol.

51, (1983), pp. 3–19; and Gerald F. Adams, David Boyce, Stephen B. Caldwell, Ronald E. Miller and T. R. Lakshmanan for permission to reprint the National Science Foundation proposal (1983) entitled 'Integration of Multiregion Models' jointly authored with them.

CHRISTINE SMITH

Introduction

Christine Smith

As mentioned at the start of Volume 1, in this and that volume we collect selected writings of Walter Isard. Isard is the founder of the interdisciplinary field of Regional Science. Aside from his entrepreneurial and organisational role in establishing and institutionalising the field, he was for almost three decades the primary leader in the development of its tools, concepts, theories and methods, and for almost two decades a dominant figure along with his students in the further development of the field. In these volumes, we attempt to present a representative collection of his refereed article contributions on diverse subjects – some of which many regional scientists and others may not be aware; we occasionally draw upon extracts from his published books and include some unpublished materials. However, a collection of such papers is unable to capture fully the concomitant breadth and depth of analysis contained in his basic writings, especially *General Theory: Social, Political, Economic and Regional* [42], *Spatial Dynamics and Optimal Space-Time Development* [58], and *Conflict Analysis and Practical Conflict Management Procedures* [62]. Many developments contained in these books are yet to be adequately appreciated by regional science, economics and other communities and may, like his 'Channels of Synthesis' section of *Methods of Regional Analysis: An Introduction to Regional Science* [28], take twenty years or more to be absorbed into the mainstream of the above communities.

Volume 1 of this collection deals primarily with Isard's writings on theory, although the first item dealt with his work on the transport-building cycle [1], [2] and [3] which represented a combination of theory, supporting empirical materials, and an attempt at application. This second volume concentrates primarily on his empirical work, both in the construction of operational models and in other processing of empirical materials in order to gain understanding of the past and the present, and in order to project into the future and to provide empirical support for urban and regional planning and policy actions. It was these empirical studies, and their usefulness for projections into the future, and for urban and regional planning and policy actions that provided the payoff that justified his demands for resources for the support of research on urban and regional problems and for the development of regional science programs, organisations and publications.

The first items in this volume of the collection relate to Isard's interest in projecting the impact of major technological developments. The first

Chapter on *Economic Implications of Aircraft* ([4], co-authored with Caroline Isard, his wife) is a natural follow-up of his historical study of previous transport developments (canals, railways, automobiles, etc.) and their impacts. His investigation of these implications led him to correctly conclude, at the time when the volume of air traffic was minor, that:

1. 'aircraft unquestionably will dominate the sphere of first-class passenger traffic, other than short distance'
2. 'short-distance passenger traffic . . . may partly fall to local and feeder airlines – But here –, as well as in the field of inter-metropolitan transit, buses and railways should retain dominance'
3. the use of aircraft for delivery of expedited first class mail and express freight
4. aircraft would not encroach upon surface modes of transport in the mass movement of heavy and bulky commodities, except perhaps in the initial phases of development of relatively isolated and backward regions with inadequate forms of surface infrastructure; and so forth.

The second Chapter is on *Some Economic Implications of Atomic Energy* [5]. Unlike the physicists who were associated with the development of the Atomic Bomb, and some leading economists like Jacob Marschak, Isard was not swept away with the notion that we would enter an era of cheap, even very cheap, nuclear power. He immediately put his finger on the basic obstacle to the realisation of such an era and the belief that bountiful growth would ensue all over the world, especially in developing countries – namely, the high capital cost of nuclear power production particularly because of the basic need to safeguard against radiation. In this article and subsequent ones (often co-authored with Vincent Whitney) he concluded that

> 'it appears unlikely that atomic energy will have a revolutionary impact upon world and national economies. We can anticipate, within broad limits, competition with coal, oil and gas as fuel in power stations, but probably not with water power at superior hydro-electric sites. The effects upon real output and industries in general should be noticeable but nothing extraordinary, although they may be of considerable consequence for certain specific industries and trades. Spatially the structure of the world economy should experience mild alterations, especially in favour of aggressive and growing economies. However, a more equitable distribution of industrial and commercial activities between backward and highly developed economies does not appear in sight' (pp. 227–8). (See also [8], [9], [12], [18], [53] and [56].)

As might be expected, an early item in this volume relates to an empirical study demonstrating the fruitfulness of formal location theory. In particular, he probed into the locational forces pertaining to the industry that was almost universally considered in the pre-Second World War era to be the most basic of all – namely the iron and steel industry. (It should be noted that in that era, to be a major power, a nation had to be a major location of iron and steel production.) Employing his early training as an economic historian, he collected and ordered diverse pieces of empirical materials in the classic article *Some Locational Factors in the Iron and Steel Industry Since the Early Nineteenth Century* [6]. This article, reproduced here as Chapter 3, traces the shift of the industry from coal to market orientation. Together with his later articles 'The Future Locational Pattern of Iron and Steel Production in the United States' ([7], co-authored with William Capron) and 'New England as a Possible Location for an Integrated Iron and Steelworks' ([10], co-authored with John Cumberland), this clearly demonstrated the applicability of Weberian location theory for empirical analysis of the location of cost sensitive, competitive economic activity such as the iron and steel industry.

The next item included in this volume also represents a logical development. Recognising the ability of Isard at conducting empirical work from his iron and steel studies, and needing research assistance for his emerging regional interests, Wassily Leontief enticed Isard to join his research team at Harvard. In addition to working with Leontief on the balanced regional input-output model (see [13]), Isard conceived of his pure interregional input-output model. In an article reproduced here as Chapter 4, namely '*Interregional and Regional Input-Output Analysis: A Model of Space Economy*' [11] he commits what must have been an unforgiveable sin in the eyes of both Kuznets (who commissioned Isard to present a paper on his ideas to the American Statistical Association, and found himself confronted by an earlier version of this article) and Leontief (who is well known to be extremely critical of economists and others who present papers on models that are not yet empirically validated) – since, as is well known, this article provides the format of a three-region, twenty-sector interregional input-output table for the United States with *no* numbers in *any* cell. However, Isard was confident that the table could and would be filled in for some part of the world – a confidence that has proved to be well founded. Isard conducted a number of other, more 'empirically sound' input-output studies at the Leontief project. See [13], [14], [16], and [30] which provide useful discussion on how to fill in the cells of an interregional input-output table given limited data availability. The most significant of these studies '*The Impact of Steel Upon the Greater New York–Philadelphia Industrial Region: A Study in Agglomeration Projection*' [17] is reproduced here as Chapter 5. This study, co-authored with Robert Kuenne, fuses comparative cost and regional input-output analysis, and as such represents his first attempt at integrating different methods of regional analysis in an empirically validated manner. Scholars interested in following up further early writings on regional and interregional input-output analysis and the comparative cost technique can consult Chapters 7 and 8 of *Methods of Regional Analysis* [28].

Note that at this time, the orientation of Isard and his colleagues was to urban problems at least as much if not more than to regional. The concern of the last study was impact on urban areas of a major industrial sector – as such it was geared to the needs of city planners and city planning commissions. Other early contributions to the analysis of economic problems of urban areas can be found in 'The Economic Base and Structure of the Urban-Metropolitan Region' ([15], co-authored with Robert Kavesh and Robert Kuenne), 'Economic Structural Interrelations of Metropolitan Regions' ([19], co-authored with Robert Kavesh), 'Municipal Costs and Revenues Resulting from Community Growth: Parts 1 and 2' ([21] and [22], co-authored with Robert Coughlin), 'Regional Science and Planning' ([33], co-authored with Thomas Reiner) and Chapters 8 and 11 (on Rent Theory and Urban Land Use) in *Location and Space Economy* [23]. Later on, however, came the challenge to construct a complete and detailed input-output table for a major metropolitan area. He and his colleagues did this for Philadelphia, becoming immersed for a number of years in collecting the survey-based data required to fill in the cells of the largest regional input-ouput table ever to have been constructed (roughly 500 × 500). Computer technology was not as sophisticated at this time, and research effort had to be devoted to working out the methods to be employed in inverting the resulting (I-A) matrix – since all computational scientists had advised him that such could not be done. We do not reproduce here any writings concerned with the mechanics of constructing and operating this model, since interested scholars can consult the extensive documentation provided in [38] and [44]. Instead we have chosen to include as Chapter 9 the results of an application of the model aimed at estimating the impact of disarmament initiatives associated with the Vietnam War on the Philadelphia region economy. The theme of '*Impact of Vietnam War Expenditures on the Philadelphia Economy: Some Initial Experiments with the Inverse of the Philadelphia Input-Output Table*' ([43], co-authored with Thomas Langford) – namely the use of economic-regional science models to provide improved estimates of the differential regional impacts of defence expenditure – is one which has been of constant concern throughout his scholarly life. For example, see [36], [39], [66], and [57]. In addition, he together with Gerald Karaska, produced an early *Atlas on United States Military Contracts and Subcontracts* [35] and published the associated sourcebooks. (See [31], [32], and [37].) These were updated somewhat in [74].

Although Isard touched upon problems of backward and developing economies in an indirect manner in the items already cited, he was not challenged to come up with specific development recommendations for such an economy until he confronted the Puerto Rican problem in the early 1950s. Neither simple comparative cost analysis on an industry-by-industry basis, nor input-output analysis by itself were able to suggest economic activities that would constitute a cluster suitable for sparking industrial development in Puerto Rico. Rather a set of activities had to be identified which were subject to important technical, marketing and other interrelations and which would thereby generate spatial juxtaposition (agglomeration) economies to offset other disadvantages of a Puerto Rico location. And so emerged the development of the technique known as industrial complex analysis and the recommendation of an oil refinery-petrochemicals-fertiliser-synthetic fibre complex for Puerto Rico. For this reason we include the item '*Industrial Complex Analysis, Agglomeration Economies and Regional Development*' ([25], co-authored with Eugene Schooler) as Chapter 6. Other writings on this newly developed technique include [20], [26], [47], ([28], ch. 9) and ([51], ch. 17). It is interesting to note that while the Economic Development Administration of Puerto Rico promoted oil refinery and petrochemical operations on the basis of his research findings, a full industrial complex was never realised in Puerto Rico. This was so since, given the nature of the tax rebate arrangements made available and the favourable access to raw materials stemming from oil refinery operations before the recent OPEC oil crisis, firms locating in Puerto Rico found it more profitable to concentrate on petrochemicals than synthetic fiber production.

Another major thrust at applied analysis by Isard and other regional scientists has been the development of the gravity model – a model which, in general, was designed to contribute towards the understanding, explaining and/or projecting of flows. Such flows could be of commodities (as in interregional trade) or of people (in journey to work and other intra-metropolitan and inter-metropolitan travel and migration) or of ideas (in communications over distances of all types). Some discussion of this model is included as another item in this volume, namely '*Gravity, Potential and Spatial Interaction Models*' ([28], co-authored with David Bramhall) as Chapter 8. Further discussion of these types of models is contained in ([27], ch. 11), 'Regional Employment and Population Forecasts via Relative Income Potential' ([24], co-authored with David Bramhall), and 'A Simple Rationale for Gravity Model Type Behaviour' [50]. By now this model is widely used in many different countries of the world and testifies to the usefulness of looking into physics and other natural sciences for techniques able to be adapted for attacking regional and urban problems.

Still another method of analysis developed by Isard and his colleague Stevens was interregional programming (see [24] and ch. 10 of [28]). Because of limitations of space, no item on Isard's work on that topic is included – but such a technique has found use in attacking urban transportation problems (e.g. the Herbert–Stevens model in the Penn Jersey Transportation Study) as well as in the development of plans for optimal resource development.

With these several methods on hand – comparative cost, interregional input-output, industrial complex, interregional linear programming and gravity models – Isard could not desist in 1959 from proposing and strongly pushing for the development of 'channels of synthesis' of these techniques. See Parts II and III of his '*The Scope and Nature of Regional Science*' [29] which is reproduced here as Chapter 7, as well as [34] and the much more detailed development in chapters 12 and 13 of [28]. Such thinking was well ahead of the times, and only took hold two decades later in the late 1970s. For example, Channel I corresponds to the CICIOP component of the integrated model described in Chapter 14 of this volume and includes a combination of comparative cost, industrial complex, linear programming and interregional input-output methods.

In the meantime, experience with the above-mentioned methods, on a totally unfused or only very partially or weakly fused basis, was accumulating. Moreover, a new set of problems – those concerned with controlling environmental pollution and ecological disturbance – had arisen and added to the more widespread acceptance of the need for such partial synthesis of approaches and/or methods of analysis. Isard and his colleagues contributed to the growing literature on those environmental problems in '*Some Notes on the Linkage of the Ecologic and Economic Systems*' [41] and '*Potentials and Problems of Economic-Ecologic Models for Management of Multi-Region Systems*' ([54], co-authored with Roger Van Zele and Phillis Kaniss) – both of which are reproduced in this volume as Chapters 10 and 11, respectively.

These items demonstrate how practical methods of regional analysis, such as input-ouput, could be extended to incorporate pollution and de-pollution coefficients and other environmental-ecologic concepts and then subsequently employed by regional planning authorities. Additional insights into his environmental research activities can be obtained from his writings on pollution coefficients [40] and [52], on activity-industrial complex analysis for environmental management [47], on a recursive programming model for handling world environmental conflicts [48], on the use of diverse regional science methods for environmental management [46], [49], [55] and his book on *Economic-Ecologic Analysis for Regional Development* ([45], co-authored with Choguill, Kissen and others).

One issue that arose in crystal-clear fashion within the last two items included in this volume is the conflict that arises between environmental and other objectives that must inevitably confront planning authorities concerned with evaluating development options available for their region. For a good description of this dilemna, see Chapters 14–17 in [55]. Isard's concern with conflict analysis and the development of practical conflict management procedures can be seen as a logical outgrowth then of his more traditional regional science modelling activities. This becomes even more apparent when one recognises the sheer volume and variety of conflicts that must come across the desk of the mayor of a large metropolitan area, the Governor of a State, the manager of a major industrial organisation and other regional decision-makers. For this reason, we include as Chapter 12 in this volume the item written with Christine Smith on '*Matching Conflict Situations and Conflict Management Procedures*' [60] and refer interested readers to the much more extensive treatment in their book on *Conflict Analysis and Practical Conflict Management Procedures: An Introduction to Peace Science* [62] wherein more than one hundred procedures are presented and their potential for use in conflict analysis and management investigated. Earlier antecedents of this type of research conducted by Isard and his associates includes chapters 4–8 in [42] and chapters 9–11 in [51]. See also [59], [63] and [68].

We next present as Chapter 13 an unpublished item, *Integration of Multiregional Models* [64] which comprises excerpts from a research proposal that was submitted to the United States National Science Foundation for financial support in 1983, but which still represents a 'state-of-the-art' paper regarding the possibilities for such integration. Co-authors of this proposal were David Boyce, Gerald Adams, Steven Caldwell, Ronald Miller and T. R. Lakshmanan. An earlier version is in ([61], co-authored with Luc Anselin) and other supplementary discussions are in [67], [71], and [73]. A basic objective of this research is to strengthen each of the six modules considered by providing to each more and higher quality exogenous inputs from the other five, and simultaneously to allow feedbacks among them. In essence by drawing upon the strengths of each module in such an integrated fashion it is argued that we can achieve the development of a more comprehensive model without sacrificing anything in terms of quality of the individual components within it. Another basic objective is to make explicit the high degree of interdependence among policies, to understand such more fully, and thereby to project better the impacts of different policies, singly and in combination. The interdependence arises from the fact that each policy, say housing, energy, transportation, industrial development, employment comes to affect the operation of any economic or other sector of a region and the region's state of affairs which then influences how other policies may affect these sectors and the region's welfare. The next item included as Chapter 15 in this volume *Policy Analysis Using an Integrated Multiregion Model* ([69], co-authored with Christine Smith) provides detail on one way in which to proceed in developing consistent sets of policy projections and hence in operating (using) the 'Conflict Management – Multipolicy Formation' module of the integrated model. For discussion of other, less developed possibilities see [67], [71] and [72].

While much scepticism was expressed at the first World Regional Science Congress (held at Harvard University in 1980) about the feasibility of constructing and operating an integrated multiregion model, we can report that work has commenced on the construction of such a model for Australia. Moreover, now that the regional science community has at least in part recognised the feasibility of such a model for a nation and being spurred on by the installation at Cornell of a super-computer facility that is being dominated in terms of usage by non-social scientists, Isard and his colleagues have moved on to the problem of constructing an integrated multiregion model for the world system. Early conceptual thinking on this subject is contained in Chapter 15 of [62]; while a later statement is contained in the article co-authored by Christine Smith on *Linked Integrated Multiregional Models at the International Level* [65] which is reproduced here as Chapter 14. In addition, a brief statement is added in Appendix 14A on *Some Notes on Current Conceptual Thinking Regarding World Models and Progress in Their Construction*. Interested readers may also wish to consult [70] and [74].

References

[1] Isard, W., 1942a. 'Transport Development and Building Cycles', *Quarterly Journal of Economics*, vol. LVII, November, pp. 90–112.

[2] Isard, W., 1942b. 'A Neglected Cycle: The Transport-Building Cycle', *Review of Economic Statistics*, vol. XXIV, November, pp. 149–58.

[3] Isard, W. and C. Isard, 1943. 'The Transport Building Cycle in Urban Development: Chicago', *Review of Economic Statistics*, vol. XXV, November, pp. 224–26.

[4] Isard, W. and C. Isard, 1945. 'Economic Implications of Aircraft', *Quarterly Journal of Economics*, vol. LIX, February, pp. 145–69.

[5] Isard, W., 1948a. 'Some Economic Implications of Atomic Energy', *Quarterly Journal of Economics*, vol. LXII, February, pp. 202–28.

[6] Isard, W., 1948b. 'Some Locational Factors in the Iron and Steel Industry Since the Early Nineteenth Century', *Journal of Political Economy*, vol. LVI, June, pp. 203–17.

[7] Isard, W. and W. M. Capron, 1949. 'The Future Locational Pattern of Iron and Steel Production in the United States', *Journal of Political Economy*, vol. LVII, April, pp. 118–33.

[8] Isard, W. and J. B. Lansing, 1949. 'Comparisons of Power Costs for Atomic and Conventional Steam Stations', *Review of Economic Statistics*, vol. XXXI, pp. 217–26.

[9] Isard, W. and V. Whitney, 1950. 'Atomic Power and the Location of Industry', *Harvard Business Review*, vol. XXVIII, March, pp. 45–54.

[10] Isard, W. and J. N. Cumberland, 1950. 'New England as a Possible Location for an Integrated Iron and Steel Works', *Economic Geography*, vol. 26, October, pp. 245–59.

[11] Isard, W., 1951. 'Interregional and Regional Input-Output Analysis: A Model of a Space Economy', *Review of Economics and Statistics*, vol. XXXIII, November, pp. 318–28.

[12] Isard, W. and V. Whitney, 1952. *Atomic Power: An Economic and Social Analysis*, (Blakiston, New York). Translation: Japanese.
[13] Isard, W., 1953a. 'Regional Commodity Balances and Interregional Commodity Flows', *Papers and Proceedings of the American Economic Association*, vol. 43, May, pp. 167–80.
[14] Isard, W., 1953b. 'Some Empirical Results and Problems of Regional Input-Output Analysis', Chapter V in Leontief, W., *et al.*, *Studies in the Structure of the American Economy*, Oxford University Press, New York, pp. 116–81.
[15] Isard, W., R. A. Kavesh and R. E. Kuenne, 1953. 'The Economic Base and Structure of the Urban-Metropolitan Region', *American Sociological Review*, vol. 18, June, pp. 317–21.
[16] Isard, W., 1953c. 'Some Emerging Concepts and Techniques for Regional Analysis', *Zeltschrift fur die Gesamte Staatswissenschaft*, vol. 109, pp. 240–50.
[17] Isard, W. and R. E. Kuenne, 1953. 'The Impact of Steel Upon the Greater New York–Philadelphia Industrial Region: A Study in Agglomeration Projection', *The Review of Economics and Statistics*, vol. 35, November, pp. 289–301.
[18] Isard, W. and R. A Kavesh, 1953. 'Some Economic Consequences of Nuclear Power', *The Annals of the American Academy of Political and Social Science*, vol. 290, November, pp. 62–6.
[19] Isard, W. and R. A. Kavesh, 1954. 'Economic Structural Interrelations of Metropolitan Regions', *American Journal of Sociology*, vol. LX, September, pp. 155–62.
[20] Isard, W. and T. Vietorisz, 1955. 'Industrial Complex Analysis and Regional Development, with Particular Reference to Puerto Rico', *Papers and Proceedings of the Regional Science Association*, vol. 1, pp. U1–U17.
[21] Isard, W. and R. Coughlin, 1956a. 'Municipal Costs and Revenues Resulting from Community Growth: Part 1 Residential Communities', *Journal of the American Institute of Planning*, vol. XXII, Summer.
[22] Isard, W. and R. Coughlin, 1956b. 'Municipal Costs and Revenues Resulting from Community Growth: Part 2 Non-Residential Communites', *Journal of the American Institute of Planning*, vol. XXII, Fall.
[23] Isard, W., 1956. *Location and Space Economy: A General Theory Related to Industrial Location, Market Areas, Land Use, Trade and Urban Structure* (MIT Press and John Wiley and Sons, New York). Translations: Japanese, Italian, Spanish.
[24] Isard, W., 1958. 'Interregional Linear Programming: An Elementary Presentation and a General Model', *Journal of Regional Science*, vol. 1, Summer, pp. 1–59.
[25] Isard, W. and E. W. Schooler, 1959. 'Industrial Complex Analysis, Agglomeration Economies and Regional Development', *Journal of Regional Science*, vol. 1, Spring, pp. 19–34.
[26] Isard, W., E. Schooler and T. Vietorisz, 1959. *Industrial Complex Analysis and Regional Development* (MIT Press and John Wiley and Sons, New York). Translation: Spanish.
[27] Isard, W. and D. F. Bramhall, 1959. 'Regional Employment and Population Forecasts Via Regional Income Potential Models', *Papers, Regional Science Association*, vol. V, pp. 25–47.
[28] Isard, W., D. F. Bramhall, G. A. P. Carrothers, J. H. Cumberland, L. N. Moses, D. O. Price and E. W. Schooler, 1960. *Methods of Regional*

Analysis: Introduction to Regional Science (MIT Press and John Wiley and Sons, New York). Translations: Polish, Russian, Japanese, Spanish and French.

[29] Isard, W., 1960. 'The Scope and Nature of Regional Science', *Papers, Regional Science Association*, vol. 6, pp. 9–34.

[30] Isard, W. and J. Cumberland (eds) 1961. *Regional Economic Planning*, OEEC/EPA, Paris. Translation: French.

[31] Isard, W. and J. Ganschow, 1962. *Awards of Prime Military Contracts by County, State and Metropolitan Areas of the United States, Fiscal Year 1960*, Regional Science Research Institute, Philadelphia, PA.

[32] Isard, W. and G. Karasaka, 1962. *Unclassified Defense Contracts: Awards by County, State and Metropolitan Areas of the United States, Fiscal Year 1962*, World Friends Research Center Inc., Philadelphia, PA.

[33] Isard, W. and T. Reiner, 1962. 'Regional Science and Planning', *Papers, Regional Science Association*, vol. 8, pp. 1–35.

[34] Isard, W. and E. Smolensky, 1963. 'Application of Input-Output Techniques to Regional Science' in T. Barna (ed.), *Structural Interdependence and Economic Development*, (Macmillan, New York).

[35] Isard, W. and G Karaska, 1963. *An Atlas on United States Military Contracts and Subcontracts*, World Friends Research Centre, Inc., Philadelphia, PA.

[36] Isard, W. and E. Schooler, 1964. 'An Economic Analysis of Local and Regional Impacts of Reduction of Military Expenditures', *Papers, Peace Research Society (International)*, vol. 1, pp. 15–44.

[37] Isard, W. and G. Karaska, 1965. *Unclassified Defence and Space Contracts: Awards by County, State and Metropolitan Area, United States, Fiscal Year 1964*, World Friends Research Centre, Inc., Philadelphia, PA.

[38] Isard, W., T. Langford Jr. and E. Romanoff, 1966–68. *Working Papers, Philadelphia Region Input-Output Study*, Regional Science Research Institute, Philadelphia, PA.

[39] Isard, W. and S. Czamanski, 1966. 'A Model for the Projection of Regional Industrial Structure, Land Use Patterns and Conversion Potentialities', *Papers, Peace Research Society (International)*, vol. 5, pp. 1–13.

[40] Isard, W., K. Bassett, C. Choguill, J. Furtado, R. Izumita, J. Kissin, E. Romanoff, R. Seyfarth and T. Tatlock 1967. 'On the Linkage of Socio-economic and Ecologic Systems', *Papers, Regional Science Association*, vol. 21, pp. 79–99.

[41] Isard, W., 1968. 'Some Notes on the Linkage of the Ecologic and Economic Systems', *Papers, Regional Science Association*, vol. 22, pp. 85–96.

[42] Isard, W., T. Smith, P. Isard, T. H. Tung and M. Dacey, 1969. *General Theory: Social, Political, Economic and Regional* (MIT Press, Cambridge, MA).

[43] Isard, W. and T. Langford, 1969. 'Impact of Vietnam War Expenditures on the Philadelphia Economy: Some Initial Experiments with the Inverse of the Philadelphia Input-Output Table', *Papers, Regional Science Association*, vol. 23, pp. 217–65.

[44] Isard, W. and T. Langford, 1971. *Regional Input-Output Study: Recollections, Reflections and Diverse Notes on the Philadelphia Experience* (MIT Press, Cambridge, MA).

[45] Isard, W., K. E. Bassett, C. L. Choguill, J. G. Furtado, R. M. Izumita, J. Kissen, R. H. Seyfarth and T. Tatlock, 1972. *Economic-Ecologic Analysis for Regional Development* (Free Press, New York).

[46] Isard, W., F. Cesario, *et al.*, 1974. *Working Papers on Environmental*

Management, Regional Science Monograph Series No. 3, Centre for Urban Development Research, Cornell University, Ithaca, NY.
[47] Isard, W., 1974a. 'Activity-Industrial Complex Analysis and Environmental Management', *Papers, Regional Science Association*, vol. 33, pp. 127–40.
[48] Isard, W., 1974b. 'World Environmental Conflicts: Some Relevant Analytical Frameworks', *Papers, Peace Science Society (International)*, vol. 23, pp. 139–53.
[49] Isard, W. and R. van Zele, 1975. 'Practical Regional Science Techniques for Environmental Management', *International Regional Science Review*, vol. 1, no. 1.
[50] Isard, W., 1975a. 'A Simple Rationale for Gravity Model Type – Behavior', *Papers, Regional Science Association*, vol. 35, pp. 25–30.
[51] Isard, W., 1975b. *Introduction to Regional Science* (Prentice-Hall, Englewood Cliffs, NJ).
[52] Isard, W., F. Cesario, and T. Reiner, 1975. *Marginal Pollution Analysis for the Management of the Urban Environment*, Regional Science Monograph Series No. 4, Center for Urban Development Research, Cornell University, Ithaca NY.
[53] Isard, W., T. A. Reiner, R. van Zele, J. Stratham, *et al.*, 1976. *Regional Economic Impacts of Nuclear Power Plants*, Regional Energy Series, Brookhaven National Laboratory.
[54] Isard, W., R. van Zele and P. Kaniss, 1976. 'Potential and Problems of Economic-Ecologic Models for Management of Multiregion Systems', *Geoforum*, vol. 7, pp. 203–13.
[55] Isard, W. and P. Kaniss, 1976. 'World Environmental Management' in M. Chatterji, (ed.), *Space, Location and Regional Development*, pp. 168–80.
[56] Isard, W., T. A. Reiner, R. van Zele, 1977. *Socio-economic Analysis of Energy Facilities, with Particular Reference to the Hartsville (Tennessee) Impact Area*, A Report for the Regional and Urban Studies Section, Oak Ridge National Laboratory.
[57] Isard, W. and P. Liossatos, 1978. 'A Model of Big Step Disarmament and Domino Effects', *Journal of Peace Science*, vol. 3, no. 2.
[58] Isard, W., P. Liossatos, *et al.*, 1979. *Spatial Dynamics and Optimal Space-Time Development* (North-Holland, Amsterdam).
[59] Isard, W. and E. Glugiewicz, 1980. 'Conflict Management within Regional Science', *Transportation and Economy*, vol. 40, December.
[60] Isard, W. and C. Smith, 1980. 'Matching Conflict Situations and Conflict Management Procedures', *Conflict Management and Peace Science*, vol. 5, no. 1, pp. 1–26.
[61] Isard, W. and L. Anselin, 1982. 'Integration of Multi-regional Models for Policy Analysis', *Environment and Planning A*, vol. 14, pp. 359–76.
[62] Isard, W. and C. Smith, 1982. *Conflict Analysis and Practical Conflict Management Procedures: An Introdution to Peace Science*, (Ballinger Publishing Co., Cambridge, MA).
[63] Isard, W. and C. Smith, 1983a. 'A Dynamical System Approach to Learning Processes in Conflict Mediation and Interaction', in W. Isard and Y. Nagao (eds), *International and Regional Conflict: Analytic Approaches* (Ballinger, Cambridge, MA), pp. 11–31.
[64] Isard, W., D. Boyce, G. Adams, S. Caldwell, R. Miller and T. R. Lakshmanan, 1983. *Integration of Multiregion Models*, Unpublished NSF Proposal.
[65] Isard, W. and C. Smith, 1983b. 'Linked Integrated Multiregion Models at the International Level', *Papers, Regional Science Association*, vol. 51, pp. 3–19.

[66] Isard, W. and C. Anderton, 1984. 'Geography of Arms Manufacture', in D. Pepper and A. Jenkins (eds), *The Geography of War and Peace*, (Basil Blackwell, London).

[67] Isard, W. and C. Smith, 1984a. 'Incorporation of Conflict and Policy Analysis in an Integrated Multiregion Model', in A. A. Andersson, W. Isard and T. Puu (eds), *Regional and Industrial Development Theories: Models and Empirical Evidence* (North-Holland, Amsterdam), pp. 109–32.

[68] Isard, W. and C. Smith, 1984b. 'Managing Conflicts Among Border and Other Impacted Regions', in L. G. Gibson and A. C. Renteria (eds) *Regional Impacts of United States–Mexico Economic Relations*, (El Colegio de Mexico, Mexico).

[69] Isard, W. and C. Smith, 1984c. 'Policy Analysis using an Integrated Multiregion Model', in M. Chatterji (ed.), *Spatial Environmental and Resource Policy in the Developing Countries*, (Gower Press, Aldershot, UK), pp. 64–78.

[70] Isard, W., 1984. 'The Possible Components of a World Development Model in Year 2000', in *Proceedings of Conference on Urban Growth and Economic Development in the Pacific Region*, Taipei, Taiwan, pp. 1–35.

[71] Isard, W. and C. Smith, 1986a. 'Estimating Economic-Demographic Linkages in an Integrated Multiregion Model', in A. Issermann (ed.), *Population Change and the Economy* (Kluwer Nijhoff, Boston, MA), pp. 159–76.

[72] Isard, W. and C. Smith, 1986b. 'Fusion of the Abstract and Practical in Conflict Analysis and Management' in J. H. P. Paelinck (ed.), *Axiomatics and Pragmatics in Conflict Analysis* (Gower, Aldershot, UK).

[73] Isard, W., 1986. 'Reflections on the Relevance of Integrated Multiregion Models: Lessons from Physics', *Regional Science and Urban Economics*, vol. 16, pp. 165–80.

[74] Isard, W., C. Smith, C. Anderton, Y. Asami, J. Bennett, B. Burton and W. Dean, 1988. *Arms Races, Arms Control and Conflict Analysis: Contributions from Peace Science and Peace Economics* (Cambridge University Press, New York).

1 Economic Implications of Aircraft[1]

Economic evolution stems from man-devised technology acting dynamically upon relatively passive geographic elements. In this paper we shall be concerned with one phase of this process: the modifications in geographic resistances to commodity and population movement which tend to result from the development of a revolutionary transport technology; and the implications of this technology for commercial-industrial structure, population distribution, and general economic prosperity.

I

To provide the necessary historical perspective we shall begin with a brief review of the impact of past transport innovations upon the United States economy, and touch upon certain relevant theoretical considerations.

During somewhat more than a century prior to the early 1930s, four major transport changes had been introduced on a considerable scale into the American economy: the canal, the railroad, the street and electric railway (trolley, subway, elevated) and the automobile.[2] The effects of these innovations can be broadly classified into two categories: those incident to reduction of the cost of commodity movement, with which the canal and railway were chiefly associated; and those which resulted from increased population mobility, in which the street and electric railway and the automobile played the principal role. In general, reduction in cost of commodity movement increases geographic specialisation, permits mass-production economies, and by differentiating between superior and inferior commercial sites and sources of material and labour, re-allocates market areas and productive activity in favour of the most efficient enterprises and strategic locales. Concentration and expansion of production ensue,[3] and we observe the emergence of huge industrial and commercial centres and the intensive exploitation of rich mineral deposits attributable in large part to construction of comprehensive canal and rail networks.

More specifically, there were four vast outbursts of canal and railroad construction in the United States during the nineteenth century,[4] each drastically lessening the cost resistance of geographic elements; and within

This chapter was written with Caroline Isard. Reprinted from *Quarterly Journal of Economics*, vol. LIX, February (1945), pp. 145–69.

the limitations of the available data we note, in one-to-one correspondence with each outburst, periods of intense activity in anthracite and bituminous coal production, pig iron production and urban construction, followed by periods of mild activity or depression in these fields. The causal flow comes in major part from the transport factor: each outburst of canal and railroad construction, through intensive and extensive enlargement of market areas and economic spheres, stimulated marked increases in industrial and mining activity (as manifest in the data on pig iron and coal production), and through consequent concentration of these activities at strategic sites, stimulated vigorous urban development (as indicated by building statistics).[5] Thus the transport factor, by affording enormous outlets for investment in its own facilities and by provoking perhaps still greater investment in those other major sectors of the American economy, was the backbone of each of the four extended periods of prosperity during the last three-quarters of the nineteenth century.

Increased population mobility achieved through the street and electric railway and the automobile permits (1) dispersion of urban population agglomerations with consequent development of suburban population and outlying territories, and (2) site selection with respect to consumer functions (an effect to be discussed in Section III). Specifically, prior to the early 1930s, the street and electric railway and the automobile were injected into the American economy in outbursts of activity,[6] and in one-to-one correspondence with these changes we observe sharp increases in non-farm residential construction. Each of these two transport changes, in turn, precipitated an intense centrifugal movement of congested urban populations, as reflected in the index of residential building,[7] and in this manner laid the basis for the formation of huge metropolitan regions. Once again the transport factor, by providing substantial investment opportunities in its own field and in the host of investment outlets associated with the peripheral development of cities and the emergence of metropolitan areas, was of crucial significance for the two extended periods of prosperity experienced during the period from the late 1890s to the early 1930s.

Two further considerations deserve comment. First, geographic characteristics as impediments to commodity and population movement are of varying import for each transport technology. When one mode of transport supplants another, the relative emphasis upon the various geographic elements may be altered. New elements may appear as major obstacles to movement, which when translated into economic terms may effect a radical change in the strategy of traffic flow, induce relocation and shift of industry and population, and modify the hierarchical order of trade routes and commercial sites. This did occur in the United States during the period under examination, when the canal technique, stressing gradients and limited water supply as obstacles, strongly differentiated between traffic arteries from the Eastern seaboard to the interior beyond the Appalachians, and when the railway, by further diminishing cost and time resistances of surface configuration, reoriented cross-country movement from circuitous waterways to much more direct East–West routes.[8]

Second, distinction should be drawn between intensive and extensive transport development of regions by technological developments affecting the cost of commodity movement. In general, intensive development perpetuates the tendencies for concentration and growth of industry and commerce at strategic sites within a given evolved region. On the other hand, the extension of transport facilities to regions formerly inaccessible or to regions which for all practical purposes have been isolated economically because of prohibitive transport costs, not only may have effects similar to intensive development through tapping unexploited markets, but, more important, may permit utilisation of new resources, institute drastic revaluation of old natural resources, and compel reconsideration of industrial and commercial locations, thereby often effecting wholesale shifts of such activities. The canal and railroad had their greatest impact upon the American economy, not so much through the substitution of the canal for the turnpike or the railway for waterways in territory already fairly well settled, but through penetration of the canal into the Appalachian interior, through access to the anthracite coal fields afforded by the canal, through the incursions of the railway into the frontier and sparsely developed regions of the Mississippi valley and the Far West. The vast industrial, commercial, and urban adaptations entailed by such thrusts as these have been the mainstays of the four extended periods of buoyant prosperity in the United States during the last three-quarters of the nineteenth century.[9]

II

Postponing application of these generalisations until Section III, let us next examine the relevant cost aspects of air transport and the role which aircraft may be expected to assume in the transport systems of the future. A multitude of uncertainties are obviously involved in such an analysis: uncertainties concerning the nature and the rate of technological progress in the aeronautical sciences,[10] in the sciences of competitive transport mechanisms, and in all other fields; uncertainties inherent in future moral, social, political and other extra-economic forces, national and international. We shall rely upon the reader's intiative to supply necessary qualifications to the analysis.

E. P. Warner has made a rough calculation of operating costs for different aircraft representative of technical progress at different times, which throw much light on the effect of technological advance upon air transport costs. The data are presented in Table 1.1.[11] Tremendous cost

Table 1.1 Costs of air transport

Aircraft	*Payload lbs*	*Direct cost in cents, capacity payload*		*Total cost (direct and indirect) in cents, capacity payload*	
		Per mile	*Per ton-mile*	*Per mile*	*Per ton-mile*
DH (1920)	600	26.4	88.1	48.3	161.0
Boeing 40 (mid-1920s)	1,200	20.2	33.7	43.6	72.7
Ford (1925)	3,200	34.1	21.3	69.3	43.3
Lockheed Vega (1929)	1,350	15.1	22.4	38.1	56.5
Boeing 247 (1933)	2,800	21.1	15.1	51.9	37.2
Douglas DC-3 (1936)	5,000	26.7	10.7	68.6	27.4
Postwar Aircraft (two to four years after war)					
a. Passenger	12,000	—	—	102	17*
b. Cargo	16,000*	—	—	76*	9.5*

Note: *See note 11.

reductions have already been achieved. Although costs per mile do not fall, ton-mile costs drop sharply, reflecting the increasing size and payload of the later models. Direct cost (comprising only the five major items) and total cost per ton-mile of the Douglas DC-3 (1936) are less than one-eighth and slightly more than one-sixth, respectively, of the corresponding figures for the DH mail plane (1920).[12]

Tremendous, too, are the cost reductions envisaged for the future. From Dr Warner's authoritative analysis are computed the minimum practicable cost data (inclusive of a profit charge of ten per cent of operating expense) of 9.5 cents and seventeen cents per ton-mile upon a capacity payload basis for first generation postwar cargo and passenger aircraft, respectively.[13] Further substantial diminutions of cost are to be anticipated from war and postwar technological advance. Dr Warner included in the design and performance of his postwar aircraft improvements effecting only one-third of possible aggregate savings in operation cost realisable in the immediate postwar period. And on the horizon are a number of outstanding innovations – the flying wind, jet propulsion, commercial glider trains, and others – any one of which, in conjunction with progress in design, fuels, and materials, may bring about a marked reduction in the cost of air transport.[14] We need not discuss this point further.

With this cost framework in mind, we may now discuss the role of aircraft in the future transport web. Obviously, this role will vary for each country and under different circumstances. We shall confine ourselves to two broad categories of circumstances: those representative of highly industrialised-commercialised areas, and those of backward, undeveloped areas, as typified by the United States and large parts of Latin America, respectively.

Aircraft unquestionably will dominate the sphere of first-class commercial passenger traffic, other than short-distance. The speed element alone would be sufficient to insure a rapidly expanding patronage (assuming increasing safety, regularity of performance and thus public acceptance[15]), but in addition a cumulatively more favorable competitive cost position for aircraft can be anticipated. A 65–70 per cent load factor applied to Warner's first-generation postwar passenger aircraft (which aircraft should characterise a large segment of domestic operations within the first postwar decade) yields an approximate passenger fare of 2.5 cents per mile,[16] a charge below prewar basic Pullman rates[17] and indicative of the keen competition which aircraft can offer to all first-class surface transport service on the cost basis alone.[18]

Short-distance passenger traffic in advanced industrialised regions may partly fall to local and feeder airlines, especially to the latter when trunk-line air connections are involved. But here the speed advantage of air travel is not so pronounced, and here, as well as in the field of urban-metropolitan transit, buses and railways should retain dominance.[19]

These forms of transportation should also retain dominance in the lower-class commercial passenger traffic of these regions, though aircraft stripped of luxurious and unessential equipment will doubtless offer some competition. Much obviously depends upon the progressiveness of railway and bus management.[20]

For backward and underdeveloped regions lacking adequate surface transport facilities and utilising primitive and obsolete mechanisms, the role of aircraft will be still more predominant in all classes of commercial passenger traffic. Not only huge time savings will be available through air service, but quite frequently cost savings as well. Nevertheless, in terms of local currencies and levels of income, air transport costs may often be burdensome; and though passenger traffic may increase relatively by tremendous strides, its absolute volume may remain comparatively small.[21]

Another important phase of the question of population movement is that of private (individual and family) aircraft. Though momentous changes in regional economic structure are implicit in such movement, little reliable and objective information and data exist upon which to formulate a prognosis. Nevertheless, certain relevant observations can be made.

Technical, rather than economic, obstacles will circumscribe future use of private aircraft. Production and operation costs seem certain in the future to be on a level permitting mass adoption.[22] It is the congestion inherent in mass commutation which must be combated; marked success in this effort is insured through the development of new techniques of navigation and traffic control, by the provision of sufficient airports, landing strips, and other aerial facilities and equipment, and through progressive city planning. Congestion, too, could be substantially alleviated, were the helicopter (much superior to the airplane in maneuverability, control of movement, safety, and landing space requirements) or a combination aircraft-automobile mechanism (capable of converging upon dense traffic areas via superhighways) widely adopted.[23] It seems reasonable to predict, very roughly, that within a quarter of a century following the close of hostilities private aircraft will be utilised as extensively as was the automobile in recent prewar years, and even more extensively in those backward regions where, as we shall note later, cargo aircraft will stimulate vigorous economic growth. Very tentatively put, the automobile should, in large part, be supplanted by aircraft in long-distance travel, but should retain dominance in local and short-distance movement, in which it will possess relative advantages in respect to cost and time, assuming the existence of roads and highways.[24]

As for cargo movement in advanced, industrialised regions, it seems reasonable to conclude that all the nations concerned will eventually come to appreciate the advantage of expedited first-class mail, and in major part employ air service without surcharge, nationally and internationally, wherever aircraft can effect quicker delivery.[25] In express traffic, too, aircraft will doubtless largely supersede surface facilities.[26] The speed element in conjunction with only moderately high charges would be sufficient to establish aircraft in a strong competitive position. In addition, aircraft can offer competition with surface facilities for express carriage from a cost standpoint. Warner's first-generation cargo aircraft, operating at 65–70 per cent capacity payload, and excluding pickup and delivery expense, permits a minimum practicable charge of somewhat above 14 cents per ton-mile. Van Zandt, peering further into the future, arrives at a corresponding charge of around 12.5 cents.[27] These figures may be compared with the 1938 basic first-class rail-express 50-mile block *haulage* tariffs of 26, 28.5 and 31 cents per hundred pounds for Eastern, Southern and Western zones, respectively.[28]

In the case of freight any encroachment of aircraft upon surface facilities will be relatively minor in advanced regions. Aircraft probably will never be able to achieve the efficiency of surface media in the mass movement of heavy and bulky commodities. Transport costs of coal shipments on the Great Lakes have been estimated as low as 0.04–0.05 cent per ton-mile (full utilisation, exclusive of terminal charges). On the Mississippi River system, over-all transport costs range chiefly from about 0.2 cent to 0.5 cent per ton-mile. By tanker, total costs are about 0.1 cent per ton-mile (50 per cent utilisation); by pipe-line, 0.38 cent per ton-mile. Non-local trucking incurs a line-haul expense in the neighbourhood of two to four cents per ton-mile. Average rail freight costs and less-than-carload service costs for the year 1932 have been estimated at 0.83 cent and 7.0 cents per ton-mile, respectively (average rail revenue per ton-mile in 1939 was 0.983 cent).[29]

The most promising opportunities for air cargo carriage are to be found in undeveloped regions now dependent upon primitive and inadequate transport facilities. Owing to the flexibility inherent in aircraft and the relatively small initial investment entailed in the establishment of a skeleton of airways, aircraft can perform a vital developmental function.[30] Already a tremendous quantity of new traffic has been developed, to say nothing of expediting existing traffic, in a number of these isolated regions, often at moderate rates.[31] Whether aircraft will retain dominance in freight carriage obviously is contingent upon the balance of traffic volume and future cost. Where moderately heavy industrialisation and commercialisation result from the stimulus afforded by aircraft, surface facilities will probably usurp freight traffic, leaving air express relatively untouched.[32]

III

In the light of the analysis developed in the preceding sections, the implications of aircraft for economic structure and the progress of society become fairly clear, and to these questions we now turn. Resemblances between the impact of aircraft and that of the street and electric railway and the automobile (resulting from increased population mobility) and that of the railroad and canal (brought about by reductions in the cost of commodity movement) will be stressed.

Participation by aircraft, especially by privately owned aircraft, in population transportation should effect vast changes in urban-metropolitan textures and patterns. The consequent increase in population mobility and greater speed and flexibility of movement will further stimulate population dispersion within urban agglomerations. Wealthier and higher-income groups will assume the lead in penetrating outlying and rural areas. As aircraft production and operation become more efficient, as physical resistances (translated into economic costs) to navigation and traffic control are overcome, moderate-income classes may be expected to participate actively in this centrifugal movement. Residences acquired in outlying territory, though amply spaced, will tend to cluster around nuclei, somewhat analogous to present-day community or neighbourhood centers, wherein will be aggregated basic retail, financial, educational, amusement and recreational services. The relative position of these satellite nuclei about the mother city will be largely determined by topographical conditions and the previous locus of population settlement.[33]

Through increased population mobility and diminished time dimension of distance, aircraft should enlarge the potential consumer hinterlands of urban centers, and thereby lead to great site differentiation and selectivity.[34] There will be fewer primary cities, but these will cater to increasingly greater populations increasingly diffused. At or about the heart of each primary city will develop increased centralisation of major distribution, transportation, financial, and perhaps governmental functions, and of select retail, choice cultural, recreational, and perhaps educational facilities. Skyscraper residential buildings may appear at the core to accommodate those of the higher-income groups who prefer to live near the center of things.[35] These developments will encroach upon central areas now designated as slum and blighted zones, driving their lower-income populations into abandoned suburbs. The latter areas, as well as those less fortunately situated middle- and large-size cities, many of whose functions will have been usurped by primary centers, may deteriorate into future slums and blighted areas.[36]

The impact of aircraft upon population movement and upon the changing structural and functional relations of city and country will open sizeable investment outlets in the postwar period. An upsurge in residential building will accompany the dispersion of urban population into outlying territories. Large developmental projects in the form of roads, community centers, public utilities, etc. will be required by the newly settled population. Primary centers will need to be reoriented and reconstructed. Such construction activities in the past have always been strong bulwarks behind economic prosperity;[37] the same may well hold for the future.

Moreover, just as the rise of the automobile stimulated the growth of a host of novel industrial and commercial enterprises, so will aircraft. On the one hand, industries to provide equipment and service for aircraft, airways, and airports will develop. These, however, already will have received a strong impetus from the war program. On the other hand, trades will emerge to satisfy the needs and desires of a more mobile population. Travel and recreation will be especially important here; aircraft, by relaxing time and geographic restrictions on movement, will give freer rein to the propensity of the American public to rest and play in distant places. In all this lie additional investment opportunities.[38]

Inherent in air cargo carriage are the emergence of new industrial and commercial areas and population agglomerations, alterations of trade channels, and outlets for new investment. Aircraft are not subject to the geographic limitations of the canal and the railroad, being barely sensitive to topography and physical boundaries. This will permit a broad reduction of transport cost in backward, undeveloped sections, and in regions accessible only with difficulty by land or sea. The geographic sphere of economic relations will be greatly extended. A revaluation of the world's resources will follow. Resources have economic significance only in reference to technology. Technological developments, such as aircraft, which bring into economic availability resources hitherto unexploited because of technical limitations, upset production functions and compel reconsideration of the most efficient locations for industry and commerce. Those areas generously endowed with strategic materials unutilised because of inadequate surface transport facilities, provided they present no such serious obstacles as climatic restrictions, diseases, extreme political instability, etc. will possess potentialities for an upsurge of industrial and commercial development. A case study of such an area, Brazil, is presented in the Appendix.[39] Some of the older industrial areas may receive a setback from the shift of industry to more favorable locations. Others, which possess decided superiority for production, may experience further concentration and growth by the extension of their market to the newly opened areas.[40] Concomitant with these shifts of industry, new trade routes and commercial sites will arise, and new agglomerations of population. Just as the canal opened up the Appalachian interior and the Pennsylvania anthracite coal fields and the railroad penetrated the Mississippi River valley region and the Far West, just as the canal and railway caused shifts and growth of industry and commerce and population redistribution within the United States, aircraft, by enlarging the world area under intercommunication, can effect marked alterations in the structure of world trade, industry and population.

Other alterations of trade routes and primary commercial centers will result from new geographic emphases. Desert, mountain, oceanic expanse and other topographic barriers to surface movement will recede in importance; atmospheric and weather characteristics, and perhaps island chains and peninsular bridge areas, will come into prominence as critical geographic detail. Accordingly, nodal sites, where sea meets land or where breaks in the earth's surface occur, lose significance for commercial activity. For international air transport, seaboard cities have no inherent advantage over inland cities. Obviously, in addition to geographic circumstances, the commercial strategy of sites for air transport will depend on (1) relative distances to other primary centres, and here distance tends to be measured along great circle courses, rather than along paths of least resistance to surface movement; and (2) the economic importance of the various sites themselves, in other words, the economic inertia and tradition of the existing pattern of industrial, commercial and population settlement.[41]

The above analysis has tremendous implications for postwar investment potentialities. If air transport is encouraged and permitted to realise its possibilities, large capital exports will be required to develop backward and unexploited areas. Intense construction activity will ensue. New industrial plant and equipment, new development and redevelopment of cities, housing and public utilities, all will be required for the new concentrations of industry and commerce and the new agglomerations of population. Substantial capital outlays will be needed for the growth, and perhaps increased centralisation, of industry and commerce in the older industrial areas; abundant investment opportunities will be provided by the shift of industrial areas, commercial sites and population within the older industrial setup. In the past, industrial, commercial and population adaptations and adjustments to transport innovation always have afforded huge investment outlets. The same relationship may well characterise the future, and thus be a strong impetus toward the achievement of postwar world prosperity.

The role of international planning and assistance in the transport development of backward regions of the world has already been referred to. Large-scale governmental assistance to, and subsidy of, transport innovations, particularly in undeveloped areas, certainly is an old and established practice. The history of the United States is replete with illustrations in the construction of its canals, railroads, and highways, and in the improvement of its waterways.[42] Similarly, to attain a comprehensive utilisation of air transport, subsidy to private enterprise and encouragement to private ownership will probably be necessary. Provision of airports, airways, equipment for traffic regulation and control, etc. seem likely to require heavy governmental assistance. We have mentioned the potentialities for postwar national progress inherent in a widespread development and adoption of aircraft as a mode of transport: the investment opportunities resulting from population movement and dispersion and the growth of huge metropolitan districts, in the rise of a host of new industries and extended recreational and travel activities, in the industrial, commercial, population and urban adaptations and structural rearrangement of the United States economy. The extent to which these investment opportunities will be realised and these national forces toward postwar prosperity set in motion will depend, in large part, upon the availability of public aid.[43]

At the same time, a broader concept of assistance must be recognised. Air transport transcends national boundaries and embraces an international order. Accordingly, assistance to the development of air transport should be on an international basis. Already various governments have subsidised private enterprise engaged in air transport; but international collaboration and cooperative aid are required for a thorough, beneficent airway development of backward countries. Mention has been made of the investment outlets inherent in the drastic reduction in transport cost permitted by aircraft, through extension of transport facilities to undeveloped and inaccessible areas and through expansion of the geographic sphere of economic relations: investment outlets available through shifts and relocations of industry and population, through alterations of trade channels and of the strategy and nodality of commercial centres. International collaboration and cooperative aid, not only in encouraging airway development of backward countries, but also in furnishing capital and technical facilities to stimulate social and economic advance, will be crucial factors. To a considerable degree they will determine the extent to which aircraft's potentialities for evoking industrial and commercial growth in backward areas and spreading prosperity through the world – for achieving postwar economic progress – will be realised.

Appendix: the case of Brazil

To illustrate more specifically the potential impact of aircraft upon undeveloped areas of the world, we examine Brazil. This country possesses a wealth of agricultural and mineral resources. Brazilian estimates have classified eighty per cent of the land as potentially productive; at present only a small fraction is in commercial use. Brazil has the largest and highest-grade iron-ore reserves known in the world. The State of Minas Gerais alone contains upwards of twelve billion tons; deposits lie close to the surface, and can be mined with ease. A vast range of other minerals are present: manganese, copper, lead, zinc, nickel, etc. Waterpower and forest resources abound. However, none of this wealth has been exploited on a significant scale.

A complex of forces has impeded Brazil's development: exploitative attitudes and feudalistic institutions of the settlers; an unstable course of settlement; a poverty-stricken, ill-educated, diseased, malnourished population; absence of aggressive political and social leadership; disunity and inadequate internal development. From an economic standpoint, there have been three principal interrelated factors operative: poor fuels, lack of capital, and inadequate transport facilities. Coal deposits, though abundant, contain considerable ash and volatile matter. It is only with difficulty that even a low-grade coking coal is manufactured. Petroleum discoveries are claimed with optimism, but they await appraisal and commercial utilisation. Lack of capital, reflecting social retardation and political instability, has precluded a program of heavy investment in capital goods – in public works as well as in agriculture, mining, and industry.

Inadequacy of transport facilities has ever plagued Brazilian progress. Geographic features – the sharply-rising gradients of the Great Escarpment south of Salvador, and the poorly-navigable, circuitous river systems (save the Amazon) which fail to converge to a natural focus – have imposed obstacles to surface transport. Equally and perhaps more important as hindrances have been human (technical) and financial limitations. Inland navigation on the whole is poor. Railroads are inadequate: poorly constructed from poor materials, poorly equipped, and poorly maintained. Highways, of recent origin and well-constructed in some regions, in general are confined to much the same areas as are the railways, namely, the coastal fringe and the southern states. The vast interior remains practically untapped by modern transport media.

Within this interior of Brazil that vicious circle of cause and effect which confronts most backward areas is present: a primitive economy cannot support the construction of surface facilities entailing heavy investment; at the same time the want of transport media impedes the much coveted economic development. Aircraft, although involving operating costs higher than those in mass shipments by surface facilities, possess an extreme flexibility and require a comparatively small outlay for airways and a minimum of equipment; in this manner they have a critical advantage as a developmental medium. Preliminary and experimental tapping of potential mineral wealth is easily and relatively inexpensively achieved with the use of aircraft. Agricultural settlement can also be stimulated. Even though aircraft may not be employed extensively in marketing heavy, bulky commodities, they can play a major role in transporting perishables, high-grade commodities, and sporadic, irregular shipments of all types of goods, as well as in providing that contact with established centres of population which is essential to expanding colonisation. Perhaps still more important is the unity which aircraft will effect. They will bind more closely Brazil's various economic regions and resources, and perhaps will provide that integrated economic structure which she has always lacked. In this manner economic growth will be encouraged. Much, however, depends upon availability of capital and technical ability, educational and hygienic progress, political attitudes and leadership, extent of postwar international collaboration and assistance, and related factors. It is conceivable that, within a favourable framework of forces, aircraft can give that impetus to the utilisation of Brazil's extensive and diverse wealth which may allow her to achieve the status of the dominant industrial reservoir (despite limitations of fuel) of a much more developed South America.[44]

Notes

1. The writers are indebted to Professors Alvin H. Hansen, Harvard University, and John U. Nef, University of Chicago, for their fruitful criticisms. Much of the material has been gathered under a fellowship granted by the Social Science Research Council.

2. The material, statistical and theoretical, is taken from two articles by W. Isard: 'Transport Development and Building Cycles', *Quarterly Journal of Economics*, November, 1942; 'A Neglected Cycle: The Transport-Building Cycle', *Review of Economic Statistics*, November, 1942.
3. For fuller exposition of these forces and the scope of their play refer to E. M. Hoover, Jr., Location Theory and the Shoe and Leather Industries (Cambridge, 1937); and W. H. Dean, Jr., *The Theory of the Geographical Location of Economic Activities*, unpublished doctoral dissertation, Harvard University Library, selections published under same title, Edward Brothers, Ann Arbor, Mich., 1938.
4. The first, extending from the mid-1820s to the early 1840s and reaching a peak in the early 1830s, consisted of canal construction; the second, from the late 1840s to the early 1860s, with a peak in the mid-1850s, almost wholly of railway construction with some canal projects in the early years; the third, from the close of the Civil War to the mid-1870s, of railroad construction, with a peak in the early 1870s; the fourth, from the late 1870s to the early 1890s, double-peaked in the early and late 1880s, of railroad construction (with an incipient stage of street and electric railway construction at the close).
5. Data covering basic industrial and mineral production do not extend far back into the nineteenth century, save for the series on anthracite and bituminous coal and pig iron, which series start in the 1820s, 1840s and 1850s, respectively. Lacking more adequate data, we view the movement of these primary series as roughly representative of industrial and mining activity. The series on building activity extends back to 1830, and for the nineteenth century depicts fairly well growth and concentration of industry, commerce, and population at primary urban centres.

   The reader is again referred to the articles by W. Isard, loc. cit., wherein are presented graphs of the series cited and a much more elaborate statistical and theoretical analysis of the relations stressed above, as well as those which follow.
6. The outbursts are timed from the late 1890s to the close of the first decade of the twentieth century and from about 1915 to the close of the 1920s, with peaks in the middle of the first decade and early 1920s, respectively.
7. Residential building, save that at the very cores of urban centers, was largely confined to suburban and peripheral areas in these years. For a partial analysis by decades of urban population movements for several primary cities during the twentieth century refer to R. D. McKenzie, *The Metropolitan Community* (New York, 1933), ch. XIII. Also see *Population of the City of New York 1890–1930*, compiled and edited by Walter Laidlaw (New York, 1932).
8. The reader unfamiliar with the details of these occurrences is referred to W. H. Dean, Jr., (Selections), pp. 42–6, and to standard texts on United States economic history.
9. See Norman J. Silberling, *The Dynamics of Business* (New York, 1943), ch. 10, for an equally strong emphasis, though with a different causal-flow analysis, upon transport innovation as a potent, dynamic stimulant to economic progress and prosperity.
10. Technical achievement in the modern era is essentially an orderly, cumulative process, continuously expanding the fund of knowledge; it thereby lends itself to rough prediction. It is in this sense we view progress in aircraft technology, progress in making explicit what is implicit in the concept of air transport, and our forecasts for the long run are evolved accordingly, though we recognise that translation of technical knowledge into practical achievement and introduction of innovations and refinements into economic life are jerky and

irregular. For further discussion of this point see A. P. Usher, *A History of Mechanical Inventions* (New York, London, 1929), chs I and II.

11. The data are assembled by E. P. Warner in two lectures: for prewar aircraft, 'Technical Development and its Effect on Air Transportation', Norwich University lecture (York, Pa., 1938); for postwar aircraft, 'Postwar Transport Aircraft', the Thirty-First Wilbur Wright Memorial Lecture, published in Aeronautical Engineering Review, vol. II, no. 10 (October, 1943), pp. 7–59.

    'To eliminate all differences except those absolutely inherent in the characteristics of the equipment itself', and thus make comparisons of prewar aircraft fair, Dr. Warner projects the planes of the past (DH, Boeing 40, and Ford) into conditions of the present (1938) and accordingly considers cost of operation for each (op. cit., pp. 33–4). Total cost is roughly arrived at by summing the five major items of direct cost, namely, fuel cost, pilot salaries, maintenance cost, depreciation and insurance and crash reserve (the sums are recorded in the table), multiplying this result by 1.2, and adding fourteen cents a mile, plus nine cents a ton-mile (of payload), thereby including indirect costs and minor direct costs. In general, total cost data 'are not far from the records of experience for the several airplanes included. For the earlier machines on the list they run somewhat lower than appears to have been the common experience at the time when they were in full service'; this could be anticipated from the elimination of differences other than in equipment (op. cit., pp. 41–2).

    Dr. Warner conceives the first postwar generation of transport aircraft (to which the data of the table relate) as comprising those models reaching the market during the period from the end of the second to the end of the fourth years after the war. The data represent minimum cost conditions (including a profit of 10 per cent of operating expenses) for aircraft constructed on the basis of dimensions and design ratios consistent with maximum economy in the present state of the art, in addition to embodying a substantial number of possible improvements whose realisation will effect an aggregate saving of 15 per cent in operating cost. For the cargo aircraft the data embrace cost of traffic solicitation and of handling and loading freight, but exclude cost of pickup from shipper and delivery to final consignee. Dr Warner computed cost data on the assumption of a 65–70 per cent load factor. For comparative purposes the writers converted the data to a capacity payload base by simple operations. The converted data are identified by asterisks, and undoubtedly are within a few per cent of what Dr Warner might have calculated, had he had a similar purpose. Obviously, the crudeness of all the data, the somewhat arbitrary and perhaps changed base of computation, and the uncertainty of prognostication warrant only rough comparisons.

12. For further evidence of the high cost of air transport operations during the early and mid-1920s, refer among other sources to *Civil Aviation*, A Report by the Joint Committee on Civil Aviation of the United States Department of Commerce and the American Engineering Council (New York, 1926); and Archibald Black, *Transport Aviation* (New York, Chicago, 1926).

    The Lockheed Vega, Boeing 247, and Douglas DC–3 are aircraft constructed primarily for passenger carriage, and Warner's data are computed accordingly. However, W. M. Sheehan, ('Airplanes for Air Freight', *Journal of Air Law and Commerce*, Vol. XI, No. 2, July, 1940) has developed for the Boeing and Douglas craft converted to cargo movement cost data which yield respectively 15.8 cents and 12.4 cents per ton-mile, capacity payload; and A. D. Lewis, (Some Economic Aspects of the Air Express Business, published by the author, Austin, Texas, 1941) has arrived at the cost figure of 12.28 cents per ton-mile for a cargo-converted Douglas DC–3, capacity payload. Though these cost data appear to have been constructed on a somewhat optimistic basis, very rough comparison with Warner's data on postwar cargo aircraft is permissible.

For an authoritative, but hypothetical, British cost estimate of efficient air transport operations during the late 1930s, see S. J. Noel-Brown, *Economics of Air Transport* (London, 1937) ch. VIII.

A partial supplement to Warner's quasi-hypothetical material is the analysis by J. Parker van Zandt (*Transportation and National Policy*, U. S. National Resources Planning Board, Washington D. C., 1942, Part II, Section I: Air Transport). The average gross operating expense per ton-mile capacity payload of the sixteen domestic air carriers has decreased steadily from approximately 50 cents in fiscal year 1935 to less than 35 cents in fiscal year 1940, and for Eastern Air Lines, one of the lowest-cost operators, has declined over the same period from about 45 cents to about 28 cents (op. cit., p. 344). Owing to incompleteness or inconsistency in reporting to the Civil Aeronautics Authority, these data are considered within 10 per cent correct. The discrepancies between Warner's computations and actual expenses of the domestic air carriers are in large part attributable to the inevitable lag in adoption of most advanced equipment for airline operation, to the inefficiencies of the various airlines, and to their provision of other than minimum-cost service. Of interest, too, though for obvious reasons not strictly comparable, are the estimates of 39 cents and 43 cents per capacity ton-mile achieved by the low-cost operators SABENA (Belgian) and KLM (Dutch), respectively, in European operations in 1937. (O. Lissitzyn, *International Air Transport and National Policy*, New York, 1942, p. 208.)

13. Other sources corroborate Warner's cost prognosis for first-generation postwar aircraft operation. Direct and total operation costs of the 44-passenger version of the Curtiss Wright Model CW-20-E are estimated at 10.3 cents and 18.5 cents capacity ton-mile, respectively. (Curtiss Wright Corporation, Airplane Division, Buffalo, N.Y., Transport Contracts Department, Operating Cost 44-Passenger Version of Model CW-20-E, Report No. TCO-13A, 20 February 1943.) Direct and total operation costs of the Curtiss–Wright commercial cargo plane, whose scheduled peace-time debut was interrupted by the war, according to estimates, hover about six cents and 10–11 cents per capacity ton-mile, respectively. (Roosevelt der Tatevasion, 'Flight Takes to the Air', *The Iron Age*, 26 February 1942, pp. 47–53. Mr Tatevasion cites from a report, Description and Estimated Performance, Curtiss–Wright Corporation, St. Louis Airplane Division). J. Parker van Zandt asserts that projection of the trend of cost during the period 1935–40 suggests gross operating expenses for domestic passenger aircraft of less than 20 cents per capacity ton-mile within a few years; and he visualises for the future cargo aircraft running at total cost of 7–8.5 cents per capacity ton-mile, exclusive of pickup and delivery expenditures (op. cit., p. 344).

On the other hand, see a more conservative report by B. A. McDonald and J. L. Drew, Air Transportation in the Immediate Postwar Period, Section IV, Curtiss–Wright Corporation, 1944.

14. For instance, Warner states that, under certain restricting conditions, if stability and control difficulties can be overcome in a true flying wing, this aircraft might have unit costs 20 per cent lower than those of his first-generation postwar passenger aircraft (Wilbur Wright lecture, p. 37).

However, one should not anticipate lowered costs through enterprise expansion. Careful analysis of airline operating statistics indicates that, contrary to railway experience, enterprises of medium size cannot expect to realise decreasing costs with expansion. Rather, constant costs are encountered 'at a scale of operation not very large'. John B. Crane, 'The Economics of Air Transportation', *Harvard Business Review*, Summer Number, 1944.

15. Warner voices the justifiable expectation that records of no more than 1.0 fatality per 100,000,000 passenger miles on a long-term average, and of 97 per cent regularity in winter and 99 per cent in summer, will be attained shortly in the United States. For fuller discussion, see Wilbur Wright lecture, pp. 55–9; also John H. Frederick, *Commercial Air Transportation* (Chicago, 1942), pp. 277–83.

16. A passenger and baggage are assumed to weigh 200 pounds. Warner's postwar passenger aircraft is designed to carry 40 passengers and two tons of mail and cargo. The payload factor applied is a maximum obtainable without impinging upon the service offered to the public. (Wilbur Wright lecture, pp. 33, 34, 37.)

17. In recent prewar years the basic Pullman charge was quite uniformly reduced to three cents per mile. Through sliding-scale and round-trip arrangements, the effective rate hovered between 2.4 cents and three cents per mile, though in the West and South a two-cent rate on 15-day round trip was established. (Kent T. Healy, *The Economics of Transportation in America*, New York, 1940, pp. 272–9). A more fundamental comparison of relative costs of travel by Pullman and aircraft involves consideration of value of time consumed, and of food, room and other supplementary expenses. Here the advantage of aircraft becomes decisive (see Frederick, op. cit., pp. 305–13).

    The encroachment of aircraft upon Pullman traffic continues, despite the fact that air rates have consistently exceeded Pullman rates by a considerable differential; however, this encroachment is difficult to measure. (For the course of passenger air rates, 1926–40, refer to Claude Puffer, *Air Transportation*, Philadelphia, 1941, p. 642). We know that a substantial fraction of business, political, and pleasure air traffic has been newly-stimulated. Nevertheless, of interest is Van Zandt's statement that 'in the last ten years, to the end of 1940, revenue passenger-miles flown on scheduled domestic airlines have increased from less than one per cent to approximately 12 per cent of Pullman passenger miles' (op. cit., p. 351). For statistics on progress of prewar domestic passenger service, consult Civil Aeronautics Authority, *Second Annual Report*, 1940 (Washington, D. C., 1941). Van Zandt conservatively predicts within the next decade or two commercial domestic air travel of six billion revenue passenger-miles, approximately a sevenfold increase over fiscal year 1940 (op. cit., p. 354). See Warner (Wilbur Wright lecture, pp. 11–13) for an estimate of roughly similar proportions.

18. Aircraft can be expected to usurp a large proportion of first and cabin class ocean travel. On intercontinental routes the time differential of air over water transport will be of still greater magnitude than air over rail travel, though simultaneously a marked cost differential of steamship rates over air rates will exist. (For a discussion of factors causing intercontinental air transport operations to be more costly than domestic, refer to Lissitzyn, op. cit., pp. 203–9.) As an example, the cost differential between aircraft and steamship fares from New York to Buenos Aires in 1940 was $291.75, whereas the time differential was 14¼ days. (Additional cases and fuller discussion are presented by William A. M. Burden, *The Struggle for Airways in Latin America*, New York, 1943, pp. 83–88, 98–107, 119.) The attraction of a sea voyage for pleasure travel, as well as the marked cost differential, should insure, however, the retention of substantial first and cabin class traffic by the steamships. On the other hand, the

creation of new business, tourist, and governmental traffic as a result of the speed offered by aircraft, should far surpass in magnitude that diverted from ocean channels. See comments and estimates of Burden, op. cit., pp. 135–6, and Warner, Wilbur Wright lecture, pp. 13–15, for American intercontinental and trans-Atlantic air traffic respectively.

Of interest are the fare reductions for Latin-American traffic recently proposed by Pan American. Postwar passenger fares would be lowered from the present schedule averaging 8.75 cents per passenger-mile to one averaging 4.25 cents with some rates as low as 3.5 cents. ('PAA Cuts Fares; Reveals Global Plans', *American Aviation*, 15 August 1944.)

19. Local and feeder air service of the airplane variety involve higher costs than long-distance service. Discussion of cost and numerous other problems of local air transport service is undertaken in an article by E. P. Warner, 'Requirements of Local Air Transport Service', *Aeronautical Engineering Review*, vol. 3, no. 2 (February, 1944), pp. 13–47. Interestingly enough, passenger traffic has been most dense over the shorter of the long runs, e.g. New York to Boston (Van Zandt, op. cit., pp. 351–3).
20. An analysis of comparative 'real business' costs of travel via bus, railway coach, and air is presented by Frederick (op. cit., pp. 308–10). It is reasonable to anticipate progress in railway and bus technologies, the embodiment of such progress in operations, and thus improvement in their competitive position. For a discussion of possible economies, see National Resources Planning Board, op. cit., pp. 2–5. The US Federal Coördinator of Transportation (*Passenger Traffic Report*, Washington, DC, 1935, Exhibit 7, Analyses 64 and 70) has calculated for the year 1933 bus revenue per passenger-mile at 1.4 cents and costs of all-coach, suburban-coach, other-coach, and reserved rail travel at 2.9 cents, 1.5 cents, 2.7 cents and 4.2 cents per passenger-mile, respectively. For discussion of elasticity of demand for lower-class rail passenger transportation, of the ability of the railways to provide such transportation profitably at greatly reduced costs, and of other relevant considerations, consult the above report, sections 20–33, 55–65, and Healy, op. cit., pp. 276–9.
21. See Burden (op. cit.) for actual and potential examples of these time and cost savings (pp. 98–101) and for general analysis of air transport in Latin America. These savings will be especially important where adverse topography obstructs surface movement. It should be remembered that aircraft, having fostered the economic development of these regions, have already been in certain cases, and may be quite extensively in the future, supplanted by superior surface transport facilities.
22. For the immediate postwar market, William D. Hall in an address before the Society of Automotive Engineers (*Flying*, vol. 34, no. 5, May 1944, pp. 28–9, 108, 112, 116) foresees light airplane models with minimum equipment priced as low as $1000, two to five-place models of medium performance, ranging from $1500 to $3000, and a high-performance, small model available at $2000. Also see J. H. Geisse, 'Suggestions for Furthering Private Flying', *Aeronautical Engineering Review*, August, 1944, wherein are presented limited cost data on private airplane operation, and William B. Stout and Franklin M. Reck, *Tomorrow We Fly* (New York, 1943).
23. The Civil Aeronautics Authority has, and the war effort will have, through training new pilots, mechanics and other aircraft personnel, and by increasing the air-consciousness and receptiveness of the public, provided a major impetus to postwar civilian flying. For statistics on the rapidly increasing, but still relatively small, reservoir of private certified pilots, 1930–42, see Ernest W.

Williams, Jr., *The Outlook for Domestic Air Transport*, Planning Pamphlet No. 21, National Planning Association (September, 1943), p. 7. The war effort, too, in its requirement of air fields, landing strips, and aerial equipment to facilitate aircraft operation, will have contributed much to the related problems of congestion and safety. To combat these problems, specialisation of airports has been suggested, with private airplane traffic serviced by a ring of plane parking airports on the peripheries of urban-metropolitan areas. (Werner Hegemann, *City Planning, Housing, Vol. III*, edited by W. W. Forster and R. C. Weinberg, New York, 1936, p. 64.)

24. It is very conceivable that private aircraft will incur sizeably greater expense in operation than the automobile; as a consequence the latter's cost appeal to lower-income groups may maintain on the highways a certain proportion of long-distance travel. In addition, lower-income groups may not be able to afford both transport means, and thus may be forced to select one of them. The alterations to the above statements, should a combination aircraft-automobile technique be evolved, are obvious. We must remember, too, that new traffic will emerge and that the aggregate of transport services obtained from private mechanisms will be tremendously increased.
25. European nations had already accepted this practice for continental mail before the present war; and England had applied it to her Empire air routes with favourable results. For arguments favoring this practice, for statistics on air mail development, and for estimates of potential air mail traffic and other relevant material for the United States, refer to Frederick, op. cit., chs 18, 19; Van Zandt, op cit.; and Francis A. Spencer, *Air Mail Payment and the Government* (Brookings Institute, Washington, DC, 1941).
26. Air cargo operations in the United States in the past have been impeded by excessive rates, as well as by lack of interest and attention and difficulties of pick-up and distribution. For a history of rates, see L. N. Selig, J. D. Michel and A. Burstein, *A Brief Study and A Plan for the Participation of the American Railroads in Air Freight Operations in the United States*, General American Transportation Corporation (Chicago, 1940), pp. 25–6; and for a relatively good discussion of air cargo development and problems of air cargo transport and for presentation of relevant statistics, refer to Frederick, op. cit., chs XX–XXI. As of 15 October 1944, American Airlines has instituted a new 'Airfreight' service inclusive of pick-up and delivery, with rates ranging down to as low as 30 cents per ton-mile ('Airfreight Rates as Much as 53 Per Cent Under Air Express', *American Aviation*, 15 October 1944.)

    Intra-European air cargo traffic in prewar years had modestly exceeded that on global and United States air routes (Lissitzyn, op. cit., table I, p. 42). Obviously, in recent years the war's impact has greatly stimulated air cargo movement, nationally and internationally, and has altered its character and distribution (Williams, op. cit., pp. 7–10, 22).
27. Van Zandt's computation is based on costs (direct and airline overhead) per capacity ton-mile of 7.5 cents and a profit of 5 per cent gross revenue. Op. cit., p. 350.
28. In a broad, rough fashion, these haulage tariffs are equivalent to 10.4 cents, 11.4 cents and 12.4 cents per ton-mile, respectively, and are subject to modest discounts progressively increasing with carriage distance. In 1939 increases of 5–10 per cent per hundred pounds were authorised by the Interstate Commerce Commission. Second-class rates are 75 per cent those of first-class, while third-class (exceptional) rates vary with traffic needs and conditions of service.

    To derive the over-all rail express charge of 1938, an allowance for rail terminal service of 30 cents per hundred pounds and another allowance (altered as of 1939) for pick-up, delivery and terminal service by the express agency of 35 cents per package less than 100 pounds, proportionally increased for packages over 100 pounds, must be added to the above haulage tariffs. Van Zandt, on the assumption that the future 'average' air shipment would weigh 25 pounds and involve 8 ton-miles of transportation, allocates 5.5 cents per ton-mile (45 cents per package-shipment) for pick-up and delivery expense; the corresponding statistics (1938–39) for typical first-class rail express are 29 pounds, 7¼ ton-miles, 4.8 cents (35 cents). In this manner Van Zandt foresees 'an average possible charge for air freight of 18 cents or less per ton-air-mile', as compared with 17.6 cents and 17.1 cents, the present (1939) average charges per ton-air-mile for rail express parcels and first-class rail express, respectively. (Van Zandt employs a ton-air-mile concept to adjust for differences in air and rail distance.) For further details see US Interstate Commerce Commission, *Reports*, Vols. 83, 89, 231, pp. 606–81, 297–323, and 471–507 respectively, and Van Zandt, op. cit., pp. 348–51. It is interesting to note here that Pan American Airlines has proposed for Latin American routes postwar rates from 20–40 cents per ton-mile for first-class commodities; 15–25 cents, second class; 10–15 cents, third- class. Pick-up and delivery service is not provided. ('PAA Cuts Fares; Reveals Global Plans', *American Aviation*, 15 August 1944, pp. 20–2.)

Obviously, costs realised on domestic air routes may exceed predictions, as a result of inefficiencies of operation, provision of luxury service (high-speed runs), extension of feeder-lines to numerous cities and communities, and other factors. At the same time, however, we can expect continuing reductions in air transport cost as a result of technological advance, though doubtless curtailment of rail and truck costs are to be anticipated from parallel technological progress and from elimination of operational inefficiencies. Also, possibility of deferred air rates for less urgent and postponable shipments must not be overlooked as a means of increasing the payload factor, thereby diminishing realisable ton-mile costs (for Latin American experience see Burden, pp. 112–13). Regarding overseas air commerce, costs, as noted previously, will be greater than on overland routes; but the tremendous time differential of aircraft over ocean vessels should more than offset its adverse cost position for express traffic.

In addition to diversion of express traffic from surface facilities, aircraft will stimulate the formation of a vast quantity of new traffic (e.g. style goods, perishable commodities). Van Zandt has estimated potential United States domestic commodity and mail air traffic at around 615 million ton-miles per annum within the next decade or two (op. cit., p. 354). For other interesting estimates see Spencer A. Larsen, Air Cargo Potential in Fresh Fruits and Vegetables, *Wayne University Studies in Air Transport*, no. 1 (Detroit, 1944).

29. Cost data are from National Resources Planning Board, op. cit., pp. 408–9, 436–42; Healy, op. cit., Chaps. X, XI; and US Federal Coördinator of Transportation, *Merchandise Traffic Report* (Washington, DC, 1934), pp. 134–5, 282–283. Brightest prospects for inroads by aircraft are in the spheres of rail less-than-carload and truck high-grade commodity traffic. For further relevant cost material on the former, see Healy, op. cit., pp. 211–12, and US Federal Coördinator of Transportation, op. cit., pp. 11–17.

Furthermore, the cost differential between water and air freight rates for transoceanic traffic probably will be enormous, though elimination by aircraft of transshipments in transport from inland point to inland point may, in some cases, outweigh this differential.

30. Total investment in Latin American-owned air transport embracing a net of approximately 20,000 miles, as of December 31, 1941, was only about $7,600,000 (Burden, op. cit., pp. 160–5). In 1939 scheduled United States domestic air operations covering a route-mileage of 36,581 involved an investment of about $40,000,000 by air carriers; and for that year approximately $84,000,000 investment in airports and airways has been allocated to foreign and domestic scheduled operations. Roughly, then, investment per scheduled domestic route-mile lies between $2000 and $3000. (US Civil Aeronautics Authority, op. cit., p. 50; Van Zandt, op. cit., pp. 337–40). In contrast, United States steam railways have a total investment of $26,502,582,158 in road and equipment, or $109,006 per mile owned (US Interstate Commerce Commission, *Statistics of Railways in the United States, 1939*, Washington, DC, 1941, p. 117). Though obviously territory serviced by railway and aircraft cannot be compared in terms of route-miles, nevertheless the relatively light investment per route-mile and the relatively short life-span of much plant and equipment confer on aircraft a most critical role for the economic growth of the regions under consideration. For a compilation of road and highway costs, chiefly in British colonial development, refer to J. Edwin Holmstrom, *Railways and Roads in Pioneer Development Overseas* (London, 1934), pp. 162–187.
31. Substantiating this statement are traffic statistics for 1938; USSR, New Guinea, Canada, Honduras and Colombia, all deficient in surface transport facilities, in order, lead other nations in tons of merchandise transported by aircraft (Lissitzyn, op. cit., pp. 49–50). The outstanding pioneer in the development of air transport in isolated regions has been TACA, operating around 5000 route-miles in Central America in 1940, carrying almost 14,000 tons of express and freight (126 per cent of United States domestic aggregate), and performing over one million ton-miles of transportation (30 per cent of United States domestic total). Characteristic of Latin American local lines have been rates averaging 80 cents to $1.00 per ton-mile for express and occasional shipments, and rates averaging 30–35 cents (and in some cases below 20 cents) per ton-mile for large, bulky and deferred shipments. The Latin American air transport industry, in contrast to those of Europe and United States, receives a large proportion of its commercial revenue from cargo carriage. But its development has been greatly impeded by insufficient subsidisation and lack of adequate technical personnel and modern equipment. See Burden, op. cit., chs V, IX and excerpts from an address by J. Parker van Zandt, *Air Transportation*, vol. 2, no. 1 (January, 1943), pp. 30–33, 37–42.
32. For instance, in Brazil, where in prewar years the existing scant net of railways was poorly constructed, equipped, and maintained, and dependent upon inferior fuels (save for imports), average receipts for 1937 and 1938 from merchandise (freight and vehicles) nevertheless approximated as low as 1.1 cent per ton-mile (Instituto de Geografia e Estatistica, Anuario Estatistica do Brasil, Ano V, 1939–40, pp. 264–5). And Holmstrom (1934) in an excellent and extensive presentation of hypothetical as well as reported rail and road operating cost data and consumer charges, chiefly in British colonial possessions, observes the greater efficiency of colonial railways over corresponding roads, and concludes that a 'typical' British colonial railway can expeditiously carry minerals, agricultural exports and fertilisers at rather under one pence per ton-mile, manufactured goods at three to six pence per ton-mile, and non-essentials at six pence to one shilling per ton-mile (Holmstrom, op. cit., chs IV–X, esp. p. 142). On the other hand, the inertia of tradition and reluctance to engage in heavy investment programs may somewhat retard

development of economic surface facilities. In that vast expanse of Latin America, as well as in numerous other regions unpenetrated by modern surface techniques, transport by primitive mechanisms is extremely expensive. (Burden, op. cit., pp. 112–13, cites 30 cents per ton-mile by mule; also see Holmstrom, op. cit., pp. 211–12.)

33. For background material, history, and trends in urban ecology, refer, among other sources, to McKenzie, op. cit., Homer Hoyt, *The Structure and Growth of Residential Neighborhoods in American Cities*, US Federal Housing Administration (Washington, DC, 1939); and Stuart A. Queen and Lewis F. Thomas, *The City; A Study of Urbanism in the United States* (New York, 1939).

34. In an analysis of census data (1940) on distribution of retail sales in cities of 10,000 or more population, the writers found strong support for the thesis that through extension of urban consumer hinterlands the automobile, and to a minor extent the street and electric railway, have induced marked selection and differentiation among these cities in Food, General Merchandise and Apparel Groups. For additional evidence, see, among other sources, P. D. Converse and R. V. Mitchell, 'The Movement of Retail Trade within a Metropolitan Area', *The Journal of Marketing*, Vol. II (July, 1937), pp. 61–7; E. Van W. Reed, 'An Analysis of the Retail Trading Relationships of Elgin, Illinois, a Satellite City', *Studies in Business Administration*, vol. IX, no. 1 (Chicago, 1938); J. A. Pfanner Jr., 'A Statistical Study of the Drawing Power of Cities for Retail Trade', *Studies in Business Administration*, vol. X, no. 3 (Chicago, 1940); and for material on rural trading areas, J. H. Kolb and R. A. Polson, 'Trends in Town-Country Relations', University of Wisconsin Agricultural Experiment Station, *Bulletin* no. 117 (Madison, 1933).

35. Regarding location of airports, growing diversification of aircraft types and aircraft flight requirements probably will confine certain types of flying activity to separate airports. Commercial transport most probably will be concentrated at a single vast municipal airport situated at short-time distance from the heart of the city. Where the airplane rather than the helicopter is employed for individual flight, rings of airports will be established around the outskirts of the city and suburban areas; these smaller airports will connect to rapid transit facilities converging at the center of the primary city. For instance, see Charles B. Donaldson, 'The Airport Situation', *Aeronautical Engineering Review*, vol. II, no. 12 (December, 1943), pp. 41–5.

36. Mention should be made of future urban-metropolitan structure from an architectural standpoint. Frank Lloyd Wright, in his *Autobiography* (New York, 1932), advocates, as an ideal, complete decentralisation of the city. He conceives a population scattered and diffused throughout the country, super-highway systems providing means of rapid intercommunication. Along these traffic arteries convenient neighbourhood distribution centers, meeting places, etc. would exist to offer all forms of decentralised but integrated services. The metropolis of the future is at most a centralised depot of some kind. Le Corbusier (*The City of Tomorrow and its Planning*, translated by Frederick Etchells, London, 1929), on the other hand, recognises the economies of specialisation and mass production in vast concentrations of industrial, commercial, financial and other activities, and accordingly emphasises functionalism. He visualises at the heart of the city huge skyscrapers set immense distances from one another, surrounded by large open spaces and parks, allocated to business and commercial purposes alone, and intricately interconnected. For residential purposes there are blocks of villas or tenement houses of relatively low height and removed from the center of the city. On the

average, nearly 90 per cent of the ground area consists of open spaces in the form of wide roadways, playing fields, gardens and parks. Combining (1) for primary cities Le Corbusier's scheme, modified somewhat to recognise limitations and the geographic sensitivities of aircraft technology, and (2) for ruralised life Wright's conception, would yield in the writers' eyes an ideal architectural model consistent with economic tendencies.

37 Refer, among other sources, to Silberling, op. cit., ch. 9; A. H. Hansen, *Fiscal Policy and Business Cycles* (New York, 1941), pp. 19–27; and C. D. Long, Jr., *Building Cycles and the Theory of Investment* (Princeton, 1940).

38. Conversion of war aircraft factories to peace-time production, and perhaps construction of new plant, will open some investment outlets, particularly for the manufacture of individual private aircraft. However, the unparalleled efficiency (traffic capacity per vehicle per year) of aircraft will severely circumscribe investment required for cargo and commercial passenger service.

For interesting remarks on the recreational potential of aircraft, see George T. Renner, *Human Geography in the Air Age* (New York, 1942), pp. 159–60, 162.

39. The developmental achievements of aircraft in Alaska, Colombia, New Guinea, and several other backward regions of the world are observed by Renner, op. cit., pp. 110–14, 162–8.

40. Within these older industrial areas, particularly the more favoured ones, there may be, in the vicinity of airports, sizeable relocation and growth of light industry and distribution trades engaged in foreign and domestic commerce.

41. The reader must bear in mind that these alterations in trade arteries and strategy of commercial sites have reference only to express, commercial passenger, and that higher-grade commodity traffic which aircraft will usurp from surface facilities or create. Ordinary freight will seek the economically least resistant surface channels, though, to be sure, some will be diverted to air lanes, when tremendous savings are available in distance and in transshipment expense (e.g. in transportation from inland point to inland point over an ocean expanse).

We must remember, too, that certain geographic details, though slight in character, may be of critical importance in determining future centers of air transport. Desirable characteristics of airport sites are approaches free of adjacent mountains and hills, relatively constant and predictable winds, good visibility and absence of fog and haze, and perhaps for heavy air cargoes, low altitude (the higher the altitude the longer the runway required).

For a discussion of air lanes and of geographic factors restrictive to air transport, consult Renner, op. cit., pp. 125–33. Also see J. Parker van Zandt, *The Geography of World Air Transport*, Brookings Institution, Washington, DC, 1944.

42. During the canal era, save for a few undertakings, states carried through the entire programme of providing canal networks, investing for those times considerable sums in these works. Subsequent inland waterway and harbour improvements and maintenance were to a still greater extent furnished by public bodies. Throughout the railway era, initiative lay with private enterprise, but subsidies by the various governments, especially for projects through undeveloped territory, were enormous. For instance, of the cost recorded to 31 December 1882, for construction and equipment of railways at that date operating Federal land-grant mileage west of the Mississippi (the major transcontinentals belong in this category) about 57 per cent is estimated as having been offset by construction aids. In the street and electric railway era, the role

of public aid was negligible, but it is significant to note that this transport medium serviced already developed population centres. Finally, public encouragement to the widespread adoption of the automobile and truck through construction of highway networks has been tremendous. In the period 1921–32 alone, approximately 22 billion dollars were expended on highways and streets by federal, state, city and local governments. Further details are available in US Office of Federal Coördinator of Transportation, *Public Aids to Transportation* (Washington, DC, 1938–40).

43. The problem of subsidy to aircraft operations is not unrelated to that of subsidy to other forms of transportation. The subsidisation of transport must be viewed as a whole, in order to avoid that social waste which results from over-encouraging two or more transport media competing in a defined field. From this standpoint, ideally the quality and character of services to be performed by each transport media should be determined at each given point of time, and subsidies meted out accordingly.

44. Among numerous other publications, source material was gathered from P. E. James, *Latin America* (New York, 1942); Brazil, Ministerio das Relacoes Exteriores, *Brazil 1939–40* (Rio de Janeiro, 1940); R. Nash, *The Conquest of Brazil* (New York, 1926); J. P. Calogeras, A History of Brazil (Chapel Hill, 1939); and M. W. Williams, *The People and Politics of Latin America* (New York, 1930).

2 Some Economic Implications of Atomic Energy[1]

Once again the world confronts the task of digesting an innovation of major proportions. Indeed this innovation may spell change of the most far-reaching and pervasive character that mankind has yet faced within international and national political, social and economic spheres. Although economic institutions and structure may be affected least of all, it is vital that whatever national and international authorities evolve be aware of the potentialities for change in this sphere, and work out operating policies in the light of these considerations to encourage constructive applications of man's newly-won ally and to minimise the economic and political frictions that may arise.

We can immediately limit our inquiry to the economic implications of atomic energy. Aside from the energy itself, there result from a fission process fissionable material capable of sustaining a chain reaction, radiations and radioactive materials.[2] We shall not discuss here the various applications of radioactive isotopes and radiations in medicine, scientific research, and industrial processes.[3] Doubtless these applications will be many and of great potential benefit to mankind; but as far as one can fore-see, each will be of relatively minor significance for the economic organisation of society, and their aggregate effect is as yet unpredictable. Our primary interest attaches to the utilisation of the energy given off in the fission process. Thus far it has not been found feasible to use this energy in any way other than in the form of heat.

Technical considerations restrict the manner in which atomic power can be utilised. The necessity of having a pile as large as the critical size, or larger, in order to sustain a chain reaction, the requirements of the cooling mechanism, and most important of all the necessity of shielding against lethal radiations, make imperative an atomic power plant of large and bulky structure. Use of light portable units, in automobiles, for example, is thus precluded. It is conceivable that rockets and passengerless aircraft might be propelled by atomic energy.[4] It is much more likely that large sea-going vessels will find it expedient to utilise atomic power, despite a large dead-weight structure. And the applications of atomic energy may be extended to such unique projects as tunelling through mountains, melting icebergs, maintaining irrigation projects.[5] But obviously the most promis-

Reprinted from *Quarterly Journal of Economics*, vol. LXII, February (1948), pp. 202–28.

ing application from the economic standpoint is in the generation of electric power with conventional turbines driven by the heat given off during the fission process. The implications of this last application are our primary concern.

I

What are the costs of generating power from atomic plants? We run into grave difficulties at the start. Not only must all of our data be hypothetical, since no nuclear plants are yet in operation, but it is not at all clear what production methods may be dictated by economic, political and military considerations. Will the nuclear fuel, the uranium isotope U-235, be concentrated by the gaseous diffusion process, the electromagnetic process, the thermal diffusion process, or by the centrifuge method?[6] Or will the nuclear fuel, plutonium (Pu-239), be prepared from primary piles of the graphite-uranium type, which may or may not concomitantly generate power, and which may or may not consume on the spot the plutonium produced? Or will both be produced? Further, will secondary reactors be utilised solely for the output of power and other non-dangerous products, or will they be used to produce nuclear fuels as well, thus effecting overall neutron economy?[7]

Political and security solutions of the atomic energy problem may well run counter to economic ones; if so, the former in all probability will override the latter. At the present moment, however, the uncertain status of any organisation for international control of atomic energy, the censorship on scientific atomic information, and the world's relative inexperience with atomic engineering and control[8] leave us very uncertain as to what methods of atomic power production will (or may) be realised in the future. In terms of political and economic feasibility, the consensus, though unofficial, seems to favour the procedure whereby power is produced from primary reactors (graphite-uranium type piles) as well as from secondary reactors using plutonium.[9]

Although we abstain from rigid judgements concerning the nature as well as the feasibility of a future net of atomic power stations, nevertheless, certain common economic features may be expected to characterise all types of nuclear power stations. Moreover we do possess two fairly reliable and authoritative estimates of the cost of generating electricity from atomic power plants.

The Thomas Report,[10] the more authoritative one, since it is based upon actual experience with the Hanford piles and the present experimental pile at Clinton laboratories, presents a very sketchy estimate in the following statement:

> . . . a large stationary nuclear power plant . . . producing 75,000 kilowatts could be built in a normal locality in the eastern United States for approximately $25,000,000. On the assumption that the plant would operate at 100 per cent capacity and that interest charges on the investment would be 3 per cent, the operating costs of the plant would be approximately 0.8c. per kilowatt hour.[11]

Recently, the estimate of capital cost has been raised to $27,000,000. The plant envisaged is basically a primary reactor; however, it would utilise in its own pile all the plutonium formed. No plutonium would be produced for use elsewhere.

A report prepared by a group of California scientists,[12] though based on less confidential information, is more satisfactory. It follows the Lilienthal report's classification of dangerous and safe activities, of primary reactors operating with uranium and plutonium, of secondary reactors operating with suitably denatured plutonium and incapable of producing any new fissionable material. A detailed cost breakdown for steam power stations based on coal, and for hypothetical atomic power stations, is presented, which is reproduced in part in Table 2.1.[13]

Careful scrutiny of this table reveals several salient points. Capital costs of secondary piles (plutonium reactors) tend to be smaller in comparison with those of coal stations, because of savings in fuel handling and storage, in boiler plant, in land and improvements, and in buildings (exclusive of piles). On the other hand, these savings are almost wholly counterbalanced by requirements for pile units and accessory facilities (such as thick shells of concrete and special insulating materials, chemical plants and the like). But it is in the construction of primary piles that huge expenditures are entailed.[14] These expenditures, being to a large extent invariable with scale, would be incurred whether the output in terms of kilowatt hours and plutonium were small or large; from an economic standpoint they dictate the erection of a primary pile plant of huge capacity to spread the resulting overhead over many units of output. The California group, compromising between efficiency and practicality, chose to consider a 500,000 kw primary pile atomic power station.

It is significant to note that the California group did not attempt to compute generating costs per kilowatt hour for the atomic power station. It confronted the insoluble theoretical problem of allocating the costs of a production process to multiple products (in the case of the primary pile: power and plutonium). Obviously, the higher (lower) the proportion of expenses of primary pile operation assigned to plutonium, the lower (higher) the kilowatt hour cost for primary pile power and the higher (lower) the kilowatt hour cost for secondary pile power. However, the group has presented some relevant computations which are given in Table 2.2.[15] The values of power per kilowatt hour assigned in the table for the

Table 2.1 Preliminary power station cost analysis[a]

	Coal	*Coal*	*Primary pile*	*Plutonium reactor*	*Plutonium reactor*
Plant comparisons					
Capacity, kw	20,000	100,000	500,000	20,000	100,000
Area, acres	2.0	10.0	(10,000)	2.0	2.0
Thermal efficiency, per cent	25	35	40	30	40
Employees, number	25	60	200	20	40
Capital costs (dollars)					
Land and improvements	60,000	600,000	(2,000,000)	60,000	100,000
Buildings (exclusive of piles)	1,000,000	3,000,000	8,500,000	700,000	2,000,000
Pile units and accessory facilities	—	—	(24,000,000)	(1,100,000)	(2,000,000)
Fuel handling and storage	—	2,000,000	200,000	—	100,000
Boiler plant	1,000,000	1,500,000	7,500,000	600,000	1,000,000
Turbine-generators and auxiliary equiment	670,000	2,400,000	10,000,000	670,000	2,400,000
Other	550,000	3,700,000	16,500,000	550,000	3,500,000
Total	3,280,000	13,200,000	68,700,000	3,680,000	11,100,000
Operating costs (dollars)					
Fuel[b]	360,000	1,000,000	100,000	270,000	1,000,000
Wages and salaries (at $4,000)	100,000	240,000	800,000	80,000	160,000
Maintenance and supplies	30,000	120,000	1,000,000	40,000	150,000
Fixed charges (15 per cent)	490,000	1,980,000	10,300,000	570,000	1,660,000
Total cost per year	980,000	3,340,000	12,200,000	960,000	2,970,000
Cost per kw hr	11 mills	7½ mills	—	—	—

Notes: [a]Costs are representative for present designs in the United States. A plant factor of 50 per cent and a utilisation factor of 90 per cent (peak load is 90 per cent of rated capacity) are used.
[b]Based on coal at $6 per ton, uranium at $10 per pound and plutonium at $20 per gram.

Table 2.2 Estimates of earned income for atomic power plants

	Primary pile unit	*Plutonium unit*	*Plutonimum unit*
Plant Comparisons			
Capacity, kilowatts	500,000	20,000	100,000
Annual power production (kw hrs)	2,190,000,000	87,600,000	438,000,000
Operating costs in dollars (estimated)	12,200,000	960,000	2,970,000
Value of power, mills per kw hr	4	11	7½
Income			
Sale of power	$8,800,000	$965,000	$3,290,000
Sale of plutonium	5,000,000 (250 kg)	—	—
Sale of tracers	—	50,000	250,000
Total	$13,800,000	$1,015,000	$3,540,000
Income above estimated costs	$1,600,000	$55,000	$570,000

several stations do not seem unreasonable, but they might very well be different if another system of cost allocation were adopted.

How do generating costs of atomic power plants and conventional power plants compare? The California report concludes from its data that 'if reasonable construction costs are assumed, it appears likely that plutonium reactors may produce power for 5 to 15 per cent less than can be done in competitive coal-fuel plants'.[16] Its hypothetical primary pile could produce electricity at 3–4 mills per kw hr (if $20 per gram is charged for plutonium), a figure which 'represents a compromise between present urban industrial power rates and those obtained in the most efficient hydroelectric projects'.[17] For rough comparative purposes we present in Table 2.3 certain estimates of the Cowles Commission for generating costs of electricity for several regions of the world.

One can draw no hard and fast conclusions as to future competitive relations between atomic and conventional power plants. To do so upon the tentative data of these three tables would be extremely hazardous. As yet we know little about how atomic piles operate at high temperatures; and we still can expect major developments in atomic power technology.[18] However, certain rather well substantiated statements can be made. For example, it does seem reasonable to expect keen competition from atomic power plants in certain localities, especially places relatively unfavorable for the expanded operation of conventional steam stations (such as Great Britain, Argentina and United States northwest of the Great Lakes). In

Table 2.3 Estimated costs of generating electricity in selected regions of the world, 1937 and 1947

	*Average generating costs in mills per kilowatt hour**	
	1937	*1947*
Argentina		
Coastal regions	9.0	16.0
Inland regions	10.0–11.0	17.0–18.0
China		
Mining regions	6.5–7.0	—
Other regions	7.0–9.0	—
Great Britain	7.25	9.5
Hungary	9.25	—
India		
Mining regions	6.5–7.5	—
Other regions	7.5–10.0	—
Soviet Union		
Ural region	6.5	—
United States (coal plants)		
At mines	5.50–6.25	5.75–6.75
Near mines or on developed waterways	6.50–7.25	7.00–8.25
Far from mines, accessible by rail	7.50–8.75	8.50–9.50

Note: *These data are computed for hypothetical, modern 100,000 kw power stations operating at a 50 per cent plant factor. For full particulars refer to the source: S. Schurr, 'Economic Aspects of Atomic Energy as a Source of Power', *Bulletin of the Atomic Scientists*, April–May, 1947, table I.

other localities atomic power plants may be called into play to supplement existing stations, to ease the strain on the power system at peak load periods. On the other hand, atomic power plants may almost wholly supplant those driven by coal and oil. But it does seem rather unlikely that in the foreseeable future nuclear power plants will prove to be more economical than plants utilising excellent sources of hydroelectric power, whose generation costs run below 2 mills per kw hr.[19] The net of power stations of the future in all probability will be one of variety, dependent to a different extent upon each type of fuel.

Despite this clouded picture, certain important relations can be discerned. Although the cost estimates may prove to be considerably in error, significant differentials in power costs will exist, not only as between stations based upon different fuels, but also between atomic power plants themselves. The values (roughly indicative, although somewhat in excess of cost) assigned by the California group to power produced by the three sizes of atomic power plants considered (500,000 kw, 100,000 kw, and 200,000 kw) were 4, 7½ and 11 mills per kw hr, respectively. It is clear that the differential in cost between the latter two nuclear power plants (both secondary reactors) is primarily attributable to scale. It would exist no matter on what basis we were to allocate cost to plutonium and power from the primary pile.[20] It is conceivable that an international and/or national atomic authority might, for diverse and unpredictable reasons, assign lower costs to plutonium and higher costs to primary pile power, so that, for example, the differential between the primary pile power and power from a 100,000 kw secondary pile would be considerably reduced or even eliminated. None the less, significant differentials of cost would exist within the atomic power system.

If we accept the scheme of the Lilienthal report, differentials in nuclear power costs may be anticipated for a second reason. Operation of the primary pile plant represents a classical case of joint products obtained in fixed proportions, i.e. the amount of power potentially available and the amount of plutonium resulting from the fission process stand in constant ratio. Furthermore, we have this technical relation: the amount of power potentially available in an initial fission process at a primary reactor will be approximately equal to the amount of power that can be produced at secondary reactors from the plutonium resulting from the initial fission process. Thus, about half of the potential power is derivable at primary reactors and half at secondary reactors.[21] Since, as has already been demonstrated, the primary pile should operate on a very large scale to gain the advantages of spreading over a broad base the fixed charges on its enormous investment in plant and equipment, it appears very reasonable to expect that lower rates would be levied for power from primary piles than for power from the average run of large secondary reactors, in order to stimulate the consumption of the tremendous power potential[22] of primary piles. From an overall economic standpoint, it is wise to avoid any waste of power whether available at secondary or primary reactors.

II

We may now inquire into some of the broader economic implications of atomic power. First, how significantly would a widespread adoption of atomic power techniques for generating electricity affect real output? Some indication is given by statistics on the aggregate costs of power and fuels to industry. For distributive and extractive industries, data are not available, but we do know that for the former the costs of fuel and power are negligible. For manufacturing industries as a whole, the Census of Manufactures estimates that for the United States cost of fuel and purchased energy accounted for 2.32 per cent of the value of product in 1939, having ranged during the preceding forty years from 1.87 per cent in 1899 to 3.23 per cent in 1921.[23] In terms of value added, the respective figures are 5.32 per cent, 4.43 per cent, and 7.78 per cent. Individual industries show great variation. In terms of value of product, those above the average for 1939 are: stone, clay and glass products (8.08); iron and steel and their products, except machinery (5.92); paper and allied products (4.18); products of petroleum and coal (2.84); chemical and allied products (2.64); and textile-mill products and other fibre manufactures (2.47). For tobacco manufactures, the lowest of the thirteen industry groups below the average, the relevant percentage is 0.17.

Clearly, since all costs may ultimately be expressed in terms of human effort and the services of natural resources, if costs of fuel and purchased energy were on the average to be cut 25 or 50 per cent by the generation of electricity at atomic power stations, the direct effect upon national output through labour freed for other pursuits, as well as the increased possibilities for leisure, would be noticeable but nothing dramatic. The direct gain could at most be expressed in terms of a few per cent.[24] More important would be the direct effect upon specific industries and the indirect effects upon national output from growth of such industries and from the overall substitution of power for labour. Much would depend upon the elasticity of this substitution, of which we know little. Historically, however, we have had a continuous succession of reductions in the generation costs of electricity. Direct operating costs per kilowatt hour have already been halved several times since 1900. It also appears from available information that fixed charges have likewise fallen significantly, although not as sharply.[25] If eventually atomic power plants are able to reduce electric generating costs by 50 per cent on the average, say from 7–8 mills to 3.5–4 mills per kw hr, there are no a priori grounds to expect, for the United States and other industrialised and commercialised nations of the world, anything more noteworthy than what has been occuring in the past, namely, an increasing resort to electricity as a source of power as well as an increasing utilisation of power in production,[26] and a consequent gradual rise in national product and welfare. In this sense the injection of atomic energy into an economy will not be revolutionary; it will be just another step forward in the developing electric power technology.

With respect to household consumption of electricity, the impact of atomic power will be far less impressive. For here distribution and, to a much lesser extent, transmission costs play a large role. One authoritative study (1939), attempting to portray a typical situation, arrives at a figure of 5 cents per kw hr as the normal cost of delivering electricity to ordinary residential users of about 40 kw hr per month; 4 cents is allocated to distribution, 1 cent to generation and transmission.[27] Obviously, a reduction of 4 mills in generation costs would not be very significant in this case, particularly in view of the great reduction in charges incident to greater residential consumption of electricity.[28]

A second group of important effects has to do with change in the spatial structure of the economy, world and national. How and to what extent will atomic energy, through its revaluation of natural and human resources, cause shifts of industry and population centres, alterations of trade routes, and changes in trade balances and arrangements;[29] how will the broad locational pattern of regions be affected?

We can immediately arrive at one conclusion. If utilisation of atomic power can induce a general lowering of power costs, it is logical to expect, through normal substitution processes, an increased application of power per laborer and a consequent increase in rates of production per laborer. Inasmuch as there is no reason to presume that consumption of food and drink per capita will be influenced by the utilisation of atomic energy, the locational pull of agricultural resources will be diminished (i.e. the proportion per worker of food and drink consumed to raw materials processed, or to finished product, or to power used, or to transport tonnage, etc. decreases). In general, differentials in agricultural productivity between various regions will be of less significance in determining the broad regional and world pattern of industry and commerce, while differentials in other locational factors, such as labor supply, transport media, proximity to market, power resources, will assume a relatively increasing importance.[30]

The diminished attraction of regions of agricultural surplus follows whether or not the reduced power costs are assumed equal throughout the world. But it is highly unrealistic and runs counter to the teachings of history to expect power costs to be uniform, or to remain so for any sizeable period, globally speaking.[31] On historical and theoretical grounds we may anticipate differentials in the future net of power stations, not only because of the various types of energy resources used in different localities and the world's unequal distribution of these resources, but also because of variations in scale of operations for plants using only one type of fuel (for example, uranium or plutonium). Of what significance could such differentials be?

Broadly speaking, we may divide into several gradually merging categories those industries using large amounts of fuels and power (for obvious reasons we ignore industries whose fuel and power consumption is relatively slight). First are industries strongly oriented to markets (such as ice manufacturing, bakeries) and to raw materials (such as mining of geographically concentrated ores.). Even though their consumption of fuel and power is considerable, other locational factors are so important that differentials in fuel and power costs between various localities are relatively minor matters. Second are industries where materials and markets are the major considerations, but where location is influenced to some extent by relative fuel and power costs – perhaps because of a wide distribution of the sources of materials, of markets, or for other reasons – as the cement, paper and compressed gas industries. Differentials in fuel and power costs could be a determining locational factor for a number of

Table 2.4 Estimates of Alcoa's postwar mill costs per pound of pig aluminum, capacity operations[a]

Cost item	*Massena, NY*	*Badin, NC*	*Average for all Alcoa plants*[b]	
	Cents	*Cents*	*Cents*	*Per cent of total cost*
Power (9 kw. hrs. per pound of metal)	1.62	1.44–1.98	1.73	20.0
Transportation:				
a. Bauxite from Dutch Guiana to Mobile, Ala[c]	0.35	0.35	0.35	4.1
b. Alumina from Mobile to reduction plants	0.85	0.52	0.69	8.0
Other costs	5.37–6.53	5.25–6.40	5.87	67.9
Total costs	8.19–9.35	7.56–9.25	8.64	100.0

Notes: [a] Adapted from US Surplus Property Board, op. cit., Appendices 15a and 15b, and from S. Schurr, op. cit., Table II.

[b] Plants located at Vancouver, Wash., Massena, NY, Niagara Falls, NY, Alcoa, Tenn., and Badin, NC. In calculating the average, when cost estimates of various items lay within a range, the arithmetic mean of the extremes was used.

[c] This cost item is included in the cost of alumina received at electrolytic reduction plants.

cases in this category. Third and fourth, respectively, are industries where fuel and power is the major locational element, but where consideration of raw materials and markets is also important (such as the glass and clay products industries) and those where fuel and power tends to be the overwhelmingly dominant locational force (as coke and carbon black manufacture, aluminum and other electroprocess industries).[32]

It would seem reasonable, a priori, to expect atomic power to have its greatest impact upon industries falling in the fourth category. To throw light upon the nature of this impact, let us examine elements of the cost structure of one of these, the aluminum industry. In Table 2.4 are presented power cost, transport cost of the bauxite ore which enters into the cost of alumina, transport cost of alumina, other costs and total cost per pound of pig aluminum for selected electrolytic reduction plants of the Aluminum Company of America; the average of all these items and their percentage of total cost for all of Alcoa's reduction plants are also given.[33] We first note the large proportion of total cost expended on power, and second that the cost of power per kilowatt hour is less than 2 mills. All alumina reduction plants are power oriented, being located at sources of hydroelectric power. As already observed, it seems unlikely that atomic power in the foreseeable future will be able to compete very successfully with hydro in such industries.[34] Consequently, their cost structure will remain unaffected and likewise their locations.

This statement must be modified to take into account the following facts: (1) that the better hydroelectric sites may be fully exploited by expansion of industries in this fourth category, and that it may be preferable to orient any further growth to nuclear power plants rather than to inferior hydro sites; (2) that revolutionary engineering advances may be achieved in the utilisation of atomic energy; and (3) that government may subsidise atomic power production by allocating the greater part of cost of nuclear power plants to military purposes. On the other hand, governments may continue to subsidise hydroelectric power by means of valley developments, as in the case of the TVA, or for military purposes, as in the case of the Canadian Shipshaw project, from which it is estimated that power will be available for the Canadian company, Aluminum Limited, at almost one-half of one mill per kilowatt hour.[35]

Eminent scientists, however, argue that even though use of atomic energy may not reduce power costs at the best power sources, it will offer great benefits to many regions of inferior power sources. These benefits will especially accrue to undeveloped regions possessing mineral and agricultural resources. However, the weakened position of agricultural resources has already been noted. Moreover, with respect to mineral resources their argument is considerably overdone, for it tends to postulate uniform atomic power costs throughout the various regions of the world. For example, a Cowles Commission report in discussing the implications of atomic energy for the location of the aluminum industry states:

> The transportation of aluminum raw materials which enter into the cost of producing aluminum results from the need to bring raw materials to cheap power. Under present circumstances, it would be even more costly to bring power to raw materials since roughly 10 lbs of coal would be required to produce sufficient power for the reduction of 1 lb. of aluminum, while only 3.4 lbs of bauxite need be shipped to power. The relationship between the weight of raw materials and the weight of fuels required per ton of aluminum could, however, be completely reversed by atomic fuels, one pound of which yields enough energy to produce 250,000 lbs of aluminum. Thus, one pound of atomic fuel could be shipped to bauxite instead of shipping close to 1 million pounds of bauxite to power.[36]

The report goes on to argue that if an integrated aluminum works could be based on cheap power in Dutch Guiana, the costs of transporting bauxite and alumina could be avoided. These costs amount to about 12 per cent of production costs (Table 2.4).[37] Such savings, though necessarily modified by perhaps greater costs of raw materials and services such as labour and marketing, 'illustrate the type of economic factors which may come into play in determining the aluminum industry's structure if cheap power can be made available, wherever needed.'[38]

This whole argument is unrealistic, for it ignores historical and empirical bases for differentials in power costs and the import of the concept of locational orientation. From Table 2.4 we note that, on the average, transport costs of bauxite and alumina per pound of pig aluminum amount to 1.04 cents for all Alcoa plants. It is immediately clear that if there should be a differential in power cost of only 1.2 mills on the average in favor of the sites at which Alcoa plants are located, as against potential Dutch Guiana sites for nuclear power development, and assuming away all other advantages that Alcoa sites might possess *vis-à-vis* Dutch Guiana sites, it will be more economical to produce at Alcoa sites. For a difference of 1.2 mills per kw hr amounts to a difference in power cost of 1.08 cents per pound of pig aluminum (9 kw hrs. being used per pound of metal). This saving would be greater than the saving in transport costs, were all production integrated at the bauxite sites in Dutch Guiana.[39] Furthermore, other considerations also place Dutch Guiana at a disadvantage – such as supplies of necessary skilled labour and of fuel resources, distance from aluminum consumption areas, except with respect to potential South American markets – so that the critical differential in power cost need be less. It should be noted, too, that from an international standpoint Alcoa electrolytic reduction sites themselves are not the best. Pig aluminum production in Canada at Arvida, Quebec, where both bauxite ore is refined and alumina reduced, incurs no transport charge for alumina, so that total transport cost before marketing is not much more than the cost of transporting bauxite to Mobile; accordingly the critical differential in kw hr cost becomes considerably smaller. In this connection we should not ignore the abnormally low cost power, estimated at almost one-half of one mill per kw hr, available for Canadian alumina operations as a result of the war subsidisation of the Shipshaw project. All this demonstrates the folly of any locational technique which pays attention only to absolute levels of cost and ignores differentials between various sites.

Thus we may conclude that the use of atomic energy will probably have little locational effect in the aluminum industry and in other industries in the fourth category, which are so strongly fuel and power oriented that they have been drawn to excellent sites of fuel and hydroelectric power. Its impact will be much greater upon the locational structure of industries falling in the second and third categories, where the pull of fuel and power and the pull of other locational factors are more on a par with one another. Here the general reduction in power cost which may be achieved in many regions and districts through nuclear fission may be effective in upsetting the balance of the locational vectors for many enterprises and in causing relocation and new growth within such regions and districts. Space is lacking for a more specific inquiry into the effects upon such industries, but we do wish to stress certain general locational considerations.

First, the differentials in power charges indicated by the available data as in all probability characterising the future net of nuclear power stations will induce more concentration and less dispersion of industry than has generally been presumed. On economic grounds alone, and in view of historic evidence, thinking in terms of primary and secondary reactors, we might anticipate a much denser agglomeration of industry about primary piles (of course beyond the range of radiation effects) where power charges from an economic standpoint would be considerably below those of secondary reactors, about which in general there would be a relatively light clustering. This would induce a much more integrated economy – one with primary reactors and their associated economic activity tending to be central cores, and a hierarchy of subordinate sites aligned with secondary reactors – undoubtedly requiring a finer articulation of production and planning.

Political and military factors will, of course, alter considerably whatever economic framework may tend to evolve. It seems unlikely that the future pattern of nuclear power plants will have very much relation to the exploitable and undeveloped mineral resources of the world. It is more logical to expect that the highly industrialised nations possessing the requisite capital, equipment and technical skill will develop and subsidise nuclear power plants within their own borders and presumably close to their industrial and commercial agglomerations[40] – especially huge plants capable of realising great economy in operation – with due modifications consistent with (1) the desirability of a geographic dispersal of plants and stocks of materials, should an international atomic authority be established, (2) solutions to the possible problems of transporting fissionable and radioactive materials and of disposing of radioactive waste materials, (3) procedures for national defence and offense,[41] and (4) other geographic limitations such as the possible necessity of having sufficient quantities of water accessible. The net effect of the total complex of forces may conceivably be in favor of greater concentration (if it is possible to measure rigorously such a phenomenon) and less scatter within the world economy.[42]

Second, any general locational analysis should take account of technical inflexibilities in production. For iron and steel output, coal has always served as raw material as well as fuel. In recent decades technical advance in the use of electricity in the iron and steel trades has been proceeding, and we find smelting of iron and production of special iron and steel alloys in electric furnaces increasing in importance in the world economy. None the less, numerous technical obstacles must be overcome before a wholesale substitution of cheap electricity from nuclear power plants for coke and coal can take place. Technology in glass manufacture, another heavy basic industry, is geared for the most part to the use of manufactured gas or natural gas. Methods designed to utilise cheap electricity are relatively expensive. And so with other processes. The need to develop new techniques to exploit cheap electricity as a fuel will be a deterrent for some time, at least, to its extensive generation from nuclear power plants.

Third, in general location analysis attention should be paid to vested interests, traditional patterns, and all other cultural and institutional elements which offer resistance to change, the more so when we are confronted by an innovation which introduces a potentially ubiquitous or mobile resource. The United States is a good example of a mature and highly industrialised economy. Here our basic industries dominated by iron and steel have concentrated about sources of fuel in the Appalachian coal fields and about the Great Lakes waterway system. Simultaneously a net of superb transport facilities has developed in these areas to serve industry. Agglomerations of population have arisen containing the labouring force of the dominant industries and of the ganglion of secondary and tertiary trades and services ancilliary to and required to service the overall population and industry. Today these highly industrialised areas possess locational advantages with respect to transport facilities, skilled labour resources, consumption and industrial markets, capital and financial resources and other factors which tend to attract new industrial growth not diverted elsewhere by strategic resources. It seems logical to expect that nuclear power plants, since they are dependent upon a mobile fuel, will be drawn to agglomerated districts enjoying such decisive locational advantages for industrial and commercial development, rather than to undeveloped areas, except in so far as the latter possess excellent sources of raw material, rare skills, potential markets and the like. The basic spatial pattern of mature and developed economies should not be very much altered by the application of atomic power.

Undeveloped regions of the world offer other types of resistances to change – cultural, political and economic. Usually they lack capital resources, skilled labour, and equipment; their transport resources or facilities or both are niggardly; their mineral and fuel resources are unexploited; their populations are often unaggressive, docile, unskilled, uneducated and live under unhygienic conditions not conducive to a high level of productivity; their political systems are frequently unstable or oppressive; their economies are unbalanced and unintegrated and composed of independent sectors or states with conflicting interests; their social customs, sentiments and emotions are frequently incompatible with industrial advance. We do not wish to imply that atomic power can not have any beneficial influence upon such regions or foster industrial development in them, particularly where there are superior mineral deposits. But we do feel that these opportunities for growth can easily be exaggerated, particularly if one concentrates upon a single factor and forgets the complex of other retarding forces.[43] Only in isolated instances have inferior fuel and power resources been the single major obstacle to development.[44]

Finally, we wish to underscore certain dynamic forces which will condition the impact of atomic energy. One is population growth. In general, a growing population implies a mobile labour supply; a stable or declining population, an immobile labour supply. A mobile labour supply usually permits a fuller and more efficient exploitation of an innovation, especially one which involves a revaluation of mineral and water resources. The United States benefited greatly from the extreme mobility of her labouring force stemming from immigration and a high reproduction rate; for it facilitated her exploitation of new resources and new technology and minimised the frictional elements in the consequent shifts and rearrangements in her internal economy. Today among the major aggressive and industrialised nations the Soviet Union promises to have the highest rate of population growth. Population projections to 1970 with 1940 as base assign an absolute decline of somewhat over 4 million to France, a fall of somewhat below 4 million to England and Wales, an insignificant increase for Germany, a rise of about 20 million for the United States and a gain of 77 million for USSR. Percentagewise, with 1940 as base, the changes are roughly: – 10, – 9, – 0.4, + 15, and + 44, respectively.[45] On grounds of labour mobility alone, the Soviet Union will probably be in a comparatively excellent position to profit from atomic energy innovations.

Further, in contrast to the advantages enumerated above, which an established locational structure comes to possess, a location analyst frequently observes the indifference to, or lag in, the adoption of new and superior production techniques which a vested and mature economy manifests. The sharp decline of the iron and steel trades in Britain relative to the rest of the world during the twentieth century was to no insignificant degree due to the inertia, complacency and passiveness of British entrepreneurship, which was content to capitalise its advantageous initial entry into the field and failed to reap the profits of a dynamic technology – this despite the fact that the British scientists and engineers were foremost in the development of iron and steel technology.[46] Again, the precipitous drop in New England textiles after 1920, which came instead of the less pronounced drop to be expected from a changed locational strategy, can be attributed to the refusal by New England manufacturers to reorient their thinking to recognise the merits of new processes and to install modern equipment.[47] On general historical grounds one may infer that aggressive and growing economies tend to exploit new technology more comprehensively than other aggressive but older and more mature economies. In this sense the Soviet Union, despite an initial handicap in scientific and engineering knowledge, may be foremost in atomic energy applications.

In conclusion, it appears unlikely that atomic energy will have a revolutionary impact upon world and national economies. We can anticipate, within broad limits, competition with coal, oil and gas as fuel in power stations, but probably not with water power at superior hydroelectric sites. The effects upon real output and industries in general should be noticeable but nothing extraordinary, although they may be of considerable consequence for certain specific industries and trades. Spatially the structure of the world economy should experience mild alterations, especially in favor of aggressive and growing economies. However, a more equitable distribution of industrial and commercial activities between backward and highly developed economies does not appear in sight, unless international authorities intervene, though instances of feverish growth of undeveloped regions may of course crop up. Similarly for various districts within any given economy, though to a lesser extent, because within heavily agglomerated areas the impetus toward decentralisation may be strengthened. Whatever rearrangements occur will probably be in those industries where the pull of fuel and power sources is on a level with the pull of other locational factors, rather than in those industries where the overwhelming dominance of fuels and power has oriented plants to superior fuel and power localities.

These tentative conclusions are based upon the presumption that the nations of the world will avert an atomic bomb race and the catastrophic destruction which would inevitably ensue. The problems of evolving an effective international atomic authority and of establishing a federation of nations of high moral integrity remain paramount in our efforts to achieve a more abundant world.

Notes

1. The writer is indebted to Professors Abbott P. Usher and Alvin H. Hansen for many fruitful suggestions, and to the Social Science Research Council for a fellowship which has made possible the carrying through of this project.
2. The reader is referred to the official report, *Atomic Energy for Military Purposes* by Henry D. Smyth, Princeton, 1945, for basic information regarding the release of atomic energy and for an understanding of the technical terms used in this article.
3. For detailed descriptions of such applications refer to 'The International Control of Atomic Energy: Scientific Information Transmitted to the United Nations Atomic Energy Commission, 14 June 1946–14 October 1946', *The United States and the United Nations Report Series*, no. 5, Washington 1947, vol. I, chs 4–6; vols. II, V.
4. For some interesting and relevant remarks see 'Atomic Power for Airplanes and Rockets', *Atomic Information*, 20 March 1947.
5. See, for example, H. E. Wimperis, *World Power and Atomic Energy*, London, 1946, pp. 58–9, where the potential effects of irrigation pumps driven by cheap atomic power are speculated upon for the Australian economy.

The recent announcement by the Los Alamos Scientific Laboratory of the successful operation of an atomic power plant using for the first time fast, instead of slow, neutrons for liberating nuclear energy from plutonium at a rate that is easily controlled represents a major step forward toward the realisation of the various possible applications mentioned. By eliminating the need of diluting (moderating) material, the bulkiness and size of the power plant can be considerably reduced. High power operation, however, is yet to be achieved. For more details see *New York Times*, 30 August 1947, pp. 1–2.

6. From present operations in the United States, the gaseous diffusion process seems to be the preferred isotope separation method. For description and explanation of these methods and those cited below, refer to Smyth, op. cit., chs VII–IX, and to 'The International Control of Atomic Energy', op. cit., vol. I, ch. 2.
7. A host of other intricate questions for various reasons remain unanswered. Many center on 'the question of whether or not enough additional fuel can be made in a power reactor to replace the original supply of nuclear fuel being consumed. If not, the world supply of nuclear fuels is measured by the amount of U-235 present in nature, extended a few-fold by such additional quantities of Pu-239 or U-233 as are generated in the consumption of U-235. On the other hand, if the regeneration of nuclear fuels can fully replace the original materials, then all of the U-238, one hundred and forty times as plentiful as U-235, and also the world's supply of thorium, which is more plentiful than uranium, constitute potential nuclear fuels.' United Nations Atomic Energy Commission, Official Records, Lake Success, NY, 1946, Part IV, 'First Report on the Scientific and Technical Aspects of the Problem of Control,' p. 27.
8. Refer to the tentative remarks and proposals for safeguards of Committee 2 of the United Nations Atomic Energy Commission in six working papers, condensed, Bulletin of the Atomic Scientists, and in Official Records, op. cit., Part V, 'First Report on Safeguards Required to Ensure the Use of Atomic Energy Only for Peaceful Purposes.' Also see *Scientific Information Transmitted to the United Nations Atomic Energy Commission*, vol. VII, 'Expert Testimony on Technological Control of Atomic Energy Activities', prepared in the office of Mr Bernard M. Baruch, United States Representative, New York, 15 December 1946.
9. This seems to be characteristic of the United States' position. For example, in *A Report on the Intenational Control of Atomic Energy*, US Department of State, 1946 (the Lilienthal Report), we find this typical statement: 'The primary production plants necessary to produce the materials required to construct *safe* power plants will in that process of production produce large quantities of power as a by-product' (p. 29). The Lilienthal Report is, of course, based on the dichotomy of 'safe' and 'dangerous' activities. Secondary reactors employing denatured U-235 or plutonium and, under the envisaged international scheme, not producing further nuclear fuel, are deemed 'safe', with certain reservations, for national operation and management with international inspection. As a matter of course, then, primary reactors (dangerous, thus calling for international management) producing plutonium as well as power would seem to be preferable to isotope separation plants (dangerous) which yield only enriched U-235, *ceteris paribus*. But the *ceteris paribus* clause contains so many important relationships, many of which are based upon classified information, that this statement becomes very tenuous.

   The scheme of the Lilienthal Report is followed in the actual experiments in atomic engineering at the various United States laboratories, which do concern the problems of generating power from primary as well as secondary reactors. For a description of these experimental activities refer to the United States Atomic Energy Commission report to Congress, 22 July 1947.

It should be noted that at present the effectiveness of denaturing is very questionable. This point is emphasised in later statements by United States officials and by Committee 2 of the United Nations Atomic Energy Commission (see the latter's working paper, 'Stockpiling, Production and Distribution of Nuclear Fuels', *Bulletin of the Atomic Scientists*, September, 1947, pp. 260–1).

10. 'The International Control of Atomic Energy', op. cit., vol. IV (Nuclear Power). Some additional comments by Charles A. Thomas appear in his article, 'Non-military Uses of Atomic Energy', *Chemical and Engineering News*, 25 September 1946, pp. 2480–3.
11. Op. cit., pp. 125–6.
12. 'Atomic Energy: Its Future in Power Production', published in *Chemical Engineering*, October, 1946, pp. 125–34. This study was prepared as a project of the Carnegie Endowment Committee on Atomic Energy under the direction of J. B. Condliffe in cooperation with the Northern California Association of Scientists. It also presents in some detail estimates of cost and potential market for fission products and by-product tracers (pp. 130–1 and Table I).
13. Ibid., p. 132.
14. The estimates of the Thomas and California reports are not as divergent as might appear at first glance. The Thomas report proceeds to state that the hypothetical atomic power plant 'is to be compared with a coal power plant which would cost $10,000,000 under the same conditions. The operating cost depends on the price of coal. The price of bituminous coal of 13,500 BTU is about $3.50 per ton at the mine and about $7 per ton delivered to the furnace of a power plant in the eastern United States. The operating cost of such a power plant would be approximately 0.65 cents per kilowatt hour, again on the assumption that the plant would operate at 100 per cent capacity and that the interest charge on the investment would be 3 per cent. Equality of operating costs between coal power plants and nuclear power plants would be reached if the coal cost $10 per ton,' (op. cit., p. 126).

    Roughly, since the capital cost of the atomic power plant exceeds that of the coal power plant by close to $17,000,000 according to the most recent estimate, and since coal handling and storage facilities are not required by the former, the cost of pile units and accessory facilities and land for the 75,000 kilowatt atomic power plant should approach $19,000,000. This is to be compared with the capital cost of the same items of $26,000,000 for the 500,000 kilowatt primary pile plant of the California report. These data seem consistent with each other, in the light of the opinion among atomic scientists that capital costs will be incurred to a large extent independently of scale of plant, i.e. 'not determined so much by the size of the vessels or equipment as by the nature of the control mechanisms required' (The California Report, p. 131). Undoubtedly the kilowatt hour cost figure of the Thomas report would be greatly improved upon for a plant of much larger scale.
15. Ibid., p. 133.
16. Ibid., p. 133.
17. Ibid., p. 132.
18. Moreover, to what extent are basic research expenses to be allocated to atomic power costs? These are not included in the data of Tables 2.1 and 2.2. How will costs of administration, management and inspection by national and interna-

tional authorities enter the picture? Will nations develop atomic energy installations for military purposes as well as to produce power; if so, on what basis will cost computations be made? These and many other considerations greatly complicate cost prognosis.

19. Refer to Appendix 8, 'Estimated Postwar Power Costs, Aluminum Reduction Plants', *Aluminum Plants and Facilities*, Report of the US Surplus Property Board to the Congress, 21 September 1945.

 Apropos of timing, J. R. Oppenheimer ('Functions of the International Agency in Research and Development', *Bulletin of the Atomic Scientists*, July, 1947, p. 174) conservatively estimates a period of five years at the most for a demonstration of usable electrical energy derived from nuclear reactors, a period of two decades at the most for the application of nucler energy to certain specific highly critical problems of energy generation, and between thirty and fifty years for atomic energy to represent a substantial addition to world power resources.

20. Theoretically, in conventional power plants using coal, oil and gas, generating cost per kw hr should fall as scale increases, at least within the range of 10,000 to 100,000 kw plants. Thus we find in Table 2.2 a substantially lower kw hr cost for the 100,000 kw than for the 20,000 kw plant. Empirically, however, this relation has not been clearly demonstrated. The best study available, 'Steam Station Cost Survey', *Electrical World*, 2 December 1939, does not show significant reductions in cost for larger scale plants after the actual costs have been adjusted to comparable plant factor and utilisation factor bases. Other factors, such as differences in age of equipment, type of fuel, cost of fuel, cost of land and site conditions obscure the relationship between kw hr cost and size of plant. However, preliminary results of a more comprehensive study of steam plants being prepared as a doctoral thesis (Harvard) by John B. Lansing indicate that kw hr cost does fall as size of plant increases, after adjustment has been made for other important variables. Also see in this connection US Federal Power Commission, *Electric Utility Cost Units and Ratios*, Washington, DC, Part I.

 In the case of plutonium reactors, on the basis of the tentative data of Table 2.2, the proportionate increase in capital costs as we proceed to larger scale plants would seem to be smaller than for coal plants. For this reason we might anticipate a more sharp decline in kw hr cost for secondary reactors than for coal power plants as scale increases.

21. The Lilienthal report, op. cit., pp. 28–9.

22. The California report states that since 'industrial regions are already equipped with large capacities for power generation, only very large population centres such as New York, Chicago, London, Paris, Leningrad, Moscow, Shanghai, Bombay, Calcutta, etc. would accommodate 200,000 kw to 1,000,000 kw all at once' (op. cit., p. 131).

23. US Bureau of the Census, *Sixteenth Census of the United States*, 'Manufactures, 1939', vol. I, Washington, DC, 1940. Also see US National Resources Planning Board, *Industrial Location and National Resources*, Washington, DC, 1942, pp. 158–60.

24. Alternative computations, arriving at the same conclusion, have been carried out by J. Marschak, 'The Economic Aspects of Atomic Power', *Bulletin of the Atomic Scientists*, 1 September 1946, pp. 8–9; and by S. H. Schurr, op. cit., p. 119.

25. Some excellent historical material is presented in J. Bauer and N. Gold, *The Electric Power Industry*, New York, 1939, chs I–V. Satisfactory figures for

overall generating costs of power for different years are not available. However, indications of fuel economy for coal stations are given by the authors in the following table in terms of average equivalent 14,000 BTU coal consumption in pounds per kw hr (p. 22):

Year	*All US plants*	*Realisable by best US Plants*	*Attainable by plant with all modern units*
1900	7.00	6.20	5.20
1910	4.75	2.70	2.10
1920	3.00	1.50	1.20
1930	1.62	1.10	0.85
1936	1.44	0.98	0.80

In the following table is summarised certain relevant information presented on the total plant performance of a metropolitan company in New York City (p. 23):

Year	*Average price of coal*	*Average weekly power-station wages*	*Operating costs* mills per kw. hr.*
1900 (estimated)	$2.70	$16	20
1910	2.65	18	8
1920	8.32	30	15
1930	4.40	33	4
1936	4.82	—	3

Note: *Excludes depreciation, general expenses, taxes and fixed charges.

Despite steeply rising costs of coal and labor, operating costs have decreased considerably during the course of the years. And doubtless if the comparative data for total cost were available, a somewhat similar pattern of reductions would be apparent.

26. Refer to US National Resources Planning Board, op. cit., pp. 157–8.
27. Bauer and Gold, op. cit., p. 75. The authors fully recognise the wide variations in generation, transmission and distribution costs, as well as in the scales of rates charged by various companies in the different geographic regions, and thus the necessity for numerous qualifications to this statement. The average revenue accruing to electric utilities per kw hr sold to residential customers in 1945 was 3.47 cents. Federal Power Commission, *Statistics of Electric Utilities in the United States*, 1945, p. X.
28. 'Hence, if average residential consumption is increased to 100 kw hr per month, the cost increment is slight and the average cost of distribution is reduced to about 2 cents per kw hr; over-all cost, including generation and transmission, would be 3 cents' (Bauer and Gold, op. cit., p. 75).
29. Unfortunately, international trade theory has treated only most superficially the effects upon trade of technological and other similar structural economic changes. In 1911 Alfred Weber ('Die Standortslehre und die Handelspolitik', *Archiv für Sozialwissenschaft und Sozialpolitik*, XXXII, pp. 667–88) bitterly

attacked, although with some confusion, the classical doctrines of international trade for failing to analyse the theoretical considerations which affect the location of industry internationally, and he maintained that problems of international trade could not be adequately handled by the classical theory. Despite Weigmann's attempt to link international trade theory with spatial analysis (*Kritischer Beitrag zur Theorie des auswärtigen Handels*, Jena, 1926) and Ohlin's efforts to incorporate international trade theory into a general localisation theory (*Interregional and International Trade*, Cambridge, Mass., 1933) and Lösch's endeavors at integration (*Die räumliche Ordnung der Wirtschaft*, Jena, 1940), international trade theory, as for example presented in the standard treatises of Haberler and Viner (*The Theory of International Trade*, London, 1936, and *Studies in the Theory of International Trade*, New York, 1937, respectively) does not meet Weber's challenge and does not encompass any generalisations to explain those realistic trade rearrangements which arise out of dynamic technological change. Viner, in fact, seems to justify this shortcoming on the ground that to explain the effects upon trade of technological improvement, such as follow from reduced transport costs, has not 'been historically the particular responsibility of the theory of international trade' (op. cit., p. 468). One is free, of course, to restrict the scope of his theory, but if the theory of international trade is to have relevance for reality it cannot arbitrarily neglect the major changes that are linked with technological development. Where it does so, one frequently obtains sterile and misleading results, such as in an article by S. E. Harris ('New England's Decline in the American Economy', *Harvard Business Review*, Spring, 1947, pp. 348–71), which should be censured no more than other works in this respect. There, to take a specific instance, he attributes the serious decline in New England's textile industry largely to the high levels of wages, *vis-à-vis* the South (ibid., p. 359). This misses the point. Marked differentials in wages for practical purposes have always existed between the South and New England. Rather, the dynamic factor is to be found in the technological development which culminated during the last quarter of the nineteenth century in high speed ring spinning and the Northrop automatic loom. These inventions caused a revaluation of labour resources; they permitted the use of the abundant and cheap unskilled labour supplies of the South and minimised the locational advantages of the established pool of highly skilled labour in New England. The devastating effects of this changed competitive position brought about by technology were wrought upon the New England textiles in the years following the collapse of 1920, despite the fact that wage rate differentials had been narrowing in the previous decades. To understand, then, the changed trading position of New England, and to prescribe alleviating measures, one must heed not merely the relative wage level, as traditional trade theory would have us do, but also underlying technical and locational forces.

It is hoped that studies now being carried on by W. F. Stolper and the author will contribute toward a more dynamic and more realistic trade theory.

30. This conclusion is pregnant with geopolitical consequences. The famed geographer, Sir Halford J. Mackinder, speaking in 1904 before the Royal Geographical Society of the great power (political and military) potentialities of the pivotal area of the world, the closed Heartland of Euro-Asia, stated that the 'spaces within the Russian Empire and Mongolia are so vast, and their potentialities in population, wheat, cotton, fuel and metals so incalculably great, that it is inevitable that a vast economic world, more or less apart, will there develop inaccessible to oceanic commerce' ('The Geographical Pivot of His-

tory', *The Geographical Journal*, vol. XXIII, April, 1904, p. 434). We do not dispute this statement, which world developments have greatly substantiated. However, we do wish to stress the point that developing power technology diminishes the relative strategy of land masses and agricultural resources, per se, and their attraction for industry and population, while it strengthens the relative strategy of regions possessing a high quality of human resources, technical equipment and knowledge, and other advantages other than agricultural. Increasing their attraction for industry and commerce, it thereby permits these regions to exploit their spatially limited land resources more intensively and to maintain a rising density of population. One should not infer that the relative strategy of the Soviet Union will tend to be unfavorably affected by atomic power technology, for in other respects, as we shall see later, the Soviet Union will be an enticing area for a comprehensive atomic power development. For several stimulating essays on geopolitical doctrines, see *Compass of the World*, edited by H. Weigert and V. Stefansson, New York, 1944.

31. The persistence throughout history of differentials and disequilibria conditions among important economic magnitudes is clearly portrayed in 'The Pattern of World Economy', an unpublished essay by Professor A. P. Usher.
32. This empirical classification follows that in US National Resources Planning Board, op. cit., ch. 7, where detailed data on fuel and power consumption of various industries are presented in several different forms.
33. The data presented by the US Surplus Property Board (op. cit.) are used in preference to the much more hypothetical and much less authoritative information available in N. H. Engle, H. E. Gregory, and R. Mosse, Aluminum, Chicago, 1945. Use of the latter source might somewhat temper the force of, but certainly not invalidate, the argument to follow.
34. The California Report estimates a generating cost of 3–4 mills per kw. hr. for a 500,000 kw. plant operating with a 50 per cent plant factor and a 90 per cent utilisation factor. Since a continuous use of electric energy characterises the operations of aluminum reduction plants, the plant factor could be very greatly improved, were such a nuclear station to supply power to electrolytic reduction mills. This would permit a considerable reduction in the cost of generating power.
35. US Surplus Property Board, op. cit., Appendix 14.
36. S. Schurr, op. cit., p. 120. The figure 3.4 in the above quotation should be corrected to read 3.8.
37. Schurr, in computing the cost of transport of bauxite from Dutch Guiana to Mobile, Alabama, included the excessive profits which Alcoa charged to its water carrier operations, these profits ranging from 45 per cent to almost 100 per cent of the estimated cost of rendering the transport services. His figures are 0.55 cents and 6.2 per cent, respectively, as contrasted with those of 0.35 and 4.1 in Table 2.4. To include such excessive profits is to distort the picture and over-emphasise the importance of transport costs. For the case where water carrier operations are integrated with several or all of the stages of pig aluminum production, as in Alcoa, the cost of operation at any stage is the relevant concern when various sites are considered for the location of that stage. For the case where water carrier operations are independent and not integrated with other stages, one can expect the highly competitive character of water-borne ore commerce to keep actual rates for transport services close to costs of operations; in this case only a minor error is committed when carrier profits are omitted.
38. S. Schurr, op. cit., p. 120.

39. Alcoa's plants at Massena and Niagara Falls, NY, incur the largest transport expense of all . . . 1.20 cents per pound of pig aluminum. Here the offsetting differential in power costs will have to be between 1.3 and 1.4 mills per kw hr. If, on the other hand, Mobile were considered as a possible location for electrolytic reduction plants, a disadvantage of less than one mill would put Mobile out of the running.
40. See in this connection an unpublished paper, 'Preliminary Estimate on Location of Atomic Energy Piles for Generation of Electricity and Plutonium', by Frederick B. Wood (University of California at Berkeley).

    The recent and *unanimous* adoption by the members of Committee 2 of the United Nations Atomic Energy Commission of the policy that an international convention rather than an operating international atomic agency should determine principles governing the rate of production and distribution among nations of nuclear fuel, and even, in certain cases, prescribe numerical quotas will set rigid limits to atomic power development. Such a policy would tend to perpetuate the existing balance of economic power, to the benefit of the more mature economies, to the detriment of growing economies and areas more enticing for atomic power development.
41. For a comprehensive survey of possible procedures see Ansley J. Coale, *The Problem of Reducing Vulnerability to Atomic Bombs*, Princeton, 1947.
42. A greater sum total of industries may well be the eventual result within regions presently possessing dense agglomerations of industry and population, but, because of various technical and safety limitations, a more diffused pattern of industry and population of more equal density throughout such regions.
43. An international authority designed to attack the deterrent factors on all fronts might be very desirable, playing a rôle similar to the United States' part in the development of the Tennessee and Columbia Valleys. With nuclear power techniques, many more regions of the world may become eligible for such development, but clearly must await concerted efforts to remove the retarding factors.
44. Note throughout the world the many good sites of hydroelectric power, for instance the Paulo Alfonso Falls of Brazil, which are not inaccessible and yet remain unutilised.

    At times even respected scientists point out undeveloped regions whose industrial growth, they contend, might be greatly stimulated by cheap power from atomic stations, but which upon closer observation we find accessible to good fuel or power resources. For example, at a meeting of the American Economic Association, January, 1947, it was remarked that the country in the vicinity of the Northern Rhodesian copper mines could very much benefit if cheap power were available from nuclear power plants. However, not too distant from these copper deposits and those of the Katanga region of the Belgian Congo are a number of coal fields: the Wankie, Lufua, Losita and Luano in Rhodesia; the Lukugo and Lualabo in Belgian Congo; (E. S. Moore, *Coal*, New York, 1940). Most of these deposits are of inferior grade, but Lord Hailey, speaking of the Wankie deposits, which are of good quality and easily worked, observed that of the estimated contents of 6,000,000,000 tons of coal 'the average annual quantity raised during the period 1904–30 was less than 500,000 tons. . . . The expansion of the Wankie field is handicapped by the fact that apart from the railways, the only consumers are the copper mines in the north' (*An African Survey*, London, 1938, p. 1495). Still more important are the tremendous resources of the Congo River system. It has been estimated to possess one-fourth of the potential water power of the world, (US Geological

Survey, *World Atlas of Commercial Geography*, Part II, p. 33), and some of its tributaries extend to the Katanga copper region, there affording a number of favourable power sites, many within access of Rhodesia. Certainly it is erroneous to attribute Rhodesia's relatively primitive economy to a dearth of fuel and power resources and to contend that atomic power (which probably will not be able to compete with hydro) will afford a major stimulus to the Rhodesian economy.

45. The data for the United States are taken from W. S. Thompson, *Population Problems*, New York, 1942, table 102, for growth under conditions of medium fertility and mortality and no net migration of foreign born. The data for other countries are from F. W. Notestein and others, *The Future Population of Europe and the Soviet Union*, Geneva, 1944, ch. II.
46. T. H. Burnham and G. O. Hoskins, *Iron and Steel in Britain, 1870–1930*, London, 1943, ch. IX and others.
47. T. R. Smith, *The Cotton Textile Industry of Fall River*, Massachusetts, New York, 1944, chs III and IV.

3 Some Locational Factors in the Iron and Steel Industry since the Early Nineteenth Century[1]

Not only many laymen but economists, as well, retain the impression that the location of coal resources is, or at least has been during the last seventy-five years, the dominant factor in determining the locational pattern of iron and steel production. And frequently inferences are drawn that nations and regions lacking coal deposits are not in a position to develop an extensive iron and steel industry and that large regional movements of coal with respect to this industry are necessarily precluded. The purpose of this essay is to demonstrate and explain important historical changes in the location of iron and steel plants in leading producing nations and to establish specifically that the locational pull of coal resources per se has not been dominant for a number of decades.

We commence with the early nineteenth century. At that time, for those processes in which coal had been substituted for charcoal as a fuel, coal deposits were, without question, locationally dominant. Ashton[2] writes of Britain: 'Drawn by a powerful magnet, the iron industry became localised on the coal fields of the Midlands, Yorkshire and Derbyshire, and South Wales'. These districts in 1820 accounted for 90 per cent of the national output.[3] Expressed in terms of Alfred Weber's framework, the weight of the coal required to smelt the ore and process the pig exceeded by a generous margin the sum of the weights of the ore consumed and the finished output, for a given unit of output.[4] Less tonnage had to be transported in bringing the ore to the coal deposits and then shipping the finished product to the market than in carrying the coal to ore deposits, even where the ore deposits happened to be near the market. In Britain, where fortuitously coal and ironstone were found together in many localities, we note little coal or ore movement associated with the iron and steel trades; but such movement as there was consisted of ore and pig and bar iron – for example, of Cumberland ores to Staffordshire, Scotland, and Wales,[5] coastal Yorkshire ores to works in Durham and Northumberland,[6] and Swedish bar iron to Sheffield.[7]

Some data are available on the consumption of coal in the production of pig iron from the middle of the eighteenth century. For that time Bell notes

Reprinted from *Journal of Political Economy*, vol. LVI, no. 3, June (1948), pp. 203–17.

that 'something like 8 to 10 tons of coal were required to make a ton of pig iron'.[8] According to Meade, in 1788 an average quantity of 7 tons of coal was employed per ton of pig iron in England and Wales; in 1798, 6 tons or less; and between 1800 and 1802, 5 tons.[9]

In the late 1820s, Nielson's hot-blast technique was successfully applied in the Clyde ironworks of Scotland. Hence consumption of coal (converted into coke) fell from 8 tons 1¼ cwt per ton of pig during the first six months of 1829 to 2 tons 13¼ cwt for the first six months of 1833. Elsewhere in Britain the savings were not so phenomenal, since the Scottish blend of coal (particularly poor for cold blast) was especially adaptable to hot-blast operations.[10] Jessop arrives at an average figure of 3½ tons of coal per ton of pig by 1840.[11] In 1851 Porter estimated that 4 tons of coal were required to smelt sufficient ore for 1 ton of pig.[12] By 1869 consumption had fallen to 3 tons.[13]

Reliable annual data for Britain became available in 1873. They are recorded for selected years in Table 3.1. By 1873 coal consumption per ton of pig iron had fallen to 2.55 tons, and by 1938 to 1.67 tons.

In the United States coal was not used successfully in the blast furnace until the late 1830s. The already advanced Welsh practices, using anthracite coal as fuel, were adopted right at the start. Therefore, in contrast to British experience, we do not find over later years such tremendous savings in coal in the American iron industries. French,[14] in his cost calculations for 1840, takes as typical a consumption of roughly 3 tons of anthracite coal per ton of pig iron. For 1850 the average figure for coal consumption must have fallen to around 2½ tons, if not less.[15] By 1879, when reliable data became available (presented in Table 3.1), the relevant figure had dropped to 2.10 tons; by 1938 it was 1.27 tons.

In contrast to this marked decline in consumption of coal per ton of pig, there has been a relatively constant consumption of iron ore. To a slight degree, a greater use of scrap, cinder, and scale in the blast furnace has reduced ore requirements. But for certain districts this has been much more than offset by the exhaustion of richer ore deposits, necessitating recourse to leaner ores. In the United States, for the earlier years, from 2½ to 3 tons of iron ore were required per ton of pig. With the discovery and exploitation of the rich Lake Superior deposits, the requirements have fallen to somewhat under 2 tons, save for the Birmingham district, where consumption of local ore hovers at about 2½ tons.[16] Great Britain's consumption of iron ore per ton of pig has differed significantly from region to region on account of the variation in the iron content of local and imported ores. The range has usually been between 2 and 3 tons, with an average close to 2½ tons over the years.[17] The rising consumption of higher-quality foreign ores has roughly balanced the greater use of leaner ores as the better domestic ores have become depleted.

Comparison of the data on iron consumption and coal consumption per

Table 3.1* Consumption of coal per ton of pig iron produced, 1873–1938

Year	Great Britain (tons)	United States (tons)
1873	2.55	—
1879	2.19	2.10
1884	2.06	—
1889	2.01	1.85
1894	2.00	—
1899	2.02	1.72
1904	2.02	1.70
1909	2.04	1.62
1914	2.06	1.57
1919	2.14	1.53
1924	2.01	1.45
1929	1.91	1.31
1934	1.75	1.28
1938	1.67	1.27

Notes: *The figures for Great Britain were calculated from data in the *Home Office Reports on Mines and Quarries* ('Parliamentary Papers' [1894–1920]) and in the annual reports, *Statistics of the Iron and Steel Industries*, of the British Iron and Steel Federation. The figures for the United States were calculated for the years 1879–1914 from data in the volumes on manufactures of the *Tenth, Eleventh, Twelfth, Thirteenth*, and *Fourteenth Census of the United States*; and for later years from data in the *Annual Statistical Report* of the American Iron and Steel Institute (1919–38).

To maintain comparability throughout the period for the United States, coke was converted into its coal equivalent, the percentage yield factors for years prior to 1919 being obtained from the respective census reports, for 1919 and later years from US Bureau of Mines, *Mineral Yearbook* (1930, 1935, 1940). In the computations from the data no distinction was made between anthracite and bituminous coal.

ton of pig iron for British and American merchant blast furnaces – furnaces which produce pig iron for sale – clearly reveals that coal resources lost their dominance[18] as a locational factor well before the middle of the nineteenth century. Similarly, economies in fuel consumption have been as great, if not greater, in the iron industry on the European Continent, where ore consumption per ton of pig has varied between 2 and 3 tons; hence we reach a similar conclusion for that area.[19]

The foregoing analysis is subject to two major qualifications. First, the data presented for coal consumption represent actual coal consumed or the coal-equivalent of the coke consumed. If the beehive process of manufacturing coke is employed, the locational considerations in utilising by-products do not enter the picture (as they do in the case of by-product ovens). Consequently, the alternatives to location at a coal site do not imply movement of coal but only of the lesser weight of coke derived from the necessary coal at beehive ovens adjacent to collieries. The beehive technique was universally employed in Britain and the United States until late in the nineteenth century and did not lose pre-eminence (50 per cent of total coke production) in either country until after The First World War. For the United States, the average coke yield per ton of coal has ranged between 60 and 70 per cent,[20] for Great Britain, about the same during the greater part of the period under consideration.[21] To illustrate: instead of an average of 1.31 tons of coal (per ton of pig) transported to blast furnaces at other than coal locations in the United States in 1929, only 0.91 tons of beehive coke would have had to be moved. Thus, owing to the widespread use of beehive coke, the locational pull of coal sources has been considerably less than may appear from the data in Table 3.1 and preceding paragraphs.[22]

Second, in making a locational decision, an entrepreneur is not so much concerned with minimising the ton-mile movement of raw materials and product as with minimising the transport expenditures incurred by such movements. The locational pull of any commodity, whether raw material or product, is stronger (weaker) the higher (lower) the ton-mile rate for its transport. If the ton-mile rate on ore is double that on coal, the attraction of coal deposits is no greater than that of ore deposits when twice as much coal is consumed. Historically, the pattern of coal, ore, and pig-iron rates has been highly irregular, not only from nation to nation but within various districts of a nation or even for alternative transport routes in any one region.[23] At most, there may have been a tendency for pig-iron rates to exceed either the coal or the ore rate for any given situation, thus enhancing the desirability of a market location.

The discussion thus far has referred to the production of pig iron itself without regard to its further processing. This excludes a very important set of locational considerations associated with the integration of the various stages of iron and steel manufacture. As Ashton observes in commenting upon the application of coal techniques and the widespread use of the steam engine during the late eighteenth century in Britain:

> Now that the whole process could be carried on with cheap mineral fuel there was no longer any reason for that separation of furnace, forge, and mill that had characterized the iron industry in the early years of the century. The typical concern of the new era was integrated in its structure: it mined the ore and coal, smelted, refined, rolled, and slit the iron into its finished form of plates and rods.[24]

Elimination of the shipping of semi-finished products, less dependence upon outside supplies, better articulation of the various production stages, fuel economies,[25] and other factors made integration desirable – both in early years when castings and wrought-iron products were predominant and in later years when steel reigned supreme. The locational strategy of integrated works for the earlier years, and to some extent for the later years, was altered in favour of coal sites, especially for malleable iron and steel products. In the production of a given weight of malleable iron or castings in the earlier years, somewhat more than that weight of pig iron was required, so that the weights of ore and coal for finished products were somewhat higher than for pig iron;[26] and for technical reasons some additional ore, scrap, and other metal may have been used.[27] But these small weight adjustments were clearly overshadowed by the large quantities of fuel consumed in processing. For the diverse finished iron and steel products of different qualities, fuel consumption after blasting operations has varied considerably, so that the construction of a consistent set of figures, even if data were available for series of years, would be an almost impossible task. However, we do have sufficient data to indicate some of the magnitudes involved.

We should distinguish between castings and wrought-iron and steel manufacture. Molding of pig iron into castings is much less prodigal in the consumption of fuel. In 1850, according to the report of the superintendent of the census, an average of 0.59 ton of coal and 7.48 bushels of coke and charcoal (or, altogether, less than 1 additional ton of fuel) were consumed in the production of 1 ton of castings from pig iron. On the other hand, an average of 1.94 tons of coal and 52.19 bushels of coke and charcoal (or more than 3 additional tons of fuel) were required to refine pig iron into 1 ton of wrought iron. Roughly, the fuel requirements of the latter process were four times those of the former.

Other data are illuminating. In 1850 the report of the Convention of Iron Masters estimated that, to make 1 ton of iron rails in Wales, the additional requirement of coal after blast-furnace operations was over 3 net tons.[28] To make a ton of bar iron on the Schuylkill River, Pennsylvania, $2\frac{1}{4}$ tons of coal were used to convert the pig.[29] And for the whole process of manufacture from ore to finished products, the report employs the rough figures of 5 tons of coal per ton of bar iron, hammered iron, hoops, and sheets; 8 tons of coal per ton of steel; and 10 tons of coal per ton of iron and steel manufactures.[30]

Clearly, then, these data indicate that, by the middle of the nineteenth century, the locational supremacy of coal sites was tottering for integrated works producing castings but the malleable iron and steel works had far from freed themselves from bondage to such sites. Of course, the data presented are, for the most part, averages. They do not reflect the wide variation in fuel requirements of individual products or the wide variation in the use of newly developing methods. An entrepreneur entering the industry or considering relocating his plant, by utilising the most recent techniques, would be able to achieve in any given year greater fuel economy than is indicated by the data. For him the locational pull of coal sites would accordingly be less forceful, and to this extent the above conclusions are modified. We should also note that, in 1850, 322,745 tons of iron castings and 278,044 tons of wrought iron were manufactured in the United States. In Great Britain iron castings did not lose ground appreciably to malleable iron and steel products until nearly the middle of the century.

Major evidence that coal sites were losing their dominance in locational decisions affecting pig iron and cruder processing works appears shortly after the middle of the century. In the early fifties John Vaughan commercially discovered and applied ore from the main seam of the great Cleveland deposit in the northeast of England. Owing both to the presence of excellent coking coal in the near-by districts of Durham and Northumberland and to the proximity of ore beds to tide-water, this deposit – despite the lean content of its ore – was to be exploited vigorously in the years to come. Vaughan 'was soon followed by other capitalists, and furnace succeeded furnace until, within a period of four to five years, the lowlying foreshore of the Tees between Middlesbrough and Eston Junction became dotted with smelting works'.[31] In 1860 almost 1½ million tons of Cleveland ironstone were raised, accounting for 18 per cent of the output of British mines; in 1870, 4 million tons of ironstone were raised, or 28 per cent of British ore mined. The pig-iron output of Cleveland and the neighbouring district of Durham, using almost exclusively Cleveland ironstone, was over ½ million tons in 1860, 15 per cent of British output; and over 1½ million tons in 1870, 27 per cent of British output.

Prior to the successful use of Cleveland ironstone, there were only thirty-eight blast furnaces in the north of England, scattered over the two counties of Durham and Northumberland, in close proximity to coal deposits. Local ironstone was used where possible, but recourse was frequently had to Whitby stone, imported from the coast of Yorkshire.[32] With the extensive exploitation of the Cleveland ore, smelting operations in the Durham coal districts increased quite rapidly – from less than 150,000 tons in 1850 to 340,921 tons in 1860, to 676,964 in 1870, and to 750,262 in 1880.[33] Most of the pig output was taken by the heavy coal-consuming wrought-iron trades.

On the other hand, pig-iron production in the Cleveland district based on the importation of south Durham coke, fortunately over a relatively short distance, skyrocketed from an insignificant figure in 1850 to 248,665 tons in 1860, to 916,970 tons in 1870, and to 1,666,156 tons in 1880. A large portion of the Cleveland pig output, in contrast to that of Durham, had always been exported to foreign countries or other British regions; in 1880, 960,581 tons (or more than half its total pig-iron output) was shipped out. Further, the Cleveland district processed less of its domestically consumed pig in the heavy fuel-consuming operations than Durham did.[34] These facts clearly demonstrate that, owing to the accumulating savings in fuel consumption, Durham coal sites lost their hold on pig-iron production itself and those iron manufactures requiring relatively small quantities of fuel. For these trades the combined locational pull of the Cleveland ironstone deposits and markets was able to vanquish the pull of the Durham coal sites.[35]

The European counterpart of this British experience is found in past interrelations of Westphalian coal and Lorraine ore. It was not until Thomas developed the basic process of producing Bessemer steel in the late 1870s that the high-phosphorous Lorraine ores became of major commercial importance. Thereafter, the Continental steelmaking based on these ores forged ahead rapidly and soon became dominant in French, Belgian, and German iron and steel trades. From the start the Lorraine fields were an orientation point for basic pig-iron production. To make 1 ton of pig, about 3 tons of Lorraine ore and 1.1–1.5 tons of coke[36] were required. Thus, where regional differentials in other costs could be neglected and where considerations of fuel economy did not dictate an integrated steelworks – i.e., did not more than counterbalance the savings in transportation costs made possible by carrying on the several stages of manufacture at different localities – it was more economical to produce pig iron at the Lorraine fields and re-work the pig at the Westphalian coal sites.[37] This was certainly true for producers supplying the huge market for steel that existed around the Ruhr coal fields.

To be sure, Lorraine ore did move to Westphalia, and some of the processes more lavish of fuel were carried on in the vicinity of Lorraine, owing to the complex pattern of markets, of alternative coal and ore sources, transport rates, industrial organisation, and differentials in other costs. None the less, the pre-First World War picture showed a marked degree of specialisation among those steel-producing regions. French Lorraine, Luxembourg, and Alsace–Lorraine concentrated on pig iron and cruder steel products. In contrast, the Ruhr Valley was more preoccupied with the later stages of steel manufacture. For instance, in 1913 Luxembourg (within the German customs union) produced 2,548,000 tons of pig iron and only 1,326,000 tons of steel. Of this output, Germany took about 70 per cent – over 1,000,000 tons of pig iron, some 250,000 tons of ingots, and over 600,000 tons (mostly semifinished) of the rolling-mill products, the great bulk going to the Ruhr.[38] It was the same with French Lorraine[39] and Alsace-Lorraine.[40] The Rhineland–Westphalian district as a whole imported 15.0 million tons of iron ore, somewhat less than one-third of which came from the Lorraine fields. The major portion of the rest came from Scandinavia and Spain. The pig-iron output of this district was 8.2 million tons, and that of steel was 10.1 million. The excess of steel over pig-iron production is obviously attributable in major part to the importation of pig iron from Luxembourg and Alsace–Lorraine.[41] Here again the material indicates that, during the pre-First World War era, for the cruder forms of production, especially where it was feasible to divorce the earlier from the later stages, coal deposits per se were no longer overwhelmingly dominant in the locational decision.[42]

Returning to the mid-nineteenth century, there is no question of the dominance of coal sites in wrought-iron and steel production. But only a few years were to transpire before Henry Bessemer proposed converting the crude product of the blast furnace into malleable iron or steel without expending any additional fuel beyond the needs of the blast. His plan was to run the molten pig iron straight from the blast furnace into a converter and there burn out the desired amount of silicon and carbon by a powerful air blast from below. Needless to say, numerous minor technical difficulties beset his path, as well as the major one of dephosphorising iron from phosphoric ores. The latter problem was solved by Thomas' and Gilchrist's development of a basic lining. The work of these men, together with the achievements of the Martin and Siemens brothers in applying the open-hearth regenerative furnace to the manufacture of steel and of Bell in experimenting with blast furnaces and analysing the chemical process taking place within them, rounded out by 1880 the series of major achievements aimed at the fuel economies subsequently realised in steel manufacture.[43]

As already mentioned, series of annual data on fuel consumption in the manufacture of various steel products do not exist; and, even if they did, it would be a very difficult task to interpret them because of the different fuel requirements and changing quality over time of steel products. None the less, it is safe to assume that tremendous savings were realised, although, on the average, the industry is far from having achieved the full fuel economies implicit in the fact that the blast-furnace and coke-oven gases yield a theoretical excess of heat sufficient to carry through the complete processing of the molten pig iron into finished steel[44] either in a Bessemer converter or in an open-hearth furnace. In 1916 Professor Bone presented a documented plan to produce a ton of finished steel sections from Cleveland ironstone and Durham coke in a fully integrated works, with an expenditure of not more than 1.6 tons of coal.[45] Actual British practice for the few firms using the 'molten pig process' fell far short of this ideal, ranging from 2 to 2.75 tons of coal per ton of finished steel sections.[46] On the Continent and in the United States, basic Bessemer works often did run on coke and blast-furnace gases and even had a surplus to sell outside the plant. Continental works were much more integrated and used substantially less fuel than did British plants.[47]

For approximate purposes the reader may conceive of the coal requirements for a ton of steel (or for comparable wrought-iron products in the earlier years before they were displaced by steel) as falling very erratically from an average of 5–8 tons in 1850 to 2 tons or less by the close of the 1930s for processes employing little or no scrap – though in individual cases the variation from these figures might be large. Thus, while fuel requirements for pig-iron production have fallen significantly over the years, they have fallen at a still greater rate for steel, using pig iron and iron ore almost entirely. The gap between the two becomes narrower and narrower over time so that, for later years, in many instances the requirements are nearly the same.

The emancipation of steelworks from coal sites was achieved early in Britain in the exploitation of the Cumberland and the Lancashire ores. Before the development of the Bessemer process these rich hematite ores of 60–65 per cent iron content were shipped extensively to Yorkshire, Staffordshire, and north and south Wales; pig-iron output was negligible. In the late 1950s the rapid growth of ironworks and steelworks based on the Bessemer process began in this region. The ore was low in phosphorous and thus was adaptable to the mass-producing Bessemer technique yielding a reliable acid steel; and the region's coastal locations were especially attractive for export trade. By 1880 Cumberland and Lancashire was the second largest pig-iron-producing region in Great Britain, with rails as the principal steel product. Its exportation of ore had fallen off to a negligible figure; in fact, on balance, the region was importing a small quantity. For the most part the enterprises imported, under a relatively heavy transport charge, south Durham coke for fuel. But a few firms did use the local coal deposits of inferior coking quality.[48]

Partial or complete divorce of steel production from coal sites was becoming more and more apparent on the Continent. For a long time the Belgian iron and steel industry had been partially dependent upon imported coal, especially from Westphalia, even though its major fuel requirements were met by the coal fields adjacent to Liége and Charleroi. Dependent on coal shipped over relatively long distances, the ore sites of Luxembourg, French Lorraine, and Alsace–Lorraine were producing large amounts of crude steel products before The First World War. The tendency, already evident before 1914, for these districts to transform more and more of their pig iron into steel and more and more of their semifinished into finished products became very marked after the Versailles Treaty had, for all practical purposes, cut off the German markets. At Stettin and other coastal locations in Germany, works have developed dependent on both Swedish ore and English coal.[49]

In the United States, too, there have been innumerable instances of estrangement from coal sites. Indeed, the relative decline of the Pittsburgh area in recent decades is to a large extent a manifestation of its gradually weakening locational pull, owing in part to the declining consumption of coal in iron and steel production. Up to the Civil War, practically all blast furnaces using coal or coke were located at coal sites – such as the Lehigh

Valley, the Pittsburgh–Wheeling district, the Mahoning (Youngstown) Valley, and the Lackawanna (Scranton) Valley.

The first major shift away from coal sites was associated with the development of Cleveland, a place admirably situated with respect to the transportation of both Lake Superior ore and Pennsylvania coal. This development was definitely linked with the fuel-saving Bessemer and open-hearth processes. The same was true of the Chicago district. After an incipient growth in the late sixties, blast-furnace operations in steel manufacture expanded at a rapid rate from the eighties on, primarily based on shipments of Connellsville coke and Lake Superior ore, with, of course, the most advanced fuel-saving plant and equipment. A striking illustration of the changing locational strategy is the shift of the Lackawanna Steel Company from Scranton to Buffalo in 1901, which gave a major impetus to the growth of Buffalo as an iron and steel center. By this change of location the company anticipated serving its eastern markets at no higher cost and being in a better position to compete for western markets. The concept of tidewater works found expression in the United States in the late 1880s at Sparrows Point below Baltimore, where foreign ores and domestic inland coal were brought together to serve the coastal and export markets.[50] These and other developments clearly testify to the continually diminishing locational pull of coal resources in iron and steel production.

There remains one other major change in the set of locational vectors. For a long time scrap has been used to supplement ore in the blast-furnace charge and to supplement pig iron in the Bessemer converter and open-hearth furnace, especially in the latter. With the growing integration of the steel industry, the gradual maturation of industrialised nations, and the ever increasing amount of obsolete iron and steel structures and discarded products, one would expect that more and more scrap would become available for consumption by the iron and steel industries. From Table 3.2 we see that the industry has taken advantage of this situation. Relevant data prior to 1909 are not available. But from 1909 to 1938 the percentage of total iron obtained from scrap by the industry in the United States increased from 35.7 to 51.3, reaching a maximum of 55.0 in 1935. The percentage of iron supplied by pig iron and iron ore fell from 64.3 in 1909 to 48.7 in 1938. Of the scrap consumed, somewhat more than half, on the average, has been 'home' scrap (i.e., material originating within the plant using the scrap) and somewhat less than half has been 'purchased'. Similar developments have taken place in other regions.

For the world as a whole this changing relation of scrap to pig iron is in large part reflected in the comparative statistics on pig-iron and steel production. After the 1870s the spread narrowed between pig iron, on the one hand, and steel and wrought iron, on the other. By 1914 the production of steel ingots and castings equaled that of pig iron. Since then, the spread between the two, owing to the continually rising consumption of

Table 3.2* Total iron supplied by type of raw material consumed in steelmaking in the United States, 1909–38 (in percentages)

Year	*Pig iron excluding ferro-alloys*	*Iron and steel scrap*			*Iron ore*
		Home	*Purchased*	*Total*	
1909	62.8	18.4	17.3	35.7	1.5
1914	58.9	20.6	18.6	39.2	1.9
1919	56.9	23.8	17.9	41.7	1.4
1929	52.2	22.5	24.8	47.3	0.5
1935	44.4	27.9	27.1	55.0	0.6
1938	48.1	27.9	23.4	51.3	0.6

Source: US bureau of Mines and Works Progress Administration, *Technology, Employment and Output per Man in Iron Mining* (Report No. E–13, Philadelphia, 1940), p. 244.

scrap, has been steadily widening in favor of steel.[51]

The significance of the increasing resort to scrap as a raw material is revealed by Table 3.3. In the United States, because of the mounting substitution of scrap for pig iron and thus for iron ore as a raw material, iron ore consumed per ton of semifinished iron and steel fell from 1.83 tons in 1900 to 1.04 tons in 1938, having been as low as 0.89 tons in 1933. Expressed in simple terms, the locational pull of iron-ore resources for steel production has, on the average, been cut almost in half since the turn of the century.

The use of scrap implies, too, a diminished consumption of fuel per ton of steel. Less fuel is required to melt the scrap than to produce an equivalent amount of pig iron. This saving, however, is partially offset by a smaller output of blast-furnace and coke-oven gases. The consumption of fuel obviously is not reduced as much as is the consumption of iron ore.[52]

Table 3.3* Ratio of iron ore consumed per ton of semifinished iron and steel in the United States, 1900–38

Year	*Ratio*	*Year*	*Ratio*
1900	1.83	1925	1.24
1905	1.66	1930	1.21
1910	1.69	1933	0.89
1915	1.46	1935	0.96
1920	1.33	1938	1.04

Source: U.S. Bureau of Mines and Works Progress Administration, op. cit., pp. 246–47.

With a reduced consumption of iron ore and coal, the attraction of the market as a production site is strengthened in relative terms. But, in a more positive manner, the mounting consumption of scrap has steadily increased the pull of the market. On the average, for the United States, somewhat less than half the scrap used has been and continues to be purchased. Although the sources of purchased scrap are geographically widespread, in general (though with many qualifications), the present pattern of purchased scrap sources largely resembles the past pattern of iron and steel markets. Location away from the market, where the market has existed for a long period of time, thus necessitates the expense of moving scrap, whereas this cost is not incurred when the industry is located at the market site itself. (To be sure, works situated at a distance from sources of scrap do use considerably less scrap than do works within surplus-scrap areas.) In this sense the locational pull of the market is measurable by the sum of the weights of finished product and purchased scrap, to the extent that the source of the latter is geographically concentrated at the market.

Undoubtedly, the use of scrap has partially accounted for a number of the shifts away from coal sites which we have cited in the steel-producing regions of the European Continent and the United States. The relative decline of Pittsburgh has, in part, been due to the persistent scarcity of scrap. Pittsburgh's productive capacity has always far exceeded its consumptive ability. Consequently, its local market supplies a smaller proportion of scrap in terms of production needs than do other areas. Chicago, on the other hand, has greater accessibility to scrap supplies. As a result, the price (including transport expense) which Chicago pays for scrap is significantly lower than is Pittsburgh's.[53] This partially explains the relatively greater growth of Chicago. The sizable development of steel production in the Detroit area in recent years is another instance where a large steel-consuming market and a large scrap-originating site have coincided.

In many instances the use of scrap has enabled small-scale steel plants to locate at or near market centres. The whole Pacific Coast is a case in point. Although its steel-ingot capacity was still small in 1939, growth had taken place at a fairly rapid rate since 1913. In 1939 the Pacific Coast had no blast furnaces. Pig iron had to be imported. However, the use of a large amount of scrap in the furnace charge (e.g., 85 per cent in 1944) enabled the steel plants producing a limited array of products to minimise the disadvantages of their poor location with respect to the raw materials – ore and coal – and to maintain their competitive position (market) on an economic basis.[54]

Political and military considerations have often motivated iron and steel developments, for example, in Italy and Japan before The Second World War. It is important to recognise that the use of domestic scrap minimises the economic penalties involved and the amount of direct and indirect subsidisation required. Japan's pre-war industry depended heavily upon imports of iron ore, coking coal, pig iron, and scrap.[55] Nevertheless, taking into consideration (1) that Japan herself was by far the major market for Far Eastern steel production, (2) that Japan's domestic supplies of non-coking coal could be used for power and heating purposes and could be mixed with northern China coal in a ratio much greater than 1:1 to produce technically satisfactory coke, and (3) that Japan was the major originating source of purchased scrap in the Far East, it was logical for Japanese sites to be the points at which to nurture Far Eastern steel production and, to a lesser extent, pig-iron production.

That coal sites per se have lost their dominance in the location of the iron and steel trades cannot be questioned. That the locational pull of ore resources relative to coal has increased continually is also an inescapable fact. That, on the average, the attraction of the market per se has been mounting relative to that of both ore and coal is also fairly clear. One wonders, then, why so much iron and steel activity remains within or near coal districts.

The main reason is that many coal districts and markets for steel have coincided. Since for some time no one of the three major locational factors has been dominant, the coincidence of any two obviously can be sufficient to determine the site of production. Coal sites have been primary market sites on several accounts. When coal was dominant in the early years of the iron and steel industry, other heavy-industry and processing trades oriented to iron and steel converged on the coal sites. Moreover, for many iron- and steel-consuming industries, coal sites have been, and in many cases remain, dominant. Historically, then, large markets have grown up in coal districts. In the absence of any new locational force, there has been, and even still is, little incentive for iron and steel industry originally oriented to coal to shift elsewhere before the coal deposits are depleted. Furthermore, there have always been and still are strong resistances to change. This can be partially explained in terms of the huge investment required and the rigidity of plant and equipment in the iron and steel industry. When a location is once established, management is naturally reluctant to relocate when large initial capital expenditures are involved, especially if the existing plant and equipment are not fully depreciated or if some obsolete plant can be used for peak loads. This same attitude toward novelty is encountered in the adoption of new techniques – where piecemeal modifications have often been used as alternatives to the wholesale scrapping of plant so often implicit in technological developments of the past.

Consequently, old locations at old coal deposits tend to persist, even when the economic basis for steel production has been undermined. Britain, of course, has afforded the classic example. During the first years of the twentieth century the excellence of location at the cheap Lincolnshire and Northamptonshire (eastern Midlands) ore sites was pointed out to British iron and steel masters. Two decades later, around 3 per cent of

British output originated in the former district, and three decades later the natural center at Northamptonshire had yet to produce steel from integrated works. In adopting new processes and organisational techniques, British entrepreneurship was overly cautious and reactionary.[56] As a consequence, throughout the twentieth century a good proportion of the British iron and steel trades has been locationally obsolete.

Finally, it is not expedient to establish at a market or an ore site a fully integrated steelworks with a capacity large enough to realise most of the economies of scale unless the new or old markets which these works are designed to serve are of sufficient size to absorb the output.[57] Inadequacy of consumptive capacity is, in fact, the limiting factor to the construction of steel capacity in a number of newly developing areas. None the less, as markets in backward and semi-developed regions do grow in terms of both effective demand and scrap supply, there is every reason to anticipate a change in the world pattern of iron and steel production in their favour especially if they have fairly easy access to ores. Lack of coal per se need not be a deterrent. The material does indicate that the locational pull of coal has fallen so considerably over the last century that interregional movement of coal has, in fact, taken place and that such movement can now take place on a much larger scale on a sound economic basis.

Notes

1. The author is indebted to Professor A. P. Usher, of Harvard University, for fruitful suggestions and to the Social Science Research Council for a fellowship which enabled him to collect the material.
2. T. S. Ashton, *Iron and Steel in the Industrial Revolution* (Manchester, England, 1924), p. 100.
3. The same districts produced 59 per cent of the national iron output in 1720 (according to a defective estimate) and 77 per cent in 1788. Dispersion and separation of the various stages of iron manufacture were necessarily characteristic before the adoption of coal techniques; forest resources for charcoal manufacture and water power were usually severely limited at any one site. The achievements of Darby, Watt, Cort, and others in the use of the new mineral fuel permitted the concentration of iron works within a narrow area (Ashton, op. cit., chs i–iv).
4. The location of coal or fuel resources, of iron-ore deposits, and of markets have been the three major regional factors in determining the geographic pattern of iron and steel production. In the earlier production stages, labour costs were of relatively minor importance, usually well below 10 per cent; in the later stages they have constituted a more substantial proportion, amounting cumulatively to as much as 20–30 per cent for the whole production process from raw material to finished steel product (for some concrete data on this point see T. H. Burnham and G. O. Hoskins, *Iron and Steel in Britain, 1870–1930* (London, 1943), pp. 150–3, 194–9, 332–5). However, it is not the absolute magnitude of labor costs that is important, locationally speaking; rather the vital considera-

tion is the comparison of differentials in labour costs among different districts with differentials in other costs. And differentials in labour costs have been small, historically, in contrast with the differentials in costs of raw materials and fuel at different localities arising from transport expense. Also see A. Predöhl, 'Die örtliche Verteilung der amerikanischen Eisen- und Stahl-Industrie', *Weltwirtschaftliches Archiv*, XXVII (April, 1928), pp. 248–50.

Other factors, such as availability of limestone and of adequate quantities of cooling water, have been of little moment in the general process of regional selection, although they have often been the determining considerations in the choice of a specific site at which to produce within a given region or district. For reasons which will be discussed later, we take the locus of purchased scrap to coincide with the locus of past markets.

5. J. H. Clapham, *An Economic History of Modern Britain: Free Trade and Steel* (Cambridge, 1932), pp. 48–9; also Ashton, op. cit., p. 49.
6. Lowthian Bell, *The Progress of the Iron Trade of Cleveland* (Middlesbrough, 1878), pp. 3–6.
7. Ashton, op. cit., pp. 54–6.
8. Op. cit., p. 4.
9. R. Meade, *The Coal and Iron Industries of the United Kingdom* (London, 1882), pp. 813, 830.
10. H. Scrivenor, *History of the Iron Trade* (London, 1854), pp. 256–60.
11. The data by regions ranged from 2 tons 16 cwt for southern Wales to 5 tons 9 cwt for Yorkshire, this discrepancy reflecting to a considerable extent the lag in adoption of hot-blast techniques (Meade, op. cit., pp. 835–6).
12. G. R. Porter, *The Progress of the Nation* (London, 1851), p. 280.
13. Meade, op. cit., p. 813.
14. B. F. French, *History of the Rise and Progress of the Iron Trade of the United States* (New York, 1858), p. 63.
15. *Documents Relating to the Manufacture of Iron in Pennsylvania* (published on behalf of the Convention of Iron Masters (Philadelphia, 1850)), pp. 89, 96.
16. For over-all averages for individual years refer to the census reports for each decade and to the annual statistical reports of the American Iron and Steel Institute.
17. For over-all averages for individual years refer to the annual statistical reports of the British Iron and Steel Federation.
18. Dominance can be rigorously defined in a locational sense. A raw material of limited geographic occurrence is dominant in a transport-oriented production process when its weight exceeds the sum of weights of all other materials that have to be transported plus the weight of the finished product, with due modifications for varying transport rates on raw materials and product.
19. For instance, in 1836 in France, where coal and coke were used in the proportion of 9 to 10, about 3 quintals of fuel were required to produce 1 quintal of iron or approximately 3.5 coal-equivalent weight units for 1 of product (Scrivenor, op. cit., p. 187). In 1930 the coal equivalent used per ton of pig was 1.56 tons (F. Benham, *The Iron and Steel Industry of Germany, France, Belgium, Luxembourg, and the Saar*, 'London and Cambridge Economic Service', Special Memorandum, no. 39, October, 1934, pp. 29–31).
20. US Bureau of Mines, *Mineral Yearbook*, annual-issues.
21. Burnham and Hoskins, op. cit., pp. 128–9.
22. On the other hand, some of the decrease in coal consumption per ton of pig iron that has occurred during the past century can be attributed to the greater recovery of coke from coal as a result of improved coking practices. For Great

Britain, coke yields rose from 60 per cent in 1910 to 70 per cent in 1930; for the United States, from 60 per cent in 1880 to 69 per cent in 1930; for Germany, from 70 per cent in 1890 to 79 per cent in 1925; for France, from 60 per cent in 1880 to 77 per cent in 1925 (ibid., p. 129). In terms of coke, then, consumption per ton of pig has declined by a somewhat smaller amount than in terms of coal equivalent for which the data have been presented.

23. For examples during the 1880s, see ibid., pp. 335–6; and reports of the Commissioners, British Iron Trade Association, *American Industrial Conditions and Competition*, ed. by J. S. Jeans (London, 1902), pp. 86–112, 362–9.
24. Ashton, op. cit., p. 100.
25. For example, the report of the US Commissioner of Corporations (*The Steel Industry*, Part III: *Cost of Production*, (Washington, 1913)) states that, for the period 1902–6, Bessemer steel plants running chiefly on hot metal (molten pig iron) seldom used 'less than 20 per cent cold metal, and seldom required less than 50 pounds of coke per [gross] ton of ingots. Plants running on cold metal entirely and using ordinary beehive coke averaged about 245 pounds of coke per ton of ingots' (p. 141).
26. Scrivenor (op. cit., p. 253), noting in 1840 the great savings in the use of pig iron per ton of iron bar resulting from Cort's innovation, states: 'At the present time the waste does not generally exceed from 6 to 7 cwt of pigs to a ton of bars, including the waste in the refinery.' In this connection also see French, op. cit., p. 63.
27. French, op. cit., pp. 136–7.
28. *Documents Relating to the Manufacture of Iron in Pennsylvania*, pp. 41–4. The breakdown was as follows: 10 cwt of coke for the refinery, 15 cwt coke for the puddling furnace and 3 cwt for the engine; 12 cwt coal for the furnace in making the top and bottom of the rail and 3 cwt for the engine; 12 cwt coal for the furnace in finishing the rails and 3 cwt for the engine.
29. Ibid., p. 91.
30. Ibid., p. 96.
31. J. S. Jeans, *Notes of Northern Industries* (London, 1878), p. 61.
32. Ibid., pp. 59–60.
33. W. G. Armstrong *et al.*, *Industrial Resources of the Tyne, Wear and Tees* (London, 1864), p. 88; and Meade, op. cit., p. 338.
34. Meade, op. cit., pp. 324 414; and Armstrong, op. cit., pp. 118–19
35. Transportationally speaking, a position on the banks of the Tees was much closer to export and British markets than was the inland location of a Durham coal site.

    The continually growing strength of the Cleveland locations is traceable to the constant experimentation with, and adoption of, improved techniques in blast-furnace operation during the 1950s and 1960s. The blast furnaces originally constructed at Middlesbrough had cubical capacities of from 4000 to 6000 feet and were no higher than 45–50 feet; the blast was heated to 600°–700° F. By 1871, which date more or less marks the end of the outburst of technical development in Cleveland furnaces, the efficacy of a furnace 80 feet in height and of 15,000 feet cubical capacity with a blast of 1000° F. was thoroughly tested and proved, though still larger furnaces were tried out and operated not unsuccessfully. These advances reduced coke consumption from 35 to 24 cwt per ton of pig (Bell, op. cit., pp. 13–14; Jeans, op. cit., pp. 64–6).
36. In view of the fact that the German and Belgian producers, who dominated production in Luxembourg and Alsace-Lorraine, were leaders in developing and practicing the science of fuel economy (*Wärmewirtschaft*) during the last

decades of the nineteenth century and the first of the twentieth, a maximum of 1.5 tons of coke per ton of pig seems reasonable. By 1913 the relevant figure was close to 1.1 tons (A. H. Brooks and M. F. LaCroix, *The Iron and Associated Industries of Lorraine, the Sarre District, Luxemburg and Belgium*, US Geological Survey, Bull. 703, pp. 96–9).

37. These weight relations were partially offset by differences in freight rates on raw materials and product. In 1880 the rate on coke was somewhat higher than that on ore. The differential between the rates widened in later decades. In 1908 the rate for 10 tons of coal and coke from Dortmund to Luxembourg was $19.20, whereas the rate for 10 tons of ore from Luxembourg to Dortmund was $12.72. In general, throughout the period, rates on pig iron were higher than those on coke (D. L. Burn, *The Economic History of Steelmaking*, Cambridge, England, 1940, pp. 155–61; Burnham and Hoskins, op. cit., pp. 335–6; and Brooks and LaCroix, op. cit., pp. 45–50).
38. F. Benham, *The Iron and Steel Industry of Germany, France, Belgium, Luxembourg, and the Saar* ('London and Cambridge Economic Service', Special Memorandum, no. 39, London, 1934), p. 39.
39. In 1913 French Lorraine produced 3.5 million tons of pig iron and 2.3 million tons of steel, shipping about 1 million tons of pig iron to other parts of France nearer to markets and/or adjacent to coal deposits. Its steel products were of the cruder type. Like Luxembourg, it relied heavily upon Westphalian coke and coal (ibid., p. 28; R. Jordan, *The French Iron and Steel Industry* ('London and Cambridge Economic Service', Special Memorandum, no. 20, London, 1926); and Brooks and LaCroix, op. cit., pp. 37–50).
40. In 1913 Alsace–Lorraine produced 3.9 million tons of pig iron, only 2.3 million tons of steel, and only 0.5 million tons of semifinished and finished products. There were shipped to Germany, chiefly to Westphalia and the Saar, 1.6 million tons of pig iron and substantial quantities of steel and semi-finished products. Coal was imported from the Saar and Westphalia (Brooks and LaCroix, op. cit., pp. 50–59; Benham, op. cit., p. 28).
41. Benham, op. cit., pp. 18–20; W. Greiling, *The German Iron and Steel Industry* ('London and Cambridge Economic Service', Special Memorandum, no. II, London, 1925); and C. M. Pepper and A. M. Thackara, *German Iron and Steel Industry* ('US Department of Commerce and Labor, Special Agent Series', no. 27), pp. 9–16.

    The situation of the Saar resembled that of the Ruhr. The former, possessing coal deposits and importing Lorraine ore, produced 2.1 million tons of steel but only 1.4 million tons of pig iron; this discrepancy in large part was balanced by importation of pig iron from Alsace–Lorraine and Luxembourg (Benham, op. cit., pp. 43–5).
42. Movement of both Lorraine ore and pig iron bears out the point that, outside of the French areas, the Lorraine sites did not have a decisive advantage in supplying pig to other Continental markets and in the export trade of the Continent. In all these instances the Ruhr sites were much closer to the market, transportationally.
43. There have been many other less important achievements, which in the aggregate have been very significant. Noteworthy contributions were made by Bunsen, Playfair, Cowper, Whitwell, and others in the utilisation of blast-furnace gases; by Thwaite, Greiner, Theissen, and others in the evolution of the large gas engine for cleaned blast-furnace gases; by a number of Belgian and German chemists and engineers in the development of the chamber coking oven, with heat and by-product recovery; by Clements, Kinney, and

McDermott in the thermal analysis of the open-hearth furnace; by a host of others in the study of gas reactions, metallography, in mechanical engineering, in the use of the dry blast and heat from molten slag, and so forth (R. A. Hadfield, *Metallurgy*, London, 1925, pp. 211–21; W. A. Bone and G. W. Himus, *Coal: Its Constitution and Uses*, London, 1936, chs xxvi–xxviii).

All these advances over time account in part for the steady reduction in average fuel consumption (see Table 3.1). However, to a large extent this steady reduction reflects the slow response of business organisation to these changes, owing to the conservatism frequently manifested by British and American entrepreneurs, to the continually expanding scope of operations which was generally found expedient, if not necessary, and to the inflexibility and long life of iron and steel plant, which often tempted entrepreneurs to defer adopting new techniques until the old facilities were fully depreciated.

44. Hadfield, op. cit., p. 215; Burnham and Hoskins, op. cit., p. 187.
45. W. A. Bone, *Coal and Its Scientific Uses* (London, 1919), pp. 409–16. The report of the British Association Fuel Economy Committee adopted 1.75 tons of good coking coal as a 'practical ideal' (Hadfield, op. cit., pp. 216–17).
46. Hadfield, op. cit., pp. 218–19. For recent years we have some relevant aggregate data on fuel economy in semifinished and finished processes. For example, in 1920 Great Britain consumed 12.8 million tons of coal in her iron and steel works, excluding blast furnaces; by 1935 this figure had fallen to 7.5 million tons. Meanwhile, production of steel ingots and castings had risen from 9.1 to 9.9 million tons, that of finished steel from 6.8 to 8.5 million tons; on the other hand, forged wrought-iron production had declined from 1.0 to 0.2 million tons and finished wrought-iron products from 0.8 to 0.2 million tons. In net terms there was a significant decrease in coal consumption in post-smelting processes (British Iron and Steel Federation, *Statistics of the Iron and Steel Industries*, 1927; Burn, op. cit., pp. 434–35).
47. Burn, op. cit., pp. 204–11, 364–66, 434–35; and Bone, op. cit., p. 416. Broadly speaking, fuel consumption in Belgium for post-smelting operations was less than one-quarter that in Britain for 1929, being as low as 3.2 cwt per ton of finished product for combined works (Burn, op cit., p. 435). In the United States, according to Burnham and Hoskins (op. cit., p. 187), the average ratio of coal used to a ton of steel product (including smelting operations) diminished from 2.01 in 1904 to 1.83 in 1914 and to 1.41 in 1927. These figures, however, in part reflect fuel savings from the increasing use of scrap, which will be noted below. In this connection also see computations by R. Hartshorne, 'Location Factors in the Iron and Steel Industry', *Economic Geography*, vol. IV, (1928), pp. 242–3.
48. Meade, op. cit., pp. 428–35, 440–59; I. Lowthian Bell, *Manufacture of Iron and Steel* (London, 1884), pp. 454–57, 619.
49. A. Delmar, *The Belgian Iron and Steel Industry* ('London and Cambridge Economic Service', Special Memorandum No. 13 [London, 1925]); Benham, op. cit., pp. 4–11, 28–43; Jordan, op. cit.; and Greiling, op. cit.

It is interesting that, from 1880 until World War I, Britain had a substantially higher ratio of pig-iron to steel production than did Germany and, for the greater part of the same period, a somewhat higher ratio than either France or Belgium. With Britain possessing cheaper and better coal than Belgium and Germany and at a great advantage in coal relative to France, one would expect her to concentrate on steel (the product consuming the greater quantity of coal) more than either of these Continental nations. Many interrelated factors explain why this was not the case. Two major reasons were the special quality and grades of British pig iron (hematite, forge, and foundry) and the tidal location of the Cleveland ironworks, which maintained for Britain a large export market. Even more important was the extreme tardiness of British producers in adopting new techniques of production and organisation, and the resulting inefficiencies in fuel utilisation. This placed Britain at a disadvantage certainly with respect to Belgium and Germany in steel production, and perhaps with respect to France for the cruder and semifinished steel products, despite Britain's superior coal position. As early as 1900, Britain found it expedient to import semifinished and heavy rolled-steel products from the Continent, where costs of production were lower. In pig-iron production, on the other hand, where perhaps the penalty of technological lag was less severe than in steel, until 1913 Britain's cost disadvantage, if any, was not serious (Burnham and Hoskins, op. cit., chs iv and vi).

50. Swank, op. cit.; V. S. Clark, *History of Manufactures in the United States* (New York, 1929), Vols. I and II; and F. Redlich, *History of American Business Leaders* (Ann Arbor, Mich., 1940), esp. pp. 132–41.
51. Burnham and Hoskins, op. cit., pp. 25–7, 272–74.
52. It is not possible to separate statistically the fuel savings from the use of scrap from those resulting from integration and technological advance. As already pointed out, some of the data cited above to illustrate reduced fuel consumption per ton of steel, effected by the latter forces, also reflect some savings arising out of the use of scrap.
53. G. McLaughlin, *Growth of American Manufacturing Areas* (Pittsburgh, 1938), ch. vii; and *Iron Age*, 5 January 1939, pp. 203–12.
54. E. T. Grether, *The Steel and Steel-using Industries of California* (Bureau of Business and Economic Research, University of California, 1946), ch. vi. In this connection also see Predöhl, op. cit., pp. 257–59.
55. During the 1930s Japan proper imported more than four-fifths of her iron supply. Her ore reserves were limited and costly to mine. Foreign ores were less expensive, despite transportation charges. Imported coking coal was supplied chiefly by northern China. Foreign supplies of pig iron came largely from British India, Manchukuo, and Korea, amounting to approximately 30 per cent of Japan's pig-iron consumption. During the period 1931–39, 8.6 million metric tons of iron and steel scrap were purchased domestically, 9.4 million self-generated, 12.8 million imported, and 26.6 million consumed. For full details see The United States Strategic Bombing Survey, *Coals and Metals in Japan's War Economy* (Washington, 1947); M. Erselcuk, 'Iron and Steel Industry in Japan', *Economic Geography*, XXIII (April, 1947), 105–29; and U.S. Tariff Commission, *Iron and Steel* (Rept. No. 128, Second Series, Washington, 1938), pp. 278–300.
56. Burn, op. cit., chs ix and xvi; Burnham and Hoskins, op. cit., ch. ix; annual statistical reports, British Iron and Steel Federation.
57. Predöhl, op. cit., pp. 286–91.

4 Interregional and Regional Input-Output Analysis: a Model of a Space Economy[1]

I

For more than four decades the construction of a general location theory has been a stimulating challenge to spatial theorists. Despite the productive efforts of Weber, Engländer, Predöhl, Ohlin, Palander, Hoover, Lösch, and others, the goals of spinning a theoretical web and specifying the design of a corresponding operational model are still remote.[2] However, the comparatively recent pioneering contributions of Leontief in developing input-output techniques for general equilibrium analysis are of consequence here. They permit an attack upon a specific set of significant problems which logically fall within the jurisdiction of a general theory of location and space-economy.

The analysis of changing interregional economic bonds and flows can be approached from several directions. The input-output method as elaborated here implicitly assumes, for the most part, unchanging spatial relations between any producer and his suppliers. Only a limited amount of industrial relocation can be tolerated with any change in the basic parameters of the economic system at any given point of time. An alternative approach has been sketched for example in an unpublished memorandum by Koopmans.[3] Such an approach would treat the space factor explicitly. It would permit major variation in the spatial extent of production for each firm and in the geographic pattern of production and resource utilisation. It would be consistent with major relocation of industry and population. It would reveal how distance actively conditions interregional economic relations.[4]

An operational model following this alternative, more desirable, approach is yet to be developed. This paper is concerned with the much more limited assignment of regional and interregional analysis within the rigid spatial framework necessitated by the postulates of input-output analysis.[5]

Reprinted from *The Review of Economics and Statistics*, vol. XXXIII, November (1951), pp. 318–28.

II

Geographic flows of commodities may be roughly attributed to two basic sets of facts. One covers the inequalities in the geographic distribution of population, income, and, in its broadest sense, resources. The other embraces the indivisibilities of production and consequent economies of large-scale operation. Lösch has elegantly demonstrated how, in the absence of inequalities, symmetrical hexagonal market areas, nets of market areas, and systems of nets would characterise a plain.[6] If one were to postulate conditions which yield approximately homogeneous market areas for all producers of a given good, and thus approximately identical structures for all regions except with respect to goods moving across regional boundaries, the application of input-output and other analytical techniques would be greatly facilitated. However, inequalities do play a dominant role in modern industrial life. Heterogeneity typifies the market areas of producers of many goods. The Birmingham pig-iron producer, because of his particular spatial relation to resources and consumers, has an entirely different market area than the Pittsburgh pig-iron producer. As a consequence, analytical techniques for interregional and regional analysis should aim at encompassing the full range of both heterogeneous and homogeneous market areas.

The most rigorous procedure possible from this standpoint would be to investigate the spatial relations between each producer and each of his customers. However, the sheer magnitude of the problem – the innumerable relations between individuals which would need consideration – makes this procedure impractical.[7] Some form of aggregation is required. Since our interest lies in geographic flows, aggregation of individuals should take place on a spatial basis, but only partially, to form meaningful regions.[8]

But how demarcate any territory into meaningful and useful regions? Economists have suggested, for example, division into areas within each of which 'exists an especially high degree of interdependence among individual incomes'.[9] Geographers, sociologists, political scientists, and others have advanced diverse criteria.[10] These criteria must be heeded by economists in segmenting the national economy, since extra-economic factors markedly influence the economic structure of any area.

No attempt will be made here to resolve this unsettled question. From an input-output standpoint, two considerations tend to be paramount. First, it is desirable to mark off regions which exhibit self-sufficiency with respect to a maximum number of like goods and services. This obviously reduces the number of goods and services which enter interarea trade, and thus reduces, too, the computational magnitude of the problem. Second, since an operational model of a space-economy rather than a purely abstract model is the goal, we are restricted to geographic divisions for which adequate data on production exist. In the United States, for example, individual states represent the finest geographical divisions for which sufficient production data are publicly available. They must constitute at present the lowest level of spatial aggregation. Census regions would represent a higher level of spatial aggregation.[11]

III

Once a fairly self-contained territory is segmented into meaningful regions, regional input-output models can be designed. Consider a general model which recognises the geographic inequalities that give rise to flows and thus encompasses the full extent of heterogeneity in existing market areas. As a necessary consequence, any given good or service produced in any region must be taken as a unique commodity, distinct from the same good or service produced in any other region. Thus, for example, if states are designated as regions, Pennsylvania brick becomes a commodity different from New York brick or California brick. In terms of a rigorous theory, this is as it should be, since Pennsylvania brick producers have a market area different in shape and extent from that of New York or California producers. In effect, then, the number of industrial categories in a territory is multiplied by the number of regions, if each region engages in each activity; by less, if this latter condition does not obtain.

Given n regions and m goods and services, the allocation of the output ${}_kX_i$ of any industry i in any region k to each industry in each region can be described by the following equation:

$$\begin{aligned} {}_kX_i - {}_{k1}x_{i1} - {}_{k1}x_{i2} - {}_{k1}x_{i3} - \ldots - {}_{k2}x_{i1} \\ - {}_{k2}x_{i2} - \ldots - {}_{kl}x_{ij} - \ldots {}_{kn}x_{im} = {}_kY_i \end{aligned} \tag{4.1}$$

where ${}_kY_i$ equals the final (bill of goods) demand for the output of industry i in region k; and ${}_{kl}x_{ij}$ equals the amount of the output of industry i in region k absorbed by industry j in region l.

More generally,

$$ {}_kX_i - \sum_{l=1}^{l=n} \sum_{j=1}^{j=m} {}_{kl}x_{ij} = {}_kY_i \tag{4.2}$$

$$i = 1, 2 \ldots m$$
$$k = 1, 2 \ldots n$$

Assuming constant production coefficients, i.e., an unchanging amount of each input required per unit output of each commodity, we derive a set of technical input coefficients (the ${}_{kl}a_{ij}$):

$$ {}_{kl}a_{ij} = \frac{{}_{kl}x_{ij}}{{}_{l}X_j} \tag{4.3} $$

$$ i, j = 1, 2 \ldots m $$
$$ k, l = 1, 2 \ldots n $$

where ${}_lX_j$ is the total output of the commodity produced by industry j in region l.

In terms of technical input coefficients and total outputs, the equations of set (4.2) become:

$$ {}_kX_i - \sum_{l=1}^{l=n} \sum_{j=1}^{j=m} {}_{kl}a_{ij}\, {}_lX_j = {}_kY_i \tag{4.4} $$

$$ i = 1, 2 \ldots m $$
$$ k = 1, 2 \ldots n $$

Elsewhere,[12] the solution of equations (4.4) for the required output of each industry of each region in terms of the items ${}_lY_j$ of the bill of goods has been developed. It is:

$$ {}_kX_i = \sum_{l=1}^{l=n} \sum_{j=1}^{j=m} {}_{kl}A_{ij}\, {}_lY_j \tag{4.5} $$

$$ i = 1, 2 \ldots m $$
$$ k = 1, 2 \ldots n $$

where ${}_{kl}A_{ij} = \frac{{}_{kl}D_{ij}}{D}$, D being the determinant of the linear system (4.4), and ${}_{kl}D_{ij}$ being the algebraic complement of the element, $- {}_{kl}a_i$, in this determinant.

Given this general model which can encompass both heterogeneity and homogeneity of market areas, what are its implications for the space-economy? Is it operational? How restrictive and meaningful are its assumptions? These questions can best be answered by first referring to a table of the form of Table 4.1, which can conveniently present the basic data required for the equations of set (4.2).

In order to bring out more pointedly the possibilities of concretely applying the model, we have chosen to construct a table for a definite number of regions and industries in an existing territory rather than for unspecified n regions and m industries in an undesignated territory. Table 4.1 refers to a twenty-industry classification[13] and a three-region division of

Table 4.1 Interregional and regional input-output table

		East	*South*	*West*		*Subtotals*
		1. Agriculture & fishing 2. Food processing . . 9. Chemicals . . . 20. Households & gov't	1. Agriculture & fishing 2. Food processing . . 9. Chemicals . . . 20. Households & gov't	1. Agriculture & fishing 2. Food processing . . 9. Chemicals . . . 20. Households & gov't	Exports (national) Total output	1. Agriculture & fishing 2. Food processing . . 9. Chemicals . . . 20. Households & gov't
East	1. Agriculture & fishing 2. Food processing ⋮ 9. Chemicals ⋮ 20. Households & gov't					
South	1. Agriculture & fishing 2. Food processing ⋮ 9. Chemicals ⋮ 20. Households & gov't					
West	1. Agriculture & fishing 2. Food processing ⋮ 9. Chemicals ⋮ 20. Households & gov't					
	Imports (national) Total Input					
Subtotals	1. Agriculture & fishing 2. Food processing ⋮ 9. Chemicals ⋮ 20. Households & gov't					

the United States. The table consists of (a) major cells, each square and heavily outlined, each containing 20^2 minor cells, each *vertically* and *horizontally* corresponding to a region, or to the subtotals category; (b) a totals row, a totals column, an import row, and an export column.

The question of data availability is discussed in a following section. For the present, assume that data are obtainable on the distribution of national imports and on the distribution of the output of each industry in each region to each industry in each region and to national exports. Or assume the equivalent, namely, that for each industry of each region and for national exports, data are obtainable on the inputs from national imports and on the inputs from each industry of each region. The first set of data would allow us to fill in the table by minor rows (horizontally), one at a time; the second set, by minor columns (vertically).

Each major cell along the major diagonal running from upper left to lower right presents the flow of goods and services from one industry to another within one and the same region. However, the cell at the extreme lower right, corresponding both vertically and horizontally to the subtotals category, presents the inter-industry flow for the nation as a whole (including national imports and exports). This in itself is a national input-output table. Any other major cell shows the flow of goods and services between the industries of a pair of regions or of a region and the nation as a whole.[14]

The subtotals major row is of particular interest. It sums the data in each minor vertical column by industry, yielding inputs into each industry of each region from each of the 20 industries as a national whole. Thus a comparison of the appropriate columns in this subtotals major row brings out the differences in production practices and consumption habits among regions.[15]

Also, it is relevant to inquire into the value of the internal interindustry structure of any region, as depicted by a major cell along the major diagonal from upper left to lower right. In general it will have considerably less value and meaning than the internal interindustry structure of the nation depicted by the major cell in the lower right corner of the table. Many industries of a region may have important relations with producers and consumers outside of the region, whereas usually fewer industries of the nation have important relations with foreign producers and consumers. Industrially, each region tends to be much less self-contained than the nation.[16]

IV

A table of the order of Table 4.1, when completely filled in, furnishes the required data for calculating the input coefficients (the ${}_{kl}a_{ij}$) according to equations (4.3) in the preceding section. When inserted into equations (4.4), these coefficients make it possible to investigate implications of changes in the bill of goods for the output of each industry in each region (the ${}_{k}X_{i}$ of equations (4.5)). *But this is so only if these coefficients can be taken to be fairly stable.* Only where coefficients are fairly stable in fact can the basic assumption of constant production coefficients be justified. The

${}_{kl}a_{ij}$ thus must reflect inherent meaningful interregional bonds, and not be purely descriptive for a given point of time.

Immediately, this requirement restricts the application of the model for projection purposes to sectors of the world within which there is, in general, stability in the relative supply prices of each output produced by several regions. When there is a significant change in the relative prices at which industry x in region d is supplied with the output of industry y in regions e and f, then there may be significant substitution between y from region e and y from region f. The coefficients ${}_{ed}a_{yx}$ and ${}_{fd}a_{yx}$ may change significantly during the period relevant for the projection. This tends to invalidate any projection analysis based on our model.

In sectors of the world-economy like the United States and many other nations, there is comparative stability in the relative prices at which any regional market of the nation is furnished a commodity produced by the several regions of the nation. The model then is, in general, applicable on an *intranational* level. This is not so on an international level, when each nation or group of nations is designated a region. For here, during the period relevant for a projection, exchange rate changes may completely upset existing market areas and supply channels.[17] The model, however, can be of value internationally where it is restricted to national (regional) economies tied to a single currency (as the sterling bloc), or to national economies which maintain stability of exchange rates among themselves.[18]

Rigorously speaking, it is not sufficient that relative supply prices be comparatively stable. Because constant production coefficients are assumed in our model, any change in an item of the bill of goods[19] *directly* calls for a proportional change in each of the inputs required in the production of that item.[20] In terms of interarea-industry relations, this requires proportional changes not only in each industry input (i.e., coal to produce Eastern steel), but also in each industry input from each region (i.e., Eastern coal to produce Eastern steel, Southern coal to produce Eastern steel). Thus if an increase in Eastern food processing is called for as an indirect result of a change in certain items in the bill of goods, this implies that there are proportionate increases in the supply of agricultural products from the East, South, and West to Eastern processors. As far as Eastern food processing is concerned, it implies no extension of the agricultural margin in terms of regions, but rather the continuance of the same pattern of relative flow intensities of agricultural products from the several regions.[21]

In effect constant production coefficients in terms of interarea-industry relations mean constant (or fixed) patterns of supply areas or channels.[22] Or, in terms of an individual industry of a particular region, they may be interpreted to mean constant resource use, region by region.[23]

Though a certain amount of unreality is introduced into the model via this assumption, the extent of such unreality should not be exaggerated. First, it should be borne in mind that the model is designed to investigate the regional and interregional repercussions of only small percentage changes in the bill of goods. This is true particularly for changes which may be expected yearly or over a short time span, and for a group of regions which in general are not operating at full capacity. The extent of the unreality obviously becomes greater as the magnitudes of the hypothesised changes increase, and as longer and longer run implications of changes are sought.

Second, in a sector like the United States where relative supply prices of producers in various regions are fairly stable, interregional flows of goods are also fairly stable. For those commodities which incur significant transport costs – e.g., brick, cement, glass – the pattern of supply areas or channels with reference to any particular point of consumption generally does not evidence much relative change with small-to-modest changes in the demand at that point. This is also true for commodities such as aircraft, where transport costs may not be important but where different attributes of the product may effectively distinguish the output of one region from another; and for commodities where cultural and institutional factors operate to fix supply channels.

It is only with reference to those commodities where transport costs are negligible, differentiation slight, and cultural and institutional factors flexible, that the postulate of constant patterns of supply areas or channels does not find an approximate reality. None the less, this explicit assumption of the interregional model is not inferior to the implicit assumption involved in most national models, whether input-output or other. Most national models implicitly postulate that any industry in a given region receives its inputs of any commodity from other regions in proportion to the output of this commodity in other regions.[24] Both the explicit and implicit assumptions are arbitrary for this group of commodities. It would seem that an assumption based on existing flow data is no worse to use for projection purposes than one based on an initial geographic distribution of output.[25] However, if the practitioner prefers, he can easily alter the relevant coefficients in the interregional model to reflect not existing flows but the initial distribution of output among regions.[26]

Finally, it is possible to allow for a certain amount of shift in supply areas, especially if the shift concerns the supply areas of items in the bill of goods. For these items in the bill of goods, their required inputs (by industry and by region) can be substituted, and thus cognizance of the shift can be taken directly, where location and comparative cost studies and other sources furnish exact quantitative information on the shift to be anticipated.

V

The assumption of fixed production coefficients for interarea industry relations is more restrictive than for interindustry relations within a closed economy. It is important to appreciate that this is, in general, well compensated by the greater degree of accuracy obtainable from working with interarea industry coefficients. As already indicated, regional differences in production techniques and in consumption patterns are registered. The fact that the inputs of coal and ore per unit output required for Eastern steel production differs from those required for Western steel production is taken into account in any operation performed with this model. In addition to its significance in regional and interregional analysis, breaking down the set of national coefficients into a much larger set of interarea coefficients is valuable for investigating the impact of a change in the bill of goods upon the *national* output of various industries. It is analogous to subdividing a given classification of goods into a finer grouping. Each permits a greater degree of accuracy in the input-output analysis because a greater range of functioning relations is recognised.

For illustration, assume an open system for the United States in 1939, with all exports treated as part of a final bill of goods, which also includes all household consumption and domestic investment. What would the cessation of exports in 1939 have implied for the United States economy in 1939? Under one set of postulates and classification of industries, it has been estimated that primary employment in the ferrous metals industry would have been reduced by 12.03 per cent, with account being taken of both direct and indirect effects.[27] However, not all regions and areas of the nation would have been affected alike. The ferrous metals industry in the coastal regions would have been expected to be affected more than that industry in the interior, certainly directly so. Since the sources from which the ferrous metals industry on the coast derives its materials are different from the sources of the materials for the industry in the interior, the outputs from all sources, too, would have been differentially affected. Also, because production practices vary from region to region, national coal and ore production would have been affected differently. For these reasons, the steel requirements of the coal and ore industries would have been different. The same reasoning applies to the whole succession of indirect differential regional repercussions. In all probability, the estimate of the reduction in primary employment in the nation's ferrous metals industry would have been more or less than 12.03 per cent (and more accurate) if differential regional impacts had been considered. This is probably true for the estimates of change in primary employment in the nation's other industries.[28]

Theoretically, then, this model can yield superior estimates of national reactions to changes in the bill of goods. More important, it can show, *within the serious limitations of our assumptions, especially the assumptions of constant resource use and a static framework,*[29] how a given autonomous impulse in one region is transmitted to other regions. It can show implicitly how the distance separating regions modifies the impact of this impulse and how the direct and indirect effects play back and forth (instantaneously) from region to region, constantly damped by spatial resistance. More specifically, it can demonstrate – always only within the framework of our postulates – how a shift of population from the East to the West, *ceteris paribus*, affects trade among all regions; affects the industrial and agricultural structure of each specific region; affects the commodity balances of payments between regions; affects output and employment in each region; and so forth.[30]

VI

The discussion thus far has skirted the practical problems of computation and of collecting and processing data. Obviously, for a regional model which contains as many industries for each region as a national (or international) model contains for the nation (or international unit), the computations are many times as extensive as those for the national (or international) model.[31] However, if it is possible to treat a national (or international) model with a 120-industry classification, then with roughly the same computational effort a three-region model with a 40-industry classification can be handled. Given the effort to be expended at computation, the use of a regional model sacrifices the superior information obtainable from a finer industrial classification. At the same time, at least equally desirable information of a regional character is acquired.[32]

It should be observed, however, that the computational requirements can be drastically cut, or a much finer industrial classification can be employed, if attention is centred primarily upon one region. In a 3-region model, for example, a 60-industry classification may be used for the region under detailed scrutiny, but perhaps only a 40-industry one for a second region, and a 20-industry one for a third region. This design would be meaningful if the 40- and 20-industry classifications were sufficient to encompass adequately the factors affecting the major interregional flows between the first and the second and third regions, respectively. In short, there is considerable flexibility in designing interregional models.

The computation problems of the model as it now stands are manageable. Not so, with the data problem.[33] For the United States, for example, appropriate data are not currently available. The Interstate Commerce Commission publishes information on state-to-state distribution of tonnage by commodity groups and on tonnage originated and terminated by groups of commodities and by geographic areas; both these sets of data are for

Class I steam railways.[34] The Army Engineer Corps publishes information on receipts and shipments of commodities by ports.[35] Both sources are inadequate for the model, since they do not provide data in terms of industry and state of origin, and industry and state of destination. Fortunately, the carload waybills furnished to the Interstate Commerce Commission by Class I railways do contain information of this character. If such information were collected for all transport media and made public, quite adequate data would be available for the operation of the model for the United States.

Pending such a development in the supply of publicly available data, estimates must be made. One method involves the use of inter-industry coefficients based upon national input-output tables. Given the available production data for each industry by regions for any base year, it is possible to obtain rough estimates of the consumption by each region of the product of each industry for the same year. This is done by deriving the requirements of each industry (including households) for producing the outputs recorded for each region. (1) Multiplying the output of each industry in a given region (say, the East) by the respective national coefficient which indicates the input of industry A (say, steel) into each industry per unit of output of each industry, (2) summing these results, and (3) adding that part of the final (bill of goods) demand for the product of industry A for which the region is responsible, yields the requirements of industry A (steel) for the given region (the East). This rough consumption figure, contrasted with the production figure of industry A in the given region, roughly determines whether the region is surplus or deficit with respect to industry A. In like manner, rough consumption estimates for each industry in each region can be derived; and surpluses and deficits with respect to each industry for each region can be calculated.

The above consumption estimates are based on the assumption that methods of production and patterns of consumption do not differ from region to region. Admittedly, they may differ significantly. As information becomes available on differences in regional practices, more accurate input coefficients can be computed, and better consumption estimates derived, for each region.

The regional production and (estimated) consumption figures provide to some extent a consistent framework for allocating the output of any industry in any region to each industry in each region. In carrying out the allocations, various location principles must be used in addition to available information on coefficients of localisation, location quotients, transport rates, distances separating surplus and deficit regions, ratios of transport cost to value of product, peculiarities of commodities such as perishability, style, etc. First, it is clear that a number of industries – for example, most of the service trades, construction, electric public utilities, transportation, communications, and households – have largely local markets. The output of these industries in any region can be distributed with only minor error to industries within the same region. The national input-output coefficients can serve as guides.[36]

In allocating the output of industries whose market areas extend beyond regional boundaries, the shipment data of the Interstate Commerce Commission and Army Engineer Corps are particularly helpful.[37] Where the unit of output is heavy and bulky and incurs relatively large transport expense, as for cement and blast furnace products, each surplus region ought to meet all its own requirements and to serve the nearest deficit regions. Or each surplus region might carve out for itself a market area from the nearest deficit regions, which would but slightly overlap the market areas of other surplus regions close enough to serve these deficit regions. Where the commodity is light, has high value, incurs a proportionately low transport expense, and tends to be nationally advertised, like standard textile products, market areas ought to overlap completely. Each surplus region ought to serve both deficit and other surplus regions.

Thus, with the use of such locational principles as the above, and of available information, the actual flow data, basic to the model, can be approximated fairly accurately for a given base year. These flow data, in the form of industry and region of origin, and industry and region of destination, permit the computation of the ${}_{kl}a_{ij}$ required for projections from the given base year.

VII

The above discussion has outlined an input-output model of a space-economy. This model permits an attack on certain problems of a space-economy, namely, changing regional and interregional structures, *but only under a set of restrictive assumptions*. Constant production coefficients, interregionally viewed, mean unchanging supply channels. The model thus forces the assumption that the spatial extent of each line of production in each region is fixed with respect to its inputs. In this respect, no spatial lengthening or shortening of any line of production can be tolerated. Viewed otherwise, transport outlays or, in physical terms, distance inputs per unit of output of any line in any region are not permitted to change. Also, as a corollary, roughly fixed price ratios are assumed, so that substitution of distance inputs for other inputs, of transport outlays for other outlays, is precluded.[38] In these respects, the underlying structure of the space-economy is held rigid.

Despite these limitations, the model is valuable because it is operational. The computations are manageable. Where data are not available, they can be estimated, especially for the United States. Thus, through projection the model can yield certain insights into the changing character of a space-economy which an anemic, non-operational model, such as Lösch's,[39] cannot. Too, the limiting assumptions of the model should be kept in mind – chiefly, a static framework (a limitation which can easily be removed with the development of dynamic input-output models) and unchanging supply channels. These, however, are frequently *implicitly* contained in many other models which aggregate on a national scale output, income, and other economic magnitudes.

The model carries through only a partial spatial aggregation.[40] It can be applied to interregional and regional analysis at an international level where stability in relative supply prices may be anticipated. On the whole, however, it is only applicable to interregional and regional analysis within a nation, and can yield as a by-product more accurate national projections when fineness of industrial classification need not be sacrificed.

The model, within its limitations, plays up three types of differential regional reaction paths: (1) that associated with the diverse industrial compositions of regions; (2) that associated with the finely interwoven and interlaced network of heterogeneous as well as homogeneous market and supply areas of a space-economy; and (3) where data are directly available and do not need to be estimated, that associated with different production practices and consumption patterns in different geographic areas. Via these three reaction paths, the impact of any regional (and thus national) change is transmitted to other regions, affecting their structure in terms of output, employment, income, etc., and affecting the flow of trade between regions.

Notes

1. The analysis in this article has greatly benefited from discussions with Guy Freutel and with Wassily W. Leontief, James S. Duesenberry, Leon Moses, and other members of the Harvard Economic Research project. However, the shortcomings of the analysis and the views expressed herein are the author's only.

   The model to be developed through spatial aggregation is basically different from one constructed by Professor Leontief through national disaggregation (in his forthcoming *Studies in the Structure of the American Economy*, ch. 4). Nevertheless, the two models should not be viewed as alternatives. Rather they are complements. The Leontief balanced regional model is particularly useful for determining regional implications of national projections; the pure interregional model, for determining national implications of regional projections. (For elaboration of this point, see W. Isard and G. Freutel, Regional and National Product Projections and Their Interrelations, sections 7 and 8, paper prepared for Conference on Research in Income and Wealth, National Bureau of Economic Research, May 1951.) Since one can proceed by degrees from one type of model to the other, it is likely that after considerable experimentation an hybrid model, involving elements of both, may prove to have the most general utility.

   The author's experience in directing the preparation and collection of the data and in carrying through the computations for the Leontief model has been of great value in setting up a feasible design for his own model.

2. For a presentation and evaluation of these efforts see W. Isard, 'The General Theory of Location and Space-Economy', *Quarterly Journal of Economics*, vol. 63 (November 1949), pp. 476–506.
3. T. Koopmans, *The Optimum Geographical Distribution of Population and Industry in the United States* (Cowles Commission for Research in Economics, January 1946).
4. Though trade theory is basically concerned with geographic flows, it has by and large ignored or set aside the space factor. This is in line with the economic theorist's customary preoccupation with the time factor.
5. Treating space implicitly or as a passive factor has also been characteristic of the better regional development studies. Colin Clark in his excellent works on economic progress in the various regions and nations of the world (*The Conditions of Economic Progress*, London, 1940, and *The Economics of 1960*, London, 1942) does not attempt at all to study the basic changes in the spatial interrelations and patterns of these nations or regions. Nor does he probe into the diverse spatial structures of nations and regions at a given point of time. To be sure, he makes spatial comparisons in the sense that he compares nations having different latitudinal and longitudinal positions. But such comparisons more accurately fall under the heading of geographic description, not spatial analysis.

   A. J. Brown (*Industrialization and Trade*, London, 1943, and *Applied Economics*, London, 1947), in ascribing a significant role to the world's uneven pattern of resources and its implications for geographic specialisation and the global locational structure of industry, takes a major step forward. But he does not transcend 'geographic positional' analysis. He does not expose the fundamental spatial scaffolding of international relations. Nor do E. F. Staley (*World Economic Development*, Montreal, 1944), the League of Nations (*Industrialization and Foreign Trade*, 1945), P. N. Rosenstein-Rodan ('Problems of Industrialisation of Eastern and South Eastern Europe', *Economic Journal*, vol. 53, June–September 1943, pp. 202–11), among others.

   K. Mandlebaum (*The Industrialisation of Backward Areas*, Oxford, 1947), too, fails to achieve spatial analysis. But, in calculating derived demand and in making appropriate adjustment for industries with relatively significant transport costs, he does accord space some explicit treatment.

   Likewise with Pei-kang Chang (*Agriculture and Industrialization*, Cambridge, Mass., 1949), despite the fact that he gives considerably more content to regional development analysis by taking over the historical-locational material of A. P. Usher (*A Dynamic Analysis of the Location of Economic Activity*, unpublished) and W. H. Dean, Jr. (*The Theory of the Geographic Location of Economic Activities*, doctoral dissertation, Harvard University, 1938).

   However, see the interesting and acute locational observations of E. Peltzer ('Industrialization of Young Countries and the Change in the International Division of Labor', *Social Research*, vol. 7 (September 1940), pp. 299–325).
6. A. Lösch, *Die räumliche Ordnung der Wirtschaft* (Jena, 1944), pt. 2. The size of the market area for each commodity would be determined by the extent to which the economies in its large-scale production would be offset by the cost of transportation.
7. Also, there is a shortcoming in confining one's attention to relations among individuals alone. Much of man's behaviour cannot be understood and explained except as conditioned by his being a member of groups of diverse sise,

structure, and composition. Aggregation of individuals to form regions can serve as only a partial automatic correction.

8. Each way of aggregation emphasises certain relations at the expense of others. Traditionally, most economic analysis – business cycle, international trade, monetary, etc. – has aggregated producers and consumers on a national basis. But, just as, for example, grouping all durable-goods production into one category conceals considerable information which would help to explain changes in their total, so the classification of producers of any industry into a national aggregate conceals considerable information concerning the spatial distribution of that industry, which information would help to explain changes in the national total. Therefore, only a partial spatial aggregation is proposed here.
9. E. M. Hoover and J. L. Fischer, 'Research in Regional Economic Growth', *Problems in the Study of Economic Growth* (Universities-National Bureau Committee on Economic Growth, New York, 1949), p. 178. Also, see W. Isard and G. Freutel, op. cit., section 6.
10. See, among others, National Resources Planning Board, *Regional Factors in National Planning* (Washington, DC, 1935), pt. IV; and 'Regional Geography', *American Geography: Inventory and Forecast* (edited by the Centennial Studies Committee of the Association of American Geographers, forthcoming).
11. States and regions formed from states are in several respects imperfect economic areas. This is especially so with respect to the flow of goods and services from and to metropolitan focal points. When sufficient data for counties and other parts of states become publicly available, it should be possible to set up superior sets of regions for the United States.
12. W. W. Leontief, 'Output, Employment, Consumption and Investment', *Quarterly Journal of Economics*, vol. 58 (February 1944), pp. 294–7.
13. As employed by W. W. Leontief, 'Exports, Imports, Domestic Output, and Employment', *Quarterly Journal of Economics*, vol. 60, February 1946, pp. 192–3.
14. Table 4.1 can be set up in an alternative fashion. Each major row and column can refer to an industry rather than a region. And each minor row or column can refer to a region rather than an industry. This arrangement of data focuses attention upon each industry rather than upon each region. It shows the relations of each industry with itself and with each of the other industries, region by region. For the study of interregional relations, however, this arrangement of data is not desirable.
15. It is possible to use another subtotals classification. Summing can be done by regions instead of by industries. This yields an arrangement of data in the subtotals major row and major column which throws light on regional balances of commodity flows and payments.
16. In general, the smaller the region for a given territory, the less self-contained. For example, if within the United States states were to be defined as regions, each region would tend to be less self-sufficient than if census regional units were defined as regions. In order to reduce to manageable proportions the task of analysing the interplay of forces among regions, it may be desirable, in designating regions, to consolidate many small areas into fewer but larger and more self-contained areas.
17. Likewise, rigid quantitative restrictions on imports and exports, and other institutional interferences with the operation of the price mechanism, may limit the usefulness of the model.

18. It is relevant to note the use of constant coefficients by international trade analysts. For example, see J. J. Polak, *Projections of the International Sector of Gross National Product*, section 3, paper prepared for Conference on Research in Income and Wealth, National Bureau of Economic Research, May 1951.

    Superficially, our model, on an international level, resembles Frisch's 'trade matrix'. Basically, this is not so. Frisch's matrix analysis emphasises the balance of the aggregates, exports, and imports, for each nation. The individual import and export items are adjusted by government authorities to be consistent with such a balance. Our model, on the other hand, assumes the uninhibited operation of the price system in determining, via comparative costs, the quantitative flows of each commodity from region to region, or nation to nation. When underlying conditions are appropriate for an application of our model, changes in aggregate exports and imports, and in the degree to which they balance, can be projected *within the framework of our assumptions* by summing the projected changes in the flows of individual commodities. See R. Frisch, 'On the Need for Forecasting a Multilateral Balance of Payments', *American Economic Review*, vol. XXXVII, September 1947, pp. 535–51; and 'Outline of System of Multicompensatory Trade', this Review, vol. XXX, November 1948, pp. 265–71.
19. The bill of goods must always be specified in terms of outputs of industries of particular regions.
20. The limitations of this assumption in terms of inter-industry relations from a national standpoint has been discussed by W. W. Leontief, *The Structure of American Economy, 1919–1929* (Cambridge, Mass., 1941), pp. 38–41, and in 'Input-Output Analysis: Discussion', *Papers and Proceedings of the American Economic Association*, vol. XXXIX, May 1949, pp. 226–40.
21. Economies of scale realisable from shifting purchasing relations with regions are already precluded by the assumption of constant production coefficients, industry by industry.
22. This assumption does, however, allow changing supply areas within any region. Before a change in the bill of goods, one state of the region may furnish 40 per cent of the supply of coal from the region to Eastern steel producers. After a change, the same state may furnish 50 per cent. As long as there is no change in the amount of coal supplied by the region per unit output of Eastern steel, such a shift in supply channels within the region is consistent with the model.
23. They also mean that, where arbitrary cross-hauling is contained in the data employed, the relative extent of cross-hauling involved in supplying the inputs required for the production of a given item does not change as the output of the item changes. But the prevalence of arbitrary (non-economic) cross-hauling in our economy can easily be exaggerated. Where transport cost is of negligible importance, and where there are significant economies in established contact and national or regional advertising, market overlapping of even slightly differentiated products not only is to be expected but has economic justification. It is only as transport cost becomes more significant, and economies of established contact and advertising less important, that cross-hauling becomes more arbitrary when market areas overlap.
24. This assumption is also employed in the Leontief regionally balanced model to determine the demands for the regional outputs of national industries.
25. Take the case of an increase in the demand for television sets by households of a given region. It would seem just as reasonable to suppose that this new demand will be supplied by the producers of the various regions in the same proportion as existing demand in the given region is supplied (as given by

current flow data), as to suppose that new demand will be supplied by various regions in proportion to their current outputs.

26. For example, if region A produces 10 per cent of the value of all television sets initially, then the coefficient relating inputs of television sets from A to households of B can be set by taking 10 per cent of the total value of television sets absorbed by households of region B and dividing by the total value of output of households in region B. Of course, a substitution of such a coefficient for one based on existing flows for any one region must be accompanied by similar substitutions for all other regions.

    Incidentally, through such substitution of coefficients a hybrid 'balanced regional-interregional' input-output model can be constructed.

27. W. W. Leontief, 'Exports, Imports, Domestic Output and Employment', *Quarterly Journal of Economics*, vol. LX (February 1946), table 2.

28. From a national standpoint, better results can be obtained by decomposing the ferrous metals industry into iron mining, operations of blast furnaces, and operations of steel works and rolling mills in order to take into account the different input requirements of each of these activities. Similarly, when the national tables are classified by regions, account is taken of the different input requirements in terms of geographic sources of supply as well as industry sources.

    Another illustration may be helpful: Suppose we wish to investigate the implications of a given increment of population in the South, *ceteris paribus*. If we were to utilise the national input-output coefficients alone, an increment of population in the South would have the same effect as a similar increment of population in any other part of the nation. The final bill of goods to be produced (where household consumption is included in this bill) would change in a given fashion. However, when we proceed on a regional basis and distinguish between the various regional patterns of consumption, then the effect of an increment of population in the South will differ from that of an increment of population elsewhere. For, *ceteris paribus*, the fuel required to meet the needs of the increment of Southern population may be less than that required to meet the needs of an equal increase of population in the East, or the needs of an equal increase of population distributed over the nation. The use of interarea industry coefficients permits greater accuracy.

    Conceivably, the use of this much larger set of coefficients might be regarded as introducing one more degree of substitution or enlarging the scope of substitution operations in input-output analysis. When the illustration cited in the text is viewed in terms of another possible set of regions, namely, census regions, it could be maintained that substitution would have taken place between Middle Atlantic steel and East North Central steel, between Mesabi iron ore and Adirondack ore, between Minnesota labor and New York labour, between coke and mechanical equipment (because, for example, Mesabi ore requires more coke in blast furnace operations while Adirondack ore must be mechanically treated and prepared before use in the blast furnace), between transport services and specific labour and equipment services (because in supplying a given demand, the product of one region may require more transport and less effort and equipment in mining and preparing the ore than the product of another region).

29. The impending development of dynamic input-output models promises to remove the limitation of a static framework. See, for example, the Leontief dynamic model (in the forthcoming *Studies in the Structure of the American Economy*, ch. 3). Once the dynamic national input-output models are tested

and put into a form suitable for operation, the interregional model can easily be adjusted to a dynamic base by introducing the appropriate matrices of stock coefficients.

30. Our model has other related values. It enables an analyst to view investment decisions and requirements with a locational perspective. A typical procedure frequently employed in evaluating investment needs is to compute national capacity and compare that capacity with the requirements to produce a given bill of goods. However, more likely than not, the geographic distribution of existing capacity does not correspond to the geographic pattern of requirements. The regional spread of electric power facilities need not reflect the regional needs for power. Hence expansion of capacity can easily take place before full utilisation of national facilities, if a region's needs outrun its capacity to produce.

    To be more concrete, the construction of a new aluminum reduction plant called for by increased demand may be possible either in region A where sufficient power facilities are available or in region B where power facilities are fully utilised. With regard to region B, national data alone would not reflect the need for new investment in power facilities. Thus, because short-run capital requirements are related to the location and relocation factors, the use of our model, together with information on regional capacities, allows more relevant comparisons and provides a sounder basis for analysing whether or not a given autonomous change will require additional capacity in industries both directly and indirectly affected.

    The model has potential value, too, for cyclical analysis. It throws considerable light upon the regional transmission of impulses, and thus upon the neglected role of space in cycle generation. However, this important matter cannot be discussed here.

31. The writer is unaware of any effort being made to apply input-output analysis to an international entity containing more than one nation without regard to national boundaries. None the less, it is useful to think of analysis at such a level in order to contrast this with analysis of an interregional character applied to such an entity.
32. This means more industrial but less spatial aggregation than in a national input-output model comparable from a computational standpoint.
33. The paragraphs which follow concern data problems for an interregional model within the boundaries of a nation. Where it is meaningful to set up an interregional model on an international level, the data problems are different. Given for each nation (region) the data of an input-output table, as well as figures on exports by industry, it would still be necessary to obtain (or estimate) data on the destination both by industry and nation of the various exports of each nation. From this set of information for any given base year, the ${}_{kl}a_{ij}$ could be computed.
34. Bureau of Transport Economics and Statistics, US Interstate Commerce Commission, *Carload Waybill Analyses, State to State Distribution of Tonnage by Commodity Groups* (Washington, DC), quarterly, and *Tons of Revenue Freight Originated and Tons Terminated in Carloads by Groups of Commodities and by Geographic Areas – Class I Steam Railways* (Washington, DC), annually.
35. US Army, Board of Engineers for Rivers and Harbors, *Commercial Statistics; Water-borne Commerce* (Washington, DC), annually.
36. Because of differences in regional practices, exact allocation according to national input-output coefficients may in a given region more than exhaust or

fail to exhaust the output of an industry with a local market. If so, this will cast some light on the nature of the differences in regional practices.

37. Here, too, some light may be shed on differences in regional practices. The derived estimates of regional consumption may need to be changed in order to be consistent with shipment data.
38. For precise definition and discussion of the concepts used in this paragraph see W. Isard, 'The Space-Economy: Distance Inputs and the Locational Equilibrium of the Firm', *Quarterly Journal of Economics*, vol. 64 (May and August 1951).
39. Lösch, op. cit., pp. 62–9.
40. The amount of spatial aggregation necessitated by computational considerations will be reduced as better highspeed computing machines are constructed and their services made available.

5 The Impact of Steel Upon the Greater New York–Philadelphia Industrial Region[1]

For a long time the inadequacy of the traditional Weberian agglomeration analysis[2] has been fully recognised. Palander, in particular, has pointed up its shortcomings.[3] More recently, a fresh and more general approach has been made by P. Sargant Florence and others.[4] This approach, primarily an attempt to identify from processed census data meaningful geographical associations in industry, of necessity does not concern itself with the development of a theoretical framework.[5]

It is the purpose of this paper to attempt some development in the theory of agglomeration and of the broad spatial clustering of all economic activities by grafting on to the sounder elements of location theory a modified regional input-output schema. Viewed from a different standpoint, we attempt an impact study, wherein the direct and indirect repercussions of the location of a basic industry in a region are evaluated.[6] In this sense, crucial elements of regional development theory are attacked. It is not our intent here to present a comprehensive analysis of the forces affecting the location of a basic industry. In other studies this has been attempted for the iron and steel industry.[7] Rather, we accept the conclusions of these studies as a starting point and attempt to appraise the direction and strength of the force generated by the basic industry upon all types of economic activity within the region. This may be defined as the 'broad agglomeration effect'.

This effect during the coming decade will be studied with reference to the Greater New York–Philadelphia urban-industrial region.[8] Some attention will be given to interregional repercussions, particularly upon the closely linked region of Puerto Rico. Throughout the study it is assumed that aggregate demand remains at a high level during the next decade, that no technological revolution in the production of steel takes place, and that no violent shocks such as a major destructive war impinge upon the economy.

This chapter was written with Robert E. Kuenne. Reprinted from *The Review of Economics and Statistics*, vol. XXXV, no. 4, November (1953), pp. 289–301.

I THE SCALE OF BASIC STEEL OPERATIONS

Past studies of the costs of assembling raw materials, production, and delivery of steel to market demonstrate that if there exists a market capable of absorbing the total output of a modern integrated steel-works, with its associated semifinishing mills, steel-producing capacity should be located in the vicinity of the market as closely as site conditions, transportation facilities, and socio-cultural factors permit. (There are of course exceptions when special conditions obtain, as when coal and ore sources coincide or are adjacent.) On the basis of these studies, it had been concluded that capacity on the eastern sea-board should be expanded. This conclusion has been borne out by the recent construction of the United States Steel Corporation's Fairless Works at Morrisville, Pennsylvania, the announcement of National Steel's intention to build at Paulsboro, New Jersey, and reports of the persistent interest of other steel companies in the Delaware Valley.

The cost rationale for the above conclusions and their subsequent fulfilment is presented succinctly in Table 5.1. Since the steel industry is a transport-oriented industry, transport costs alone need be presented.[9] The advantage of a Trenton location is clear-cut and marked.

From the standpoint of sound regional planning, the mere specification that an industry should be situated in an area on the basis of a cost advantage is insufficient; the scale of its operation must also be given. For such a market-oriented industry as steel, this entails the estimation of future demand for steel in the area. *Iron Age* estimated the consumption of steel by metal-working industry alone in the New York–Trenton–Philadelphia industrial area in 1948 as 5.1 million tons.[10] (In 1948, for the nation as a whole, the metal-working industry received 46 million of the 66 million

Table 5.1 Transportation costs on ore, coal, and finished products for selected producing locations serving New York City* (in dollars)

Location	*Transportation costs on:* *Ore*	*Coal*	*Finished products*	*Total*
New London (hypothetical)	3.68	5.42	8.80	17.90
Pittsburgh	5.55	1.56	12.40	19.51
Cleveland	3.16	3.85	14.00	21.01
Sparrows Point	3.68	4.26	8.40	16.34
Buffalo	3.16	4.27	11.60	19.03
Bethlehem	5.56	5.06	5.80	16.42
Trenton	3.68	4.65	4.80	13.13

Note: *The costs on iron and coal are calculated per net ton of steel.
Source: Isard and Cumberland, op. cit., p. 257.

tons of steel shipped by steel mills.) The *Iron Age* estimation procedure, however, is subject to serious criticism.[11] The much more reliable Census data record consumption by metal-fabricating industries of 4.0 million tons for 1947 after appropriate adjustment of state data to correspond with our area.[12]

If the adjusted Census figure is blown up by 43 per cent to account crudely for consumption by nonmetal-working industry, if, in addition, allowance is made for a share of the coastwise trade and export markets of the United States,[13] and if the expected growth of effective demand by steel consumers in the area is considered, an optimist might predict steel consumption of at least 9 to 11 million tons annually by the early 1960s. On this basis he might contend that an almost equivalent amount of new capacity could be justified. Less sanguine observers, cognizant of the intense competition from Sparrows Point, Bethlehem, and Buffalo, might with good reason restrict their estimates of new capacity by 1962 to 5 or 6 million tons.[14]

For purposes of this article, a range of estimates would unduly complicate computations, as is easily seen later. The use of an average also is undesirable, since in terms of planning capital outlay and development of the greater metropolitan regions within the area it is more desirable to have firm estimates of minimum expansion effects than infirm estimates of most likely effects.[15] Hence, here, as throughout this paper, we attribute expansion to an industry in our area only when a sound basis exists for such a prospect. Our minimum estimate for new basic steel capacity for this area, exclusive of steel capacity induced by the new steel demand incident to the agglomeration effect in its broadest sense, is 3.0 million tons. This new steel capacity, in effect a displacement of extra-areal capacity, estimated as of autumn 1951, was based on the announced intention of the United States Steel Corporation to construct 1.8 million tons of capacity, and on our considered judgement – particularly in the light of National Steel's proposal to erect 1 million tons of capacity – that at least 1.2 million tons of additional new capacity was justified.[16]

From here on, the problem decomposes into two major parts: (1) to estimate the resulting direct expansion of existing and influx of new steel-fabricating activities within the area, and (2) to quantify the direct and indirect repercussions of the new steel production and associated new steel-fabricating operations upon all activities in the area.

II DIRECTLY INDUCED STEEL FABRICATION

Several attempts have been made to identify the forces dictating the juxtaposition of industries representing consecutive stages in a production process or related in some other vertical or horizontal manner.[17] The approach of these attempts becomes more valid and useful when location theory and historical trend analysis supplement the location quotient and coefficient of geographical association upon which these studies heavily depend. In associating steel-fabricating activity with basic steel capacity, we have treated individually the steel-fabricating industries listed below in Table 5.4 by three-digit Census classification. For each such activity: (1) we compare, for each of 69 industrial areas as defined by *Iron Age*, the number of production workers engaged in it and the capacity to produce the major basic steel items consumed by it;[18] (2) we observe the rates of growth of the given activity during the period 1939–47 in those areas where capacity for producing the major steel items consumed by the activity increased considerably more rapidly than for the nation as a whole, and we compare such rates with those of other areas;[19] and (3) we consider particular features and characteristics of the production process in the given activity, with special reference to transportation costs on raw materials and the finished product, economies of scale, industrial organisation, and other location factors whose effect might easily and objectively be identified.

To illustrate the procedure, industry number 343, Heating Apparatus (Except Electric) and Plumbers' Supplies (SIC code) is analysed below. This industry produces the following major products: cast iron heating boilers; cast iron radiators; residential steel heating boilers; warm air furnaces; gas, coal, wood, and kerosene domestic heating stoves and cooking stoves, ranges, and appliances (except electric); water heaters (except electric); steam and hot water unit heaters; floor furnaces; tanks for water heaters; oil burners, domestic and industrial; plumbing fixture fittings and trim; cast iron and steel kitchen sinks; cast iron and steel bathtubs. In Table 5.2 are recorded for 69 industrial areas and for the balance of each state not included in the industrial areas capacity for hot-rolled sheet and strip production in 1948 (column 1) and *Iron Age* estimates of production workers in 1948 (column 2). It should be noted that of the steel consumed by industry 343 in 1947, 86 per cent consisted of sheet and strip.[20]

Careful scrutiny of the table reveals two types of relationships. The fairly large concentrations of employment in the New York–Newark–Jersey City area (5876), Bridgeport–New Haven (2853), Hartford–New Britain (1988), and other southern New England districts, indicate a strong market pull. This, too, is borne out when the disproportionately small capacity in the combined San Francisco–Oakland–Los Angeles area is contrasted with the large amount of employment in industry 343 (12,123).

Less clear-cut cases of strong market pulls are Philadelphia–Camden (3909 employed), and the entire state of Wisconsin (13,002), including Milwaukee (3768) and Sheboygan (5500). However, each one of the areas already mentioned has some sheet and strip capacity or, aside from New

Table 5.2 Production employment in industry 343 and hot-rolled sheet and strip capacity, 1948, by selected areas*

Industrial area	*Capacity (in tons)*	*Employ-ment*	*Industrial area*	*Capacity (in tons)*	*Employ-ment*
Chicago–Gary	3,879,700	12,959	Minneapolis–St Paul	—	1,073
Youngstown	2,638,350	650	Seattle	—	943
Pittsburgh	2,614,520	3,437	Fall River–New Bedford	—	935
Detroit	2,451,000	3,875	Boston	—	885
Baltimore	1,195,000	2,031	Moline	—	717
Buffalo	1,155,380	2,540	Worcester	—	640
Cincinnati-Covington	950,000	3,960	Dallas–Ft Worth	—	611
Balance of Ohio	793,000	791	Rochester	—	610
Birmingham	725,200	2,210	Utica	—	610
Cleveland	651,000	7,510	Ft Wayne	—	565
Sharon–New Castle	600,000	1,110	Syracuse	—	565
Balance of Kentucky	374,000	300	San Diego	—	500
St Louis	328,500	5,361	Akron	—	500
Canton	314,000	625	Benton Harbour	—	435
San Francisco–Oakland	238,926	2,918	Balance of Alabama	—	410
Mansfield–Marion	205,000	2,040	Balance of Texas	—	381
Los Angeles	113,284	9,205	Springfield–Holyoke	—	361
Bridgeport–New Haven	150,000	2,853	Trenton, NJ	—	330
Philadelphia–Camden	105,100	3,909	Allentown–Bethlehem	—	275
Anderson–Muncie	95,000	855	Balance of Missouri	—	220
Columbus	61,500	2,715	Balance of Michigan	—	210
Indianapolis	55,000	1,477	Balance of Georgia	—	200
Atlanta	25,500	620	Toledo	—	186
New Castle, Ind.	25,000	0	Flint	—	175
Kansas City	23,500	426	Balance of Indiana	—	175
Houston	9,000	68	Balance of Washington	—	150
Balance of Minnesota	7,120	785	Evansville, Ind.	—	109
Louisville	—	6,820	Balance of Connecticut	—	80
New York–Newark–Jersey City	—	5,876	Dayton	—	65
Sheboygan	—	5,500	Albany	—	50
Milwaukee	—	3,768	Binghamton–Elmira	—	50
Balance of Wisconsin	—	3,734	Balance of California	—	40
Lansing	—	2,830	Balance of Maryland	—	25
Hartford–New Britain	—	1,988	Balance of New York	—	0
Peoria	—	1,875	Balance of Pennsylvania	—	—
Erie	—	1,760	Balance of New Jersey	—	—
South Bend	—	1,599	Hagerstown	—	—
Rockford	—	1,540	Danville, Ill.	—	—
Lancaster–York	—	1,450	Madison, Wis.	—	—
Springfield, Ill.	—	1,376	Balance of Kansas	—	—
Wichita	—	1,255	Mobile	—	—
Reading	—	1,206	New Orleans	—	—
Balance of Illinois	—	1,167	Balance of Louisiana	—	—

Table 5.2 *continued*

Industrial area	*Capacity (in tons)*	*Employ-ment*	*Industrial area*	*Capacity (in tons)*	*Employ-ment*
Grand Rapids	—	1,097	Beaumont	—	—
Scranton–Wilkes–Barre	—	1,095		—	—

Note: *Heating apparatus (except electric) and plumbers' supplies.

England points, is no more than about 200 miles from a major sheet- and strip-producing point. This testifies to the pull of sheet- and strip-producing points which is even more forcibly illustrated by Chicago-Gary (12,959 employed); Cleveland (7510); Louisville, somewhat over 100 miles from Cincinnati-Covington, (6820); Cincinnati-Covington (3960); Detroit (3875); and Pittsburgh (3437). In short, both market and sheet- and strip-producing points exert significant pulls upon the location of the industry. This suggests that with the addition of new sheet and strip capacity at the Fairless Works and elsewhere in the Delaware Valley, there should be a net positive attraction for industry 343, especially since, relative to the other major population focal points, such as Chicago–Gary and San Francisco–Oakland–Los Angeles, our area is deficient in its per capita representation of this industry.

Table 5.3 Production employment in industry 343

	1939	*1947*	*Per cent rate of growth, 1939–47*
United States	78,330	126,725	61.8
Maryland	1,128	2,499	121.5
California	4,652	9,373	101.5

A study of relevant historical material is also illuminating. In Table 5.3 are presented Census data on the rates of growth of industry 343 for the United States and for Maryland and California, 1939–47. (Because of data shortcomings, a comparison over a longer historical period is not justified.) Maryland and California were during this period among the fastest growing states in terms of sheet and strip capacity,[21] and they embrace or are peripheral to major market areas. The growth rates for California and Maryland well exceed that for the nation. This suggests that in any given large market area growth of industry 343 is at least partially related to the existence and growth of sheet and strip capacity in that area. This adds further support to the thesis that installation of sheet and strip capacity in the Delaware Valley will induce additional employment in this industry, particularly since the existing Maryland capacity can be considered peripheral to our area.

Theoretical considerations reinforce the empirical suggestions. The products of industry 343 tend to be heavy and bulky and to bear high transport costs relative to their value if not produced at the market; similarly, transport costs on the raw material steel are high, although to a lesser degree, if production does not take place at a steel source. A situation which involves the coincidence of both market and raw material, as in our area, strongly argues for significant expansion of employment in this industry. Thus, since this industry is a relatively fast growing one, as indicated in Table 5.3, it seems logical that a decreasing proportion of the area's demand for the products of this industry will be met from plants in the Western Pennsylvania–Ohio area, and an increasing proportion from plants within the area.

Having established on three counts that the coexistence of market and capacity in our area warrants expansion in industry 343 during the coming decade, we confront the problem of estimating the specific magnitude of this expansion. No neat estimating formula exists. Judgment based on knowledge of the area and conversations with informed persons must be resorted to. In line with such knowledge and conversations and in view of the fact that in 1948 our area had 12,968 production workers whereas the Chicago–Gary area had 12,959 (exclusive of the very substantial employment in the 200-mile area about Chicago–Gary), we judge that 20 per cent represents a firm minimum rate of increase in production workers in industry 343 incident to our area's increased ability to compete as a result of new steel capacity. This 20 per cent increase represents an increment of 2594 workers.

Similarly, for every other three-digit steel-fabricating activity, an estimate of numbers of new production workers was derived.[22] These estimates are listed in Table 5.4.[23] The total number of new production workers is 61,326 which, when divided by the 9166 production workers to be associated with the 3.0 million tons of new steel capacity, yields a ratio of 6.7. Despite its serious shortcomings, a comparison of this ratio with the 8.9 national ratio of production workers in steel-fabricating activities to production workers in steel manufacture suggests the conservative nature of our estimates, especially since of all the areas in the United States our area represents the most concentrated market.[24]

In setting forth these estimates, we do not claim that each will be realised. It is well recognised that unpredictable events and inadequate knowledge of the interaction of basic underlying forces as they affect an individual industry may have led us astray in individual instances. However, as a whole, we feel that these estimates represent a set of firm minimum figures upon which subsequent analysis may rest.[25]

Table 5.4 Estimates of new production employees in the area by steel-fabricating activity

Industry number	*Title*	*Per cent rate of growth*	*New production employees*
341	Tin cans and other tinware	10.0	923
342	Cutlery, hand tools, and general hardware	—	—
343	Heating apparatus (except electric) and plumbers' supplies	20.0	2,593
344	Fabricated structural metal products	4.6	1,132
346	Metal stamping, coating, and engraving	10.0	2,408
347	Lighting fixtures	2.5	1,911
348	Fabricated wire products	10.0	670
349	Miscellaneous fabricated metal products	17.5	2,695
351	Engines and turbines	8.0	723
352	Agricultural machinery and tractors	10.0	294
353	Construction and mining machinery and equipment	10.0	252
354	Metal-working machinery	13.5	2,137
355	Special-industry machinery (except metal-working machinery)	16.0	5,204
356	General industrial machinery and equipment	25.0	5,566
357	Office and store machines and devices	10.0	1,225
358	Service-industry and household machines	20.0	4,512
359	Miscellaneous machinery parts	25.0	5,666
36	Electrical machinery, equipment, and supplies	5.0	8,025
371	Motor vehicles and motor-vehicle equipment	20.0	7,331
372	Aircraft and parts	3.0	871
373	Ship and boat building and repairing	5.0	1,989
374	Railroad equipment	3.0	364
375	Motorcycles, bicycles, and parts	—	—
379	Transportation equipment (n.e.c)	7.0	32
38	Professional, scientific, and controlling instruments; photographic, etc.	—	—
39	Miscellaneous manufacturing industries	10.5	4,803
Total			61,326

III REPERCUSSIONS OF INITIAL ACTIVITIES

To quantify the direct and indirect repercussions of the operation of the above new steel-fabricating industries as well as of the new steel capacity, a modified regional input-output schema is employed. The entire set of inputs absorbed directly by the steel and steel-fabricating activities constitutes, in a sense, the 'bill of goods' that must be furnished to our area. This bill of goods is constructed: (1) by multiplying the cents' worth of every input required per dollar output of a given steel-fabricating activity by the dollar volume of that activity corresponding to the estimated new employment;[26] (2) by multiplying the cents' worth of every input per dollar output of steel by the dollar volume of new steel output corresponding to our estimate of new steel capacity; and (3) by summing horizontally for each of the 45 different industrial inputs the requirements by steel and each fabricating activity. The totals for each of the 45 inputs are recorded in column 1 of Table 5.5.[27]

In designing our study we made an important modification of the traditional input-output matrix. Since in regional analysis it is important to catch the local multiplier effect resulting from the generation of new income, households have been introduced into the structural matrix as an industry. Hence, the labor and other household services required by the new activities are recorded.[28]

After the bill of goods had been constructed, it was necessary to determine the fraction of the total amount of each input minimally expected to originate in the area. (It is clear that of the bill of goods requirements some will be produced within the area, while others, such as coal, will come from outside the area.) This step was necessary since in this study the expansionary effect upon the economy of the area is maintained by only the outputs which are produced within the area. It is not to be denied that the area may expand as a result of new demands for its products by population elsewhere, for example as a result of new demands by the population in the Appalachian coalfields consequent to additional demand for coal within the area. However, the inability to apply an interregional model[29] which would quantify these feedbacks compels us to neglect this indirect expansionary effect. This is not inconsistent with our accumulated bias toward underestimation, and hence is fully in line with the objectives of this article.

The determination of the above fraction for each industrial activity was based on the small amount of location theory directly relevant, on the accumulated production and derived consumption data for our area, on data of flows of commodities into and out of the area, and, finally, upon judgements of informed persons within the area. These fractions are recorded in percentage terms in column 2 of Table 5.5. Again many subjective decisions had to be made, and in each case we attributed to the area the production of only that fraction of the inputs required of an activity for which we felt there was a firm basis. For example, because the new industry will draw upon existing agricultural labour in the area, because some of the best agricultural land will be diverted into industrial, commercial, and residential uses, and in view of the past relative decline of agriculture in the area, it was felt that there was no firm basis for projecting an expansion of agriculture incident to steel development. Hence, a percentage of zero is recorded for agriculture, even though the likelihood exists that there will be some expansion. Similarly, in the case of chemicals, a 45 per cent figure represents an understatement in view of the rapid expansion of, and the natural advantages for, chemical production in the area. Observe, too, that the highest percentage figure associated with any activity is 95, even though it is likely that for certain service industries the percentage figure will be higher.[30]

Applying the percentages of column 2 to the corresponding items of column 1 yields column 3, which records (as firm minimums) the first round expansion of each industrial activity in the area which we anticipate will be required to maintain the new steel capacity and its associated fabricators in reasonably full operation. It should be noted that in column 3 of Table 5.5 the requirements of steel and of the diverse outputs of steel-fabricators are considered to represent demand which is over and above the demands for steel and fabricated steel products which initially justified the erection of new basic steel and steel-fabricating capacities; these industries must produce beyond their initial capacity. In effect, we have shaped the analysis so as to leave these activities within the structural matrix.

To produce the first round of expansions requires in turn a whole series of inputs. One can derive this series of inputs: (1) by multiplying the cents' worth of each input per dollar output of a given activity by the dollar volume of output of that activity recorded as the first round expansion; and (2) by summing horizontally over all activities the requirements so derived for inputs from each industrial category. However, not all of these second round input requirements will originate in the area. Multiplying these second round input requirements by the corresponding percentages in column 2 of Table 5.5 yields firm minimum estimates of the second round expansions for the area, which are recorded in column 4. In turn, third round input requirements and third round expansions (column 5) are derived, and so forth.[31] Since the rounds converge rather rapidly, it was necessary for our problem to compute only the first six rounds of expansions with rough extrapolations to cover the infinite number of succeeding rounds. By industrial category, the sum of the first six rounds plus extrapolations are listed in column 6 of Table 5.5.[32]

For each industrial category (except the household sector), the value of output per worker in 1947 was used to obtain a crude estimate of new employment[33] corresponding to the sum of the round expansions of

Table 5.5 Direct and indirect repercussions of new basic steel capacity

	Input requirements of initial steel and steel-fabricating activities (in $ thousand) (1)	*Minimum percentage of input requirements to be produced in area (2)*	*First round expansions in area (in $ thousand) (3)*	*Second round expansions in area (in $ thousand) (4)*	*Third round expansions in area (in $ thousand) (5)*	*Sum of round expansions in area (in $ thousand) (6)*	*Total new employees corresponding to round expansions (7)*	*Total new employees in initial steel and steel-fabricating activities (8)*	*Over-all total of new employees (9)*
1. Agriculture & fisheries	50.0	0	0.0	0.	0.	0.	0		0
2. Food & kindred products	294.6	60	176.8	17,660.	8,249.	42,492.	1,833		1,833
3. Tobacco manufactures	0.0	0	0.0	0.	0.	0.	0		0
4. Textile mill products	3,864.7	10	386.5	406.	39.	1,280.	142		142
5. Apparel	1,285.6	75	964.2	10,124.	3,461.	21,155.	2,302		2,302
6. Lumber & wood products	5,610.7	5	280.5	93.	36.	450.	64		64
7. Furniture & fixtures	1,753.4	33	578.6	802.	198.	2,000.	234		234
8. Paper & allied products	4,818.7	40	1,927.5	1,674.	1,297.	6,574.	426		426
9. Printing & publishing	425.5	90	383.0	5,929.	3,014.	14,617.	1,667		1,667
10. Chemicals	10,626.4	45	4,781.9	3,599.	1,630.	12,077.	601		601
11. Products of petroleum & coal	10,936.6	25	2,734.2	2,547.	1,118.	7,634.	228		228
12. Rubber products	8,381.5	15	1,257.2	355.	102.	1,879.	169		169
13. Leather & leather products	647.7	20	129.5	679.	194.	1,371.	150		150
14. Stone, clay, & glass products	9,031.7	15	1,354.8	441.	139.	2,083.	268		268
15. Iron & steel	121,170.5	50	60,585.3	13,566.	2,965.	78,335.	6,093	11,666	17,759
16. Nonferrous metals	33,997.4	20	6,799.5	1,667.	381.	9,063.	505		505
17. Plumbing & heating supplies	3,192.4	25	798.1	248.	50.	1,189.	118	3,640	3,758
18. Fabricated structural metal prod.	3,480.7	40	1,392.3	312.	33.	1,809.	151	1,420	1,571
19. Other fabricated metal products	31,770.9	40	12,708.4	2,146.	561.	16,121.	1,537	10,060	11,597
20. Agric'l, mining, & const. machinery	3,651.3	5	182.6	46.	11.	251.	22	707	729

21. Metal-working machinery	7,389.1	25	1,847.3	270.	43.	2,210.	289	2,705	2,994
22. Other machinery (except electric)	28,463.6	40	11,385.4	2,675.	551.	15,384.	1,486	28,607	30,093
23. Motors & generators	11,265.9	20	2,253.2	226.	42.	2,560.	301		
24. Radios	4,562.2	30	1,368.7	428.	101.	2,026.	192	10,392	12,312
25. Other electrical machinery	21,773.9	50	10,887.0	2,011.	432.	13,903.	1,427		
26. Motor vehicles	50,530.8	10	5,053.1	742.	260.	6,421.	389	8,770	9,159
27. Other transportation equipment	2,605.5	20	521.1	276.	69.	958.	117	4,605	4,722
28. Professional & scientific equipment	3,221.4	50	1,610.7	801.	287.	3,123.	416		416
29. Miscellaneous manufacturing	5,116.8	60	3,070.1	2,888.	982.	8,418.	845	6,108	6,953
30. Coal, gas, & electric power	7,767.0	50	3,883.5	1,843.	2,693.	11,079.	1,100		1,100
31. Railroad transportation	13,575.8	75	10,181.9	6,010.	2,390.	21,532.	3,308		3,308
32. Ocean transportation	457.3	75	343.0	331.	170.	1,021.	110		110
33. Other transportation	4,179.4	95	3,970.4	8,422.	2,836.	19,694.	2,394		2,394
34. Trade	13,969.8	95	13,271.3	36.585.	11.855.	83.642.	13,874		13,874
35. Communications	1,790.7	90	1,611.6	2,409.	1,283.	7,305.	1,191		1,191
36. Finance & insurance	3,086.2	90	2,777.6	9,472.	5,062.	25,252.	2,329		2,329
37. Rental	3,018.8	95	2,867.9	26,222.	9,603.	55,680.	909		909
38. Business services	5,338.5	95	5,071.6	2,385.	2,406.	13,384.	1,305		1,305
39. Personal & repair services	396.9	95	377.1	14.399.	5,088.	24,212.	4,443		4,443
40. Medical, educ., & nonprofit organisations	000.0	90	0.0	9,811.	2,160.	17,271.	4,370		4,370
41. Amusements	000.0	90	0.0	3,677.	1,066.	6,591.	1,100		1,100
42. Scrap & miscellaneous industries	8,388.2	50	4,194.1	2,054.	727.	7,411.	771		771
43. Undistributed	103,638.6	50	51,819.3	5,875.	6,019.	69,236.	7,208		7,208
44. Eating and drinking places	000.0	95	0.0	16,916.	3,903.	29,551.	3,705		3,705
45. Households	348,218.0	82	285,590.4	63,002.	80,894.	509,578.			
Total	903,807.7		521,377.2	282,024.	164,400.	1,177,822.	70,089	88,680	158.769

column 6. These estimates include nonproduction employment. They are listed in column 7 of Table 5.5, and total 70,089 workers. To them must be added the employees in the initial new capacity of steel[34] and steel-fabricating activities (column 8, Table 5.5). This yields column 9 of Table 5.5, and a total of 158,769 workers.[35] Finally, new employees expected to be engaged in construction (8574)[36] and government activity (12,884)[37] must be included. The over-all new employment thus amounts to 180,228 workers.

It should be emphasised that the above total is a gross figure and relates to change in employment only in activities which expand as a result of new steel capacity. Theoretically, one should subtract from this figure the decline in employment in activities which are forced to contract as a result of labour shortages, high wages, and the like. However, contractions are likely to be more severe in the shortrun than in the longrun when the new employment openings will have had an opportunity to exert a positive effect upon labour migration into the area. Since this study is concerned with conditions a decade hence, at which time we do not feel that such contractions will be of major proportions, and since we have already allowed for some contractions when we weighed expanding sectors against declining sectors for an industrial category in deriving the percentages of column 2 of Table 5.5, we retain the gross figure of 180,228 as an over-all firm minimum estimate of new employment. The reader may prefer to make a small downward adjustment.

In terms of the population engaged in the labour force (roughly 43 per cent) which characterised the area in 1950,[38] the total new population resulting, on the basis of the above gross employment figure, is 419,000. To the extent that the new wage jobs replace existing wage jobs and cause the decline of the industries associated with the latter, and to the extent that the above per cent increases in the future, the net increase in population supported by the new industrial stratum will be smaller.

IV BASIC QUALIFICATIONS

In appraising the above results, we need not discuss the qualifications springing from the use of constant production coefficients. They have been fully discussed elsewhere.[39] However, the inclusion of constant consumption coefficients in the structural matrix requires some discussion. To the extent that the new income generated does accrue as income to new job-holders (either from in-migration or natural population growth)[40] and does not increase wages of existing job-holders, and to the extent that the age and sex composition, ethnic and racial backgrounds, savings habits and other socio-cultural characteristics of the new job-holders and their families are similar to those of the population as a whole in 1947, the use of constant coefficients is not so tenuous a procedure as might appear at first sight. Also, it might be argued that in the light of the inadequacies of existing consumption studies, no better alternative exists.

In addition, we have abstracted from regional production coefficients and have assumed that production practices (as well as consumption habits) for the area are identical with those for the nation as a whole. This assumption is necessary since only national coefficients exist. However, from our knowledge of the resource characteristics of our area and its industrial composition, it is considered unlikely that the use of national coefficients entails serious distortions.[41] Nor, because of obvious difficulties, have these coefficients been adjusted industry by industry for increases in labour productivity or for changes in technology on the horizon; hence, our results reflect the 1947 productivity-technological structure.[42]

Since our projection is made for a decade hence, by which time the construction and equipping of our new plants and facilities will have been completed, we exclude input requirements to build up this capital stock. Our analysis is aimed at the delineation of current intersectoral flows to be expected in the early 1960s.[43]

One major problem which is not readily apparent needs to be bared. To be of practical use all types of input-output analysis require some leakage factor or device which reflects social frictions and the retarding effects of a whole array of noneconomic factors, as well as resource limitations and other economic resistance to expansion. In the typical input-output study, this is implicitly obtained by excluding sufficiently important sectors from the structural matrix (households, government, foreign trade, inventory, and capital formation) and placing them in the bill of goods. In our analysis, the restoration of households to the structural matrix without compensating changes would yield a leakage factor insufficient to guarantee a realistic convergence by rounds. In other words, the sum of the rounds would yield an expansion effect which far exceeds the upper limit of a range of reasonable levels. The system would be insufficiently anchored.[44]

The counterbalancing factor in our analysis is the leakage introduced through the omission of interregional feedback effects. In other words, no expansionary impulse comes from the increased income in other areas incident to their supply of inputs to our area. Though neglect of the interregional feedback effects is a shortcoming which calls for improved models,[45] none the less in our crude framework it provides anchorage and serves the desirable purpose of compensating, if not overcompensating, for the elimination of the customary household leakage. This point is strongly supported by the fact that if the bill of goods of the 1947 BLS national interindustry study were increased by amounts corresponding to the outputs of the new steel and steel-fabricating capacities (which form the bill of goods in the problem of this paper) the resulting expansions in all activities of the nation which would be required would yield a dollar total somewhat greater than that obtained in this paper for the subsequent expansions of activities in our area. Put otherwise, for every new employee in the initial steel and steel-fabricating plants, we estimate that on the average another 0.79 persons would be employed in the subsequent expansions of activities in our area. In comparison, where the same bill of goods increase is injected into the BLS national model a somewhat greater number of persons would be employed in the required expansions of all activities in the nation for every employee corresponding to this bill of goods increase.

Notes

1. This study was done partly under the auspices of the Social Science Research Center, the University of Puerto Rico, and of the Center of Urban and Regional Studies, the Massachusetts Institute of Technology. The basic projections were completed in February 1952 for investment use by the firm of Minot, De Blois and Maddison, who generously financed the work in the initial stage.
2. C. J. Friedrich trans., *Alfred Weber's Theory of the Location of Industries* (Chicago, 1929), chs 5–7.
3. T. Palander, *Beiträge zur Standortstheorie* (Uppsala, 1935), ch. 8; see also E. M. Hoover, Jr., *Location Theory and the Shoe and Leather Industries* (Cambridge, Mass., 1937), pp. 89–99; and O. Engländer, 'Kritisches und Positives zu einer allgemeinen reinen Lehre vom Standort', *Zeitschrift für Volkswirtschaft und Sozialpolitik*, Neue Folge, vol. V (1926). For a defense of Weber's agglomeration theory, see E. Niederhauser, 'Die Standortstheorie Alfred Webers', *Staatswissenschaftliche Studien*, Bd. vol. XIV (1944), pp. 181–7.
4. National Resources Planning Board, *Industrial Location and National Resources* (Washington, 1943), ch. 5; and P. S. Florence, *Investment, Location, and Size of Plant* (Cambridge, England, 1948), esp. ch. IV.
5. Florence's interpretations may be construed by some as implicit theorising. In particular, objection may be made to a conversion of his coefficient of geographical association into a general indicator of economic spatial dependence between industries. It should be noted that such a conversion can be questioned in view of the statistical shortcomings of the coefficient. See Robert E. Kuenne, *The Use of Input-Output Techniques for the Estimation of Employment in the Delaware Valley* (PhD dissertation, Harvard University, 1953), ch. 2.
6. In many respects our impact study for a region resembles the pioneering, but neglected, study of Barfod's for a city. (Børge Barfod, *Local Economic Effects of a Large-Scale Industrial Undertaking* (Copenhagen, 1938). Because of the larger quantities of data currently available, we have been able to adopt a more comprehensive approach.

   Among other studies, somewhat related to ours, are: M. C. Daly, 'An Approximation to a Geographical Multiplier', *Economic Journal*, vol. L, June–September 1940, pp. 248–58; Federal Reserve Bank of Kansas City, 'The Employment Multiplier in Wichita', *Monthly Review*, vol. 37 (September 1952); E. Ullman, 'The Basic-Service Ratio and the Areal Support of Cities', forthcoming.

7. W. Isard 'Some Locational Factors in the Iron and Steel Industry Since the Early Nineteenth Century', *Journal of Political Economy*, vol. LVI, June 1948, pp. 203–17; W. Isard and W. Capron, 'The Future Locational Pattern of Iron and Steel Production in the United States', *Journal of Political Economy*, vol. LVII, April 1949, pp. 118–33; W. Isard and J. Cumberland, 'New England as a Possible Location for an Integrated Iron and Steel Works', *Economic Geography*, 26 (October 1950); John H. Cumberland, *The Locational Structure of the East Coast Steel Industry With Emphasis on an Integrated New England Steel Mill* (PhD dissertation, Harvard University, 1951). For the analysis of locational factors in another basic industry, see John Krutilla, *The Structure of Costs and Regional Advantage in Primary Aluminum Production* (PhD dissertation, Harvard University, 1952).
8. This region includes the following counties: Fairfield and New Haven in Connecticut; Kings, Queens, Richmond, New York, Bronx, Westchester, Orange, Putnam, Rockland, Suffolk, and Nassau in New York; Bergen, Passaic, Morris, Essex, Hudson, Union, Middlesex, Monmouth, Ocean, Burlington, Atlantic, Salem, Gloucester, Camden, Mercer, Hunterdon, Somerset, Warren, and Sussex in New Jersey; and Chester, Delaware, Philadelphia, Montgomery, and Bucks in Pennsylvania.

   Ideally, the area should be defined in such a way as to include all of those counties and parts of counties which together comprise a continuous spatial entity and which will undergo noticeable change as a result of the steel development. Practically, because of data inadequacies, only whole counties can be treated. From a consideration of the nature and direction of transportation facilities, industrial and residential land use, population distribution, and similar characteristics, the above listed counties seem to form the most appropriate spatial entity for our problem, although undoubtedly some contiguous areas will be directly affected, too.
9. See Isard, op. cit., pp. 203–04; and Isard and Capron, op. cit., pp. 120–3.
10. O. Johnson, *Steel Consumption in 1948: A Report by Iron Age to the Metal-working Industry*. The New York–Trenton–Philadelphia industrial area, considerably smaller than our Greater New York–Philadelphia urban-industrial region, includes the following counties: Fairfield and New Haven in Connecticut; Bronx, Kings, Nassau, New York, Queens, Richmond, and Westchester in New York; Bergen, Burlington, Camden, Essex, Gloucester, Hudson, Mercer, Middlesex, Passaic, and Union in New Jersey; and Bucks, Chester, Delaware, Montgomery, and Philadelphia in Pennsylvania.
11. For Connecticut the *Census of Manufactures*, 1947, vol. I, pp. 244–5, reports consumption by metal-workers of 509,210 tons as against 1,194,694 tons in *Iron Age*; for New York, 2,109,010 tons as against 3,232,356 tons; for New Jersey, 1,130,845 tons as against 1,684,699 tons; and for Pennsylvania, 5,600,224 tons as against 5,498,654 tons.
12. The adjustment procedure was as follows: for those sectors of our area in Connecticut, New York, and Pennsylvania, the consumption for each (as given in *Iron Age*) was taken as a percentage of the appropriate *Iron Age* state figure. These percentages were multiplied by the appropriate Census data for the state, and the products were added to the total consumption listed in the *Census* for New Jersey.
13. Exports from the port districts of New York and Philadelphia of iron and steel mill products and semi-manufactures were 2.6 million tons and 451,800 tons respectively in 1947, and 1.6 million tons and 283,000 tons respectively in 1948. US Army, Board of Engineers for Rivers and Harbors, *Commercial Statistics; Water-borne Commerce* (Washington, DC); and Annual Reports, pt. 2, 1947,

pp. 231, 285–366, and pt. 2, 1948, pp. 255, 315–97.

14. In view of such competition we postulate that only 50 per cent of the new steel demand arising as a result of this regional development will be supplied by producers in the area. For discussion of this point see page 106 and column 2, Table 5.5 below.
15. It should also be noted that a secondary objective of this study was to isolate and pin-point those localities in our area which would experience highest rates of population growth and physical development and hence would be most attractive for nonspeculative real estate investment, *ceteris paribus*. For this objective, too, firm minimum estimates are more desirable insofar as they reduce the element of risk.
16. It is interesting to note that the United States Steel Company in discussing the operation of the Fairless Works stated: 'Initial operation of the plant starting in 1952 will require delivery of 3,300,000 gross tons of coal and foreign ore annually by water. . . . The rolling mills which will be installed at the Fairless Works in 1951–52 will have sufficient capacity to consume the output of probable future open hearths and blast furnaces requiring an additional annual tonnage of 3,700,000 gross tons of iron ore and coal'. (*Statement by United States Steel Company at Public Hearing by US Army Engineers March 14, 1951, Trenton, New Jersey*, mimeographed, pp. 2–3.) These statements, in addition to others, suggest a capacity well beyond the 1.8 million tons currently under construction.
17. Florence, op. cit.; and National Resources Planning Board, op. cit., ch. 5.
18. Though the quality of the Iron Age data is poor, they are used since there are no other data on three-digit steel-fabricating activities for such a fine breakdown of the United States.

    Because of the aforementioned reasons and those cited in the source quoted in footnote 5, we have not computed the coefficient of geographical association. Rather, we have carefully inspected these two sets of data.
19. Moreover, because costs of relocation are significant, consideration was given to the rate at which each activity has been expanding nationally. It was felt that, *ceteris paribus*, greater employment could be anticipated for our area in the faster growing than in the slower growing industries.
20. *Census of Manufactures*, 1947, vol. I, p. 238.
21. Hot-rolled sheet and strip capacity grew from 498,000 tons in 1938 to 1,195,000 in 1948 for Maryland; and for the same years from 85,100 tons to 352,210 tons for California. These represent respectively a more than doubling and tripling effect for these states. Data from American Iron and Steel Institute, *Annual Reports*.
22. The lack of any clear-cut patterns in the three-digit industries of the two-digit categories, industry groups 36, 38, and 39, strongly argued for the use of the two-digit aggregates of these three-digit industries.
23. It should be noted that expansion in metal furniture trades has been omitted. Difficulties of separating metal furniture from nonmetal furniture manufacture, both in the above steps and the steps to follow, argued for excluding this activity, although about 2500 new production workers in metal furniture trades might be anticipated. This contributes to the minimal nature of our projections. It still further balances any overestimates that may have crept into the analysis, especially in the industry groups, 354–59, where a reexamination of procedure after computations were completed indicated that we had made less conservative estimates for these three-digit industries than for other three-digit industries.

24. For curiosity's sake alone, a multiple linear correlation between production workers and four explanatory variables was made for the 69 industrial areas of Table 5.2 (balances of states were excluded). The four variables were: (1) market proximity, as given by an index measuring distances of any given area from the eleven largest market centres in the United States; (2) distance of that area from a region of major steel capacity (arbitrarily defined as one capable of producing more than 1,250,000 tons annually); (3) population (1950) of that area; and (4) steel capacity of the given area.

    A multiple correlation coefficient of .82 was obtained. However, only 11 of the estimates derived from the estimating equation were within plus or minus ten per cent of the actual employment data. Injection of new values for the variables to account for new steel capacity in the New York–Trenton–Philadelphia industrial area led to an estimate of 45,000 new production workers in metal-fabricating activities for this area. However, no significance should be attributed to this finding, since extreme items have seriously distorted the influence of steel capacity upon fabricating employment, and since, in any case, multiple correlation analysis has little meaning in this particular context. For a more detailed presentation of the correlation analysis, see Kuenne, op. cit., ch. 2.
25. To the extent that the input structures of these individual activities tend to resemble one another, errors of over-and under-estimation tend to cancel themselves when we proceed to quantify subsequent repercussions. In the limit, when input structures are identical, the industries may be aggregated into one industry without loss of accuracy.
26. These cents' worth of inputs are simply the technical coefficients in the Leontief input-output model. The coefficients for a 50 × 50 matrix were developed for 1947 by the Bureau of Labor Statistics. See table 5 in W. D. Evans and M. Hoffenberg, 'The Interindustry Relations Study for 1947', *The Review of Economics and Statistics*, vol. XXXIV, May 1952.

    The dollar volume corresponding to new employment in steel and each fabricating activity was crudely obtained by multiplying the dollar volume per production worker for steel and for each activity, as derived by the 1947 *Census of Manufactures* for the United States as a whole, by the anticipated new employment. This derivation of input requirements assumes linear homogeneous production functions. In addition it should be borne in mind that the production worker data of the *Census* and *Iron Age* diverge. Hence, unavoidable error is introduced, since new employment estimates of three-digit categories are stated in terms of *Iron Age* production workers while the dollar volume expansion of each activity is based on *Census* production worker data.
27. An error of computation crept into the construction of this bill of goods. Production worker employment in industry 343 was overstated by 477. In view of the crudeness of the analysis and the smallness of the error, recomputation of the round effects did not seem to be warranted.
28. New construction was removed from the Bureau of Labor Statistics structural matrix because its inclusion is questionable. Given the data currently available, any separation of construction into that on current and that on capital account involves subjective judgements. As a consequence, results obtained on the requirements of construction when construction is included in the structural matrix tend not to be superior to those obtained by the crude rule-of-thumb method which will be employed later.
29. See W. Isard, 'Interregional and Regional Input-Output Analysis: A Model of a Space-Economy', *The Review of Economics and Statistics*, vol. XXXIII,

November 1951. Experimentation with an interregional model is currently being undertaken.

30. In the case of households a figure of 82 per cent was used in order to allow for some outflow to persons residing in other areas and to recognise leakages through savings and other channels.

    A useful check on the percentages was furnished by data derived for 1939 in connection with regional input-output work of the Harvard Economic Research Project (see W. Isard, 'Some Empirical Results and Problems of Regional Input-Output Analysis', ch. 5 in W. Leontief, *et al.*, *Studies in the Structure of the American Economy*, New York, 1953). In this work, for each of a group of industries somewhat similar to ours, estimates were made of the consumption of outputs by states. Then, subtracting for any given activity the consumption estimate from the actual production for a given state yields a surplus or deficit which can be expressed as a percentage of total consumption of that activity in that state. This percentage gives some indication of the degree to which a state supplies its own needs for the outputs of any given industry. The percentages derived for 1939 for the states of Connecticut, New York, New Jersey, and Pennsylvania tend to have either smaller deficits than our percentages would yield or actual surpluses.

31. Under certain circumstances, different sets of percentage reduction factors can and should be used from round to round to estimate expansions. See W. Isard and M. J. Peck, 'Location Theory and International and Interregional Trade Theory', *Quarterly Journal of Economics*, forthcoming.

32. These, of course, represent the dollar value of each industry's product in 1947 prices.

33. Data were obtained from the *Census of Manufactures*, 1947; *Census of Business*, 1948; *Minerals Yearbook*, 1947; *Survey of Current Business*, July 1949; various issues of the *Monthly Labor Review*; and from the Harvard Economic Research Project.

34. The 11,666 estimate of number of total workers for the 3.0 million tons of initial new capacity is obtained from data available on the Fairless Steel Works. See Pennsylvania State Planning Commission, *Boom in Bucks County* (mimeographed), p. 1.

35. Column 9 of Table 5.5, when converted into dollar figures, approximates the column vector X, where $X = (I-RA) - {}^{1}Y$. In the term on the right-hand side of the equation, Y represents the bill of goods column vector whose elements are the outputs of the initial steel and steel-fabricating activities. It is multiplied by the inversion of the matrix $(I - RA)$ where A is the coefficient (structural) matrix constructed to include households, and where R is the diagonal matrix whose elements are the set of percentages of column 2, Table 5.5.

36. We estimated construction employees: (1) by dividing the total national labour force in 1947 less government and construction workers into the total of construction workers for the nation; and (2) by multiplying this ratio by the total of the estimates in column 9.

37. We estimated government employees: (1) by dividing the total national labour force in 1947 less all government employees (state, local, and federal) into the total of state and local government employees in 1947 for the nation; and (2) by multiplying this ratio by the total of the estimates in column 9 plus new construction employees. This calculation, which excludes an allowance for federal employment, is in keeping with the probable less-than-proportional expansion of government employees, and also in line with our 'firm minimum' objective.

38. According to the *Census of Population*, 1950 (general characteristics reports by states), the percentages for New Jersey, for the New York–Northeastern New Jersey metropolitan area, and for the Philadelphia metropolitan area are 43.5, 44.0, and 42.2 respectively.
39. Among others, see 'Input-Output Analysis: Discussion', *Papers and Proceedings of the American Economic Association*, vol XXXIX, May 1949, pp. 226–40; and W. Leontief, *The Structure of American Economy, 1919–1929* (Cambridge, Mass., 1941), pp. 38–41.
40. This is not at all unlikely since our problem concerns a long-term adjustment. In the longrun labor from other regions can be attracted to these industries in the area, and growth of the area's population can take place. As a consequence the new income is likely to be absorbed by new job-holders.
41. These distortions are certainly not likely to be any greater than those arising in general input-output analysis from insufficient disaggregation of industrial categories into component activities, from insufficient attention to varying product-mix of firms encompassed by any one component industry, and from differential efficiencies among such firms. The latter, too, pervade our analysis.
42. The reader may care to adjust our employment estimates by a net labor saving coefficient. See W. Leontief, *et al.*, op. cit., pp. 27–35, for some analysis of changes in coefficients in past periods.
43. Included, however, is an allowance for 'average' new construction activity. This activity plus replacement of depreciated facilities engages the labor of the new employees in construction.
44. This point is neatly borne out by operations with the Leontief balanced regional model. (Leontief, *et al.*, op. cit., chs. 4 and 5.) In the Leontief model, households are removed from the bill of goods and put into the structural matrix without any compensating changes. Hence, the use of the resulting inverted matrix gives the following unrealistic results: an increase of one dollar in the bill of goods requirements for the output of agriculture and fishing necessitates an expansion of $7.59 in outputs for all industries in the nation; a similar increase in the bill of goods requirements for food-processing necessitates expansions amounting to $6.64 for all industries; and the same increase for iron and steel foundry products leads to a total expansion of $7.03 for all industries; and so forth. Clearly, these expansionary effects are extreme, and point up the inadequate leakage of the Leontief model. On the other hand, when households are retained in the bill of goods, similar one dollar increases in the bill of goods requirements for the output of each of these three activities necessitate respective expansions of $1.73, $2.14 and $1.89 in the outputs for all industries.

    These statements, however, do not constitute a criticism of the Leontief balanced regional model. The Leontief model was designed to obtain the percentage change of the output of each activity from its 1939 level when the bill of goods demand for any given activity is changed in order to make comparisons among these percentage changes. It was not designed to obtain absolute changes. The data resulting from various operations with this model are presented in percentage terms, and not in dollars.
45. Ideally, as improved models are developed to encompass this interregional feedback effect, those elements of each industrial category which do not respond readily in approximately proportionate manner to general impulses and consequent changes in their outputs and the outputs of related industries should be removed from the structural matrix. For example, that set of inputs which households require but which are relatively unrelated to income and

must be considered as functionally related to other variables should be taken out of the matrix and transferred to the bill of goods. Thus, the construction of better models necessitates the inclusion of more realistic response patterns in the matrix.

6 Industrial Complex Analysis, Agglomeration Economies, and Regional Development[1]

I INTRODUCTION

This article is a sequel to a paper on industrial complex analysis[2] which reported on the first phase of a research project with reference to Puerto Rico. In addition to reporting on the second phase of this project, this article attempts to attack some of the difficult conceptual problems which still remain in the general field of agglomeration theory and interindustry relations.

II THE DIMENSIONS OF THE PROBLEM

One distinguishing characteristic of modern industrial production is the existence of numerous groups or combinations of activities, the activities of each group being subject to important technical, marketing, or other interrelations. (In this article an activity is defined as a single productive process.) Such a group of activities may comprise the successive stages in the manufacture of an end-product; or the joint production of two or more commodities from a single class of raw materials; or the production of two or more intermediate commodities and their combination to form other intermediates and end-products; or various combinations of these. (In fact, one may define a *major* activity group in terms of a set of smaller activity groups).

If a locational study is undertaken as part of an economic evaluation of such a group of activities, it must, to be meaningful, analyse the group as a whole as well as each of its several activities. Analysis which considers each activity alone, as though the activity were independent of the others, is inadequate. The necessity for joint as well as individual consideration is apparent when the activities represent successive stages in the production of a single commodity or commodity class. For example, the location pattern of ingot steel production cannot be ascertained without a knowl-

This chapter was written with Eugene W. Schooler. Reprinted from *Journal of Regional Science*, vol. 1, Spring (1959), pp. 19-34.

edge of the location of pig iron production. But the location pattern for pig iron production cannot be established unless the location of its market, i.e., the production of ingot steel, is known. Furthermore, these activities are also influenced by the location of ore and coal deposits as well as steel rolling and finishing operations and the markets for fabricated steel products. Thus an adequate location study must consider the whole combination of activities.

Joint analysis is no less necessary when the products of the individual activities have multiple uses, or when the group of activities include more than one industry type. Acetylene and hydrogen cyanide produced from refinery gases can be used to manufacture acrylonitrile, which can then be transformed into orlon fiber and ultimately into textile fabrics. In this group of activities three industries are represented: petroleum refining, chemicals, and textiles. Furthermore, most of the intermediate commodities have alternative uses, and thus alternative markets. But with regard to this particular production sequence, none of the activities could be considered a priori as having an independent production location pattern. A joint locational evaluation is necessary. It is quite possible that such analysis would indicate multiple locations for some of the activities, as a consequence of the multiple uses for the commodities.

In this connection, perhaps the objective of a locational analysis may be to assess the chances of a *newly developing* region to share in the growth of a given industrial activity. If the activity in question is part of a group of interrelated activities such as those described, the analyst should not base his evaluation on the assumption that the location pattern of all the other activities in the group will be unaffected, i.e., may be taken as fixed. Such a procedure will often yield misleading results, and will generally tend to understate the locational advantage of the region under study. An adequate analysis needs to consider possible locational changes in *all* the interrelated activities. If for example it is felt that a newly developing region might be a favourable location for automobile production or assembly, account should also be taken of the possibility that such activities as glass, steel, and aluminum production might also locate in the same region and thus favourably influence a locational decision on an automobile assembly works. Or perhaps a region is interested in acquiring a meat packing industry. Any attempt at a detailed evaluation should consider the possibilities and influences of changes of the locational patterns of leather production and shoe manufacture, e.g., that leather production and shoe manufacture may also be profitably located in the region when the group of activities is considered as a whole.

Account should be taken at this juncture of the possible objection that since *all* economic activities are interrelated – that everything depends on everything else – the analyst could, strictly speaking, justify nothing less than the evaluation of a general spatial equilibrium in each locational study. Actually, the very selection of the group of activities to be evaluated involves, in one sense, a consideration of all economic activities, with the decision that the interrelations of most of them with the activity or activities under investigation are of such a small order that they can be justifiably be ignored (particularly when research resources are limited).

III THE INDUSTRIAL COMPLEX CONCEPT AND ITS ANALYSIS

Associated with the type of regional analysis envisioned in the above paragraphs is the concept of an industrial complex. An industrial complex may be defined as one or more activities occurring at a given location and belonging to a group of activities which are subject to important production, marketing, or other interrelations. (The case of one activity represents an extreme case of an industrial complex; this case is included in the definition simply for reasons of convenience in exposition.)

Having defined an industrial complex, we may now proceed to analysis. We could develop in the rest of this paper the step-by-step procedure used over the last four years in arriving at conclusions regarding a set of specific industrial complexes. But, now that our study of these complexes is complete, we realise that many of the steps in our procedure were haphazard and that this reflects the hit-or-miss approach characteristic of much basic research. After considerable thought on procedure we better see what is an efficient set of steps. Therefore, instead of presenting the procedure actually used we shall attempt to develop a logical framework for industrial complex analysis, with a minimum of deviation from standard location approaches and terminology. This means that the data which we shall present will not be as meaningful as they would be in connection with our outmoded procedures.

It now seems that industrial complexes can be attacked in a fashion quite similar to the Weberian approach to the location study of a particular industry. There is however, the one major difference, namely, that an industrial complex is typically not a single activity or industry but a set of locationally interdependent activities which are hooked up by a matrix of structural relations. For convenience sake, this matrix may be conceived of as an internal input-output matrix; but it is more, for it allows variable coefficients of production as well as process substitution (i.e. alternative processes) in the linear programming sense.

Now, following modern Weberian doctrine, we assume a fixed market for the products of the industrial complex. (Later, we will check the 'first-run' results against the assumed pattern of markets, make changes in our assumptions if there are major inconsistencies, and proceed in an iterative fashion.) Next we assume as given the geographic positions of sources of each of the raw materials. (At the start we can assume either one or more sources of a particular raw material. If there are more than one source of a given raw material we can choose that source which initially seems best; and, after the first-run results are obtained and even modified, consider, with the use of standard market area analysis, whether or not a second or third or fourth, etc. source is more desirable.)

Next we specify a reasonable internal structure for a reasonable industrial complex. For example, Figure 6.1 depicts a reasonable internal structure for the Dacron A complex, which as we shall see, is reasonable too. This complex has been described in the previous paper.[3] It is not necessary to go into details for this paper. Suffice is to indicate that the complex has an oil refinery which produces the customary refinery products. Several of these products are gases. One of these gases, namely ethylene, is processed along with imported paraxylene, ultimately to become Dacron staple (upper part of the figure). Other gases, such as hydrogen and methane, are processed into ammonia, from which comes fertiliser (lower part of the figure).

Once a reasonable internal structure of a reasonable industrial complex is determined, we calculate total input requirements and output yields. (That is, we obtain the inputs of raw materials and labour, and the output, which are assumed in the Weberian locational polygon analysis.) Such total input requirements and output yields are calculated via an input-output table such as Table 6.1, (and via other means when input-output relations are non-linear). This table is described in the first paper. We do not have the space to repeat this description here.

To illustrate total input requirements for a set of reasonable industrial complexes, we present Table 6.2. In Table 6.2, negative signs indicate inputs; positive signs, outputs. Column 1 of Table 6.2 presents most of the total input requirements and output yields for the Dacron A complex.

Given the above assumptions, and the original and derived materials for any industrial complex, we seek the optimal transport point. That is, we construct a Weberian locational polygon, and via the use of geometry, or a Varignon physical model, or the Palander–Hoover isodapane approach, or a simple dollars-and-cents cost calculation, determine the point of minimum transport cost. Unfortunately we cannot present here all the empirical materials which would fully develop a case. However we can give some illustrations with respect to the refinery-petrochemical-synthetic fibre complexes mentioned above. For these complexes, the Gulf Coast turns out to be the point of minimum transport cost.[4] If we now take any other site of production, say Puerto Rico, we should expect total transport cost to be higher at this latter location. The extent of this difference (disadvantage) is approximately indicated in column 1 of Table 6.3 for selected reasonable industrial complexes.

Still following Weber, once the point of minimum transport cost is

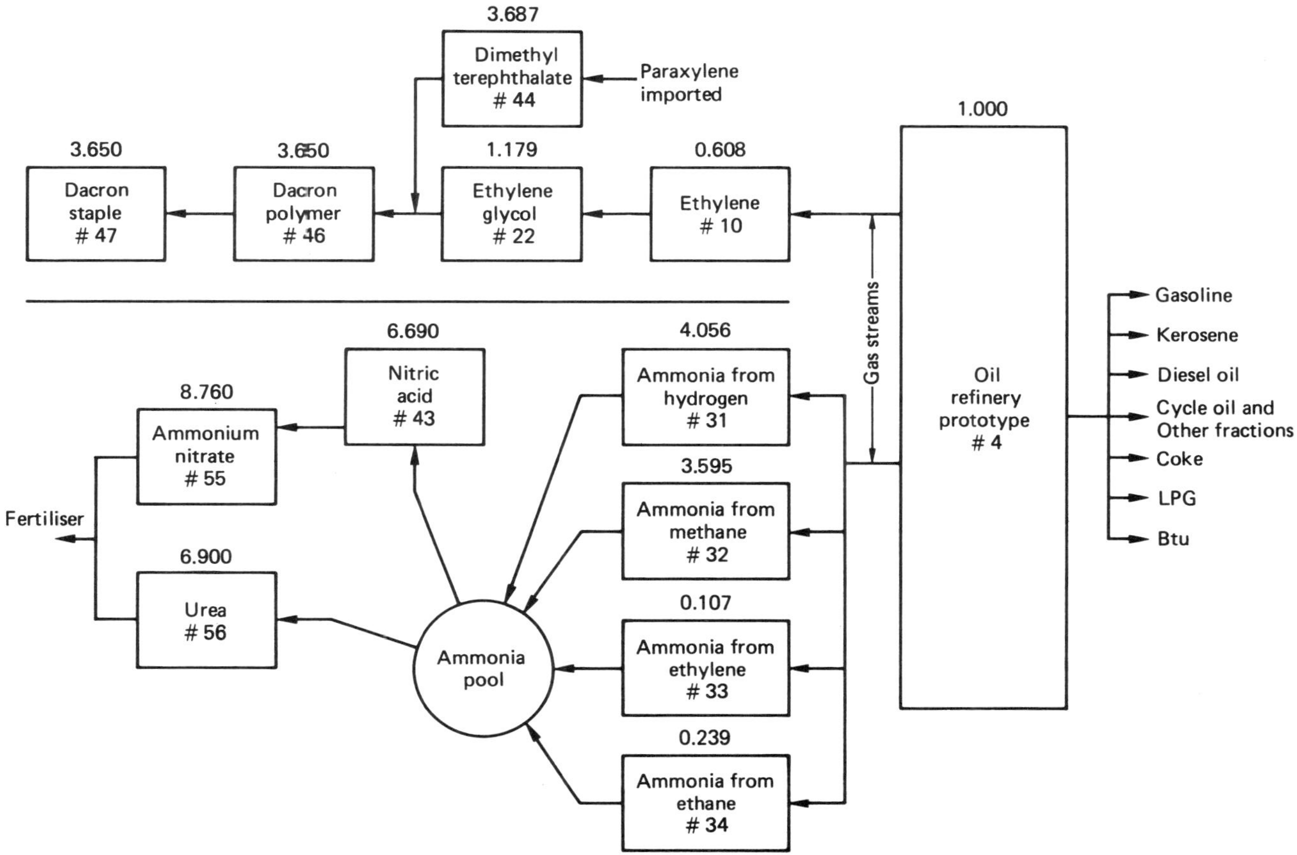

Figure 6.1 'Dacron A' programme

19. Propane MMlb	+ 2.150	+ 5.080	− 5.080												
20. Butylenes MMlb															
21. Butanes MMlb															
22. Pure ethylene MMlb			+16.100	− 8.300											
23. Pure ethane MMlb			+30.190												
24. Steam MMMlb	− 0.801	− 1.402	− 0.148	− 0.103		− 0.023	− 0.023	− 0.023		− 0.030	− 0.060	− 0.500	− 0.007	− 0.028	− 0.555
25. Power MMKWH	− 2.511	− 3.999	− 0.194	− 0.800	− 4.640	− 5.600	− 5.600	− 5.600	− 1.200	− 5.200	− 2.500	−12.000	− 0.170	− 0.340	−16.000
26. Fuel 10×MMMBTU	−139.000	−242.000		− 2.010		− 0.450	− 0.450	− 0.450		− 2.800	− 1.000			− 2.250	
34. Nitrogen MMlb				+68.000											− 2.200
35. Ethylene glycol MMlb				+10.000							− 3.230				
39. Ammonia MMlb					+10.000	+10.000	+10.000	+10.000	− 2.860				− 2.380	− 5.800	
40. HCN MMlb															
41. Acrylonitrile MMlb															
42. Methanol MMlb										− 4.000	+ 3.350				
43. Sulphur MMlb															
44. Sulphuric acid MMlb															
45. Nitric acid MMlb									−10.000				− 7.630		
46. Paraxylene MMlb										− 6.800					
47. Dimethyl Terephthalate MMlb										+10.000	−10.000				
48. Dacron polymer MMlb											+10.000	−10.000			
49. Dacron staple MMlb												+10.000			
59. Ammonium nitrate MMlb													+10.000		
60. Urea MMlb														+10.000	
61. Carbon dioxide MMlb														− 7.500	
74. Nylon salt															−10.000
76. Nylon filament															+10.000

19. Propane MMlb	+ 2.150	+ 5.080	− 5.080												
20. Butylenes MMlb															
21. Butanes MMlb															
22. Pure ethylene MMlb			+16.100	− 8.300											
23. Pure ethane MMlb			+30.190												
24. Steam MMMlb	− 0.801	− 1.402	− 0.148	− 0.103		− 0.023	− 0.023	− 0.023		− 0.030	− 0.060	− 0.500	− 0.007	− 0.028	− 0.555
25. Power MMKWH	− 2.511	− 3.999	− 0.194	− 0.800	− 4.640	− 5.600	− 5.600	− 5.600	− 1.200	− 5.200	− 2.500	−12.000	− 0.170	− 0.340	−16.000
26. Fuel 10×MMMBTU	−139.000	−242.000		− 2.010		− 0.450	− 0.450	− 0.450		− 2.800	− 1.000			− 2.250	
34. Nitrogen MMlb				+68.000											
35. Ethylene glycol MMlb				+10.000							− 3.230				− 2.200
39. Ammonia MMlb					+10.000	+10.000	+10.000	+10.000	− 2.860				− 2.380	− 5.800	
40. HCN MMlb															
41. Acrylonitrile MMlb															
42. Methanol MMlb										− 4.000	+ 3.350				
43. Sulphur MMlb															
44. Sulphuric acid MMlb															
45. Nitric acid MMlb									−10.000				− 7.630		
46. Paraxylene MMlb										− 6.800					
47. Dimethyl Terephthalate MMlb										+10.000	−10.000				
48. Dacron polymer MMlb											+10.000	−10.000			
49. Dacron staple MMlb												+10.000			
59. Ammonium nitrate MMlb													+10.000		
60. Urea MMlb														+10.000	
61. Carbon dioxide MMlb														− 7.500	
74. Nylon salt															−10.000
76. Nylon filament															+10.000

31. Limestone MMlb		− 14.185		− 14.185			− 29.927	
32. Lime (hydrated) MMlb		− 0.005		− 0.005				
33. Ethylene Oxide MMlb					+ 0.013		+ 0.003	
34. Nitrogen MMlb	+ 80.172		+ 80.172		+265.948			
35. Ethylene glycol MMlb	+ 0.001	+ 0.001	+ 0.001	+ 0.001				
36. Ethylene dichloride MMlb							− 0.003	
37. Acetylene MMlb						+ 0.002		+ 0.002
38. Vinyl chloride MMlb							+ 0.005	+ 0.005
39. Ammonia MMlb	− 0.032	− 0.024	− 0.018	+ 0.004	− 0.003	− 0.020	+ 0.001	+ 0.017
40. HCN MMlb					+ 0.002	+ 0.002	+ 0.002	+ 0.002
41. Acrylonitrile MMlb					+ 0.005	+ 0.005		
42. Methanol MMlb	− 2.522	− 2.522	− 2.522	− 2.522				
43. Sulphur MMlb			− 1.777	− 1.777				
44. Sulphuric acid MMlb			− 0.002	− 0.002				
45. Nitric acid MMlb	+ 0.061	+ 0.061	+ 0.065	+ 0.065	+ 0.060	+ 0.060	+ 0.061	− 0.020
46. Paraxylene MMlb	− 25.072	− 25.072	− 21.753	− 21.754				
47. Dimethyl terephthalate MMlb	+ 0.005	+ 0.005	+ 0.005	+ 0.005				
48. Dacron polymer MMlb								
49. Dacron staple MMlb	+ 36.500	+ 36.500	+ 36.500	+ 36.500				
50. Dacron filament MMlb								
51. Dynel polymer MMlb								
52. Acetone MMlb							− 10.950	− 10.950
53. Dynel staple MMlb							+ 36.500	+ 36.500
54. Dynel filament MMlb								
55. Orlon polymer MMlb								
56. Dimethyl formamide MMlb					− 8.395	− 8.395		
57. Orlon staple MMlb					+ 36.500	+ 36.500		
58. Orlon filament MMlb								
59. Ammonium nitrate MMlb	+ 87.600	+ 87.600	+ 87.600	+ 87.600	+ 87.680	+ 87.680	+ 87.600	− 87.680
60. Urea MMlb	+ 69.000	+ 69.000	+ 69.000	+ 69.000	+ 68.960	+ 68.960	+ 69.000	+ 68.960
61. Carbon dioxide MMlb	− 51.750	− 51.750	− 51.750	− 51.750	− 51.720	− 51.720	− 51.750	− 51.720
Textile labour MMmhr	− 5.284	− 5.359	− 5.283	− 5.358	− 5.315	− 5.280	− 5.396	− 5.334
Chem-pet. labour MMmhr	− 2.229	− 2.473	− 2.226	− 2.469	− 2.329	− 2.217	− 2.590	− 2.390

31. Limestone MMlb		− 14.185		− 14.185			− 29.927	
32. Lime (hydrated) MMlb		− 0.005		− 0.005				
33. Ethylene Oxide MMlb					+ 0.013		+ 0.003	
34. Nitrogen MMlb	+ 80.172		+ 80.172		+265.948			
35. Ethylene glycol MMlb	+ 0.001	+ 0.001	+ 0.001	+ 0.001				
36. Ethylene dichloride MMlb							− 0.003	
37. Acetylene MMlb						+ 0.002		+ 0.002
38. Vinyl chloride MMlb							+ 0.005	+ 0.005
39. Ammonia MMlb	− 0.032	− 0.024	− 0.018	+ 0.004	− 0.003	− 0.020	+ 0.001	+ 0.017
40. HCN MMlb					+ 0.002	+ 0.002	+ 0.002	+ 0.002
41. Acrylonitrile MMlb					+ 0.005	+ 0.005		
42. Methanol MMlb	− 2.522	− 2.522	− 2.522	− 2.522				
43. Sulphur MMlb			− 1.777	− 1.777				
44. Sulphuric acid MMlb			− 0.002	− 0.002				
45. Nitric acid MMlb	+ 0.061	+ 0.061	+ 0.065	+ 0.065	+ 0.060	+ 0.060	+ 0.061	− 0.020
46. Paraxylene MMlb	− 25.072	− 25.072	− 21.753	− 21.754				
47. Dimethyl terephthalate MMlb	+ 0.005	+ 0.005	+ 0.005	+ 0.005				
48. Dacron polymer MMlb								
49. Dacron staple MMlb	+ 35.500	+ 36.500	+ 36.500	+ 36.500				
50. Dacron filament MMlb								
51. Dynel polymer MMlb								
52. Acetone MMlb							− 10.950	− 10.950
53. Dynel staple MMlb							+ 36.500	+ 36.500
54. Dynel filament MMlb								
55. Orlon polymer MMlb								
56. Dimethyl formamide MMlb					− 8.395	− 8.395		
57. Orlon staple MMlb					+ 36.500	+ 36.500		
58. Orlon filament MMlb								
59. Ammonium nitrate MMlb	+ 87.600	+ 87.600	+ 87.600	+ 87.600	+ 87.680	+ 87.680	+ 87.600	− 87.680
60. Urea MMlb	+ 69.000	+ 69.000	+ 69.000	+ 69.000	+ 68.960	+ 68.960	+ 69.000	+ 68.960
61. Carbon dioxide MMlb	− 51.750	− 51.750	− 51.750	− 51.750	− 51.720	− 51.720	− 51.750	− 51.720
Textile labour MMmhr	− 5.284	− 5.359	− 5.283	− 5.358	− 5.315	− 5.280	− 5.396	− 5.334
Chem-pet. labour MMmhr	− 2.229	− 2.473	− 2.226	− 2.469	− 2.329	− 2.217	− 2.590	− 2.390

costs in attracting such labour. Such costs may be approximated at around 2.5 million dollars as indicated in column 3. Subtracting from Puerto Rico's advantage on textile labour account both the disadvantage on transport account and the disadvantage on chemical labour account yields column 4. This column does indicate the economic feasibility of deviating at least selected industrial complexes from the sites of an optimal pattern on the Mainland to a cheap labor site, namely Puerto Rico. But perhaps it would be still more economic to consider deviating only part of these selected industrial complexes and not the whole. This matter could logically be considered at this point. However, for exposition purposes, we prefer to illustrate this step with empirical materials at a later point.

An industrial complex analysis must proceed further. Modern Weberian doctrine would introduce agglomeration economies – more specifically: scale economies, localisation economies, and urbanisation economies, and certainly no later than this point modern Weberian doctrine would permit substitution among factor inputs and product outputs. Let us first treat substitution.

When we consider the location of a complex at two different sites, say a resource site and a market site, it is wrong at least theoretically to postulate that the internal structure of the complex will be identical at both sites. For at the resource site, the delivered price of the resource will be lower than the delivered price of the resource when location is at the market. Therefore we can expect that relative to the optimal internal structure at the market, location at the resource site will tend to involve substitution of the resource for other inputs. Or, if a complex either in whole or part deviates to a cheap labour point, we anticipate the general substitution of the cheap factor, labour, for other factors as well as a host of subsidiary substitutions. Typically in industrial complexes, such substitution, i.e., change in factor proportions, involves process change, since small substitutions along a continuous transformation function are in actuality infrequent. This is why in the original input-output matrix all feasible production processes must be included. (And in this sense the input-output matrix is identical with the activity matrix of linear programming.) We shall illustrate process change and substitution later.

IV AGGLOMERATION AND SPATIAL JUXTAPOSITION ECONOMIES

Now we may introduce the first major type of agglomeration economy, namely scale economies, as modern Weberian theory would have us do. Such economies differ from location to location. For example, major industrial activity existing at one site may make possible economies of scale not possible at a second site where no industrial activity is present. This relation is clearly demonstrated in our refinery-petrochemical-synthetic fibre complex. The Gulf Coast is a major industrial area. Puerto Rico is at most a minor industrial area. Therefore, practically all petrochemical activities can be operated at scales and achieve scale economies in the Gulf Coast not possible for Puerto Rico. Hence realistic analysis must account for differences in scale economies, as we do in column 5 of Table 6.3. But bear in mind that scale economies also involve process change, and thus influence in a significant way factor proportions and the internal structure of industrial complexes at any site. To illustrate, because the market for nylon salt is relatively small in Puerto Rico, a process, namely an all-adipic acid process, which achieves relatively low unit costs at relatively small outputs is employed in Puerto Rico. In contrast, because the market for nylon salt is relatively large at the Gulf Coast, a process, namely an adipic acid-adiponitrille process, which does not achieve low unit costs at small outputs but is best for achieving low unit costs at high outputs, is used at the Gulf Coast. Hence in column 5 of Table 6.3 we cover both scale and process differences. In this column we assume that the Gulf Coast plant has only moderate scale economy advantages over Puerto Rico plants. In Table 6.4 we consider the case where Gulf Coast plants have maximum scale economies relative to Puerto Rico plants.

In addition to scale economies, agglomeration economies include localisation economies, i.e., economies that occur when like plants congregate at one site. These economies are quite important for certain complexes such as the garment trades complex. In the refinery-petrochemical-synthetic fibre complexes, however, they tend not to exist since scale economies are so major that one rather than two or more plants of the same type tends to evolve.

Finally, Weberian agglomeration analysis for an industrial complex would encompass urbanisation economies, or what might be more appropriately termed urbanisation-regionalisation economies. These economies emerge when unlike plants are spatially juxtaposed rather than geographically separated. And as with individual industry analysis, these economies are difficult to identify in many cases. To be more concrete we may enumerate some of these economies. In an integrated steel works, heat economy achieved in the transferal of molten pig iron to open hearth furnaces is such an economy. Or in refinery-petrochemical-synthetic fiber complexes, the economy in the use of optimum size power and steam plants is such an economy. This latter economy would not be realised if we were to separate geographically two or more activities, where each had to produce its own power but did not require the output of an optimum steam plant. (Such a situation does not occur in the complexes examined.)

Much more difficult to estimate in industrial complex analysis as well as in single industry analysis, are other types of urbanisation-regionalisation economies. Among these are: (1) administrative economies and other

relation is clearly demonstrated in our refinery-petrochemical-synthetic fibre complex. The Gulf Coast is a major industrial area. Puerto Rico is at most a minor industrial area. Therefore, practically all petrochemical activities can be operated at scales and achieve scale economies in the Gulf Coast not possible for Puerto Rico. Hence realistic analysis must account for differences in scale economies, as we do in column 5 of Table 6.3. But bear in mind that scale economies also involve process change, and thus influence in a significant way factor proportions and the internal structure of industrial complexes at any site. To illustrate, because the market for nylon salt is relatively small in Puerto Rico, a process, namely an all-adipic acid process, which achieves relatively low unit costs at relatively small outputs is employed in Puerto Rico. In contrast, because the market for nylon salt is relatively large at the Gulf Coast, a process, namely an adipic acid-adiponitrille process, which does not achieve low unit costs at small outputs but is best for achieving low unit costs at high outputs, is used at the Gulf Coast. Hence in column 5 of Table 6.3 we cover both scale and process differences. In this column we assume that the Gulf Coast plant has only moderate scale economy advantages over Puerto Rico plants. In Table 6.4 we consider the case where Gulf Coast plants have maximum scale economies relative to Puerto Rico plants.

In addition to scale economies, agglomeration economies include localisation economies, i.e., economies that occur when like plants congregate at one site. These economies are quite important for certain complexes such as the garment trades complex. In the refinery-petrochemical-synthetic fibre complexes, however, they tend not to exist since scale economies are so major that one rather than two or more plants of the same type tends to evolve.

Finally, Weberian agglomeration analysis for an industrial complex would encompass urbanisation economies, or what might be more appropriately termed urbanisation-regionalisation economies. These economies emerge when unlike plants are spatially juxtaposed rather than geographically separated. And as with individual industry analysis, these economies are difficult to identify in many cases. To be more concrete we may enumerate some of these economies. In an integrated steel works, heat economy achieved in the transferal of molten pig iron to open hearth furnaces is such an economy. Or in refinery-petrochemical-synthetic fiber complexes, the economy in the use of optimum size power and steam plants is such an economy. This latter economy would not be realised if we were to separate geographically two or more activities, where each had to produce its own power but did not require the output of an optimum steam plant. (Such a situation does not occur in the complexes examined.)

Much more difficult to estimate in industrial complex analysis as well as in single industry analysis, are other types of urbanisation-regionalisation economies. Among these are: (1) administrative economies and other

Table 6.4 Overall advantages of selected staple fibre programmes, in $M/yr, by mainland scale economies assumption and under each of three assumptions of wage rate differentials

Programme	$\Delta CL = +100^1$ $\Delta TL = -75^2$ Mainland scale economies			$\Delta CL = 0^1$ $\Delta TL = -75^2$ Mainland scale economies			$\Delta CL = +50^1$ $\Delta TL = -37.5^2$ Mainland scale economies		
	Max. (1)	Mod. (2)	Min. (3)	Max. (4)	Mod. (5)	Min. (6)	Max. (7)	Mod. (8)	Min. (9)
Orlon A	−2515	−1731	−1145	−219	565	1151	−3356	−2572	−1986
Orlon B	−1994	−1042	−332	335	1287	1997	−2823	−1871	−1161
Orlon J	−569	383	1093	1648	2600	3310	−1441	−489	221
Dynel A	−3070	−2221	−1267	−490	359	1313	−3798	−2949	−1995
Dynel F	−1122	−272	−681	1268	2118	3071	−1927	−1077	−124
Dacron A	−790	311	1471	1439	2540	3700	−1657	−556	604
Dacron D	−2427	−1326	−166	42	1143	2303	−3202	−2101	−941
Nylon A	−5003	−1569	−1235	−2379	1055	1389	−5718	−2284	−1950
Nylon G	−2841	592	927	−566	2867	3202	−3691	−258	77
Nylon K	−4148	−585	−250	−1607	1956	2291	−4895	−1332	−997
Short Programme A	1563	1563	1563	2944	2944	2944	367	367	367
Short Programme D	1395	1785	2074	2959	3349	3638	248	638	927
Short Programme E	−612	−222	67	−187	203	492	−498	−108	181
Short Programme F	983	1373	1662	2789	3179	3468	−99	−291	580

Notes: 1. ΔCL is wage rate differential (Puerto Rico minus mainland) on chemical-petroleum labour, in ¢/mh.
2. ΔTL is wage rate differential (Puerto Rico minus mainland) on textile labour, in ¢/mh.

We recognise of course that any attempt at identifying the general magnitude of spatial juxtaposition economies (such as welfare effects) many of which are intangible, is a very difficult step.[9] Certainly it is as difficult, if not more so, for industrial complex analysis than for single industry analysis. Yet this difficulty may not always constitute a serious obstacle to analysis. For example, in certain cases the interrelations of activities might be such that spatial juxtaposition of all the activities leads to substantial economies which could not be achieved in any but a full combination. If so, a realistic locational or regional analysis need only evaluate for each potential location a complex of the full group of activities and need not consider split location patterns. Furthermore, in such a situation it can often be further postulated that the economies of spatial juxtaposition are approximately equal in magnitude at all feasible locations. Thus it is is sufficient for the analyst to pursue a locational comparison of the full complex, using only the relatively objective data on regional differences in production and transport costs; such data are more easily obtainable.[10,11]

resource use, industrial location, and regional development, industrial complex analysis is a useful technique. It can identify and evaluate profitable situations and activity combinations which cannot be accurately assessed either by simple individual industry comparative cost studies, or by strictly linear interindustry techniques.

Yet there are limits to the use of the industrial complex approach. Obviously it should not be used where the technical and production interrelations among activities are relatively insignificant. Further, its effectiveness may be limited in situations where some of the activity interrelations result in economies of spatial juxtaposition which are largely non-quantifiable. Also, because of the magnitude of the computations involved, the analyst cannot expect to treat all the interrelations in every situation. Many times he must be content to evaluate the major ones, assuming optimistically that the less important ones do not lead to significant interregional cost differences.

Nevertheless, despite its limitations, industrial complex analysis can contribute importantly to the understanding of regional growth and development, especially when employed in conjunction with other regional science techniques.

Notes

1. The authors are respectively, Chairman of the Department of Regional Science and Lecturer in Industry at the University of Pennsylvania.
2. W. Isard and T. Vietorisz, 'Industrial Complex Analysis and Regional Development', *Papers and Proceedings of the Regional Science Association*, vol. 1, 1955, pp. U1–U17.
3. W. Isard and T. Vietorisz, loc. cit.
4. An alternative minimum transport cost pattern for some complexes involves location in the South of the activity producing the fibre from the chemical intermediates and the location at the Gulf Coast of all other activities.
5. For full treatment of the point, see Isard, Schooler, and Vietorisz, *Industrial Complex Analysis and Regional Development*, (MIT Press and John Wiley and Sons, New York, 1959).
6. All the full programmes comprise refinery, petrochemical, fertiliser, and synthetic fibre activities. The major differences among the full programmes are in the types of fiber produced and in the petrochemical intermediates from which these fibres are derived. All the Orlon programmes involve the production of the intermediate acrylonitrile, but Orlon A and Orlon B derive the acrylonitrile from hydrogen cyanide and ethylene oxide (oxidation process), while Orlon J derives acrylonitrile from hydrogen cyanide and acetylene.

   All Dynel programmes require the production of vinyl chloride and acrylonitrile as intermediates for dynel fiber. In Dynel A, both these intermediates are based on ethylene and chlorine (along with HCN in the case of the acrylonitrile); in Dynel F the vinyl chloride is derived from acetylene and hydrogen chloride (ex chlorine) and the acrylonitrile comes from acetylene and HCN.

Dacron fibre requires the intermediates ethylene glycol and dimethyl terephthalate. In Dacron A the ethylene glycol is derived from ethylene by the oxidation process and the DMT is produced by the air oxidation of paraxylene. In contrast, Dacron D calls for ethylene glycol from ethylene and chlorine via the chlorhydrin process, and DMT from paraxylene via nitric acid oxidation.

Nylon fibre is derived from the intermediates hexamethylene diamine and adipic acid. Nylon A requires the production of HMD from adiponitrile, which in turn is derived from furfural. The adipic acid in this programme is produced by the air oxidation of cyclohexane. In the Nylon G programme, the HMD is derived from adipic acid, which thus is the sole intermediate and is derived entirely from the air oxidation of cyclohexane. In contrast, Nylon K calls for the production of both HMD and adipic acid from adiponitrile, all of which is derived from butadiene.

7. Short Programme E consists of refinery and fertiliser activities alone.
8. The first column of Table 6.4 is applicable under conditions of maximum mainland scale economies for petrochemical activities. These conditions obtain when the activities of a mainland complex can be operated at scales considered optimal from an engineering-economic standpoint. For example, these optimum scales for the Dacron A activities are listed below in column (a).

	Mainland scales		*(MM lb/yr)*	*Puerto Rico scales*
	Max.	*Mod.*	*Min.*	
	(a)	*(b)*	*(c)*	*(d)*
Ethylene Glycol	100	50	12	12
DMT	100	50	40	40
Ethylene	200	50	10	10
Amonia	135	100	100	100
Nitric acid	200	100	70	70
Ammon. nitrate	200	100	90	90
Urea	200	100	70	70

The second column of Table 6.4 refers to conditions of moderate mainland scale economies. Such conditions are represented by activity scales on the mainland which are common at the present time. For Dacron A these activity scales are recorded in the above table in column (b).

Column 3 of Table 6.4 refers to conditions of minimum scale economies on the mainland. Essentially such conditions obtain if mainland market and other conditions warrant complexes with most activities of the same scales as those postulated for the Puerto Rico complexes. For Dacron A; these mainland activity scales are recorded in column (c) above, and the Puerto Rico scales are recorded in column (d).

In Table 6.4, columns 4 through 6 and 7 through 9 repeat the maximum, moderate, minimum order of mainland scale economies assumptions, but the data of these columns were calculated on the basis of two more alternative assumptions as to the Puerto Rico mainland wage rate differential.

9. Some of the spatial integration economies are of course much more amenable to calculation than others. For example, the heat economies and power economies mentioned in the examples given in the text are rather easily identified or estimated. Also, savings in clerical labour and other indirect costs as well as scale advantages in steam and power production, all resulting from spatial integration, can in certain cases be fairly accurately estimated. And

even such advantages as better organisation of production and more effective quality control can conceivably be estimated quantitatively. The principal problem in this connection – one which it may optimistically be hoped is not insuperable – is in persuading interested parties who possess the required data to make the relevant calculations, or at least to make the data available.

The most difficult economies to estimate are of course the positive welfare effects. However, it is possible that certain indirect effects of alternative size complexes upon community or regional growth can be compared by some sort of regional multiplier approach, perhaps utilising input-output concepts. If such analysis could be accomplished successfully, the results could be used as a partial basis for decisions by regional governments concerning a variable tax-subsidy program applicable to a full complex on the one hand and to various possible smaller sub-complexes on the other.

10. It might be objected in this connection that such a procedure disregards possible differences in the magnitude of external economies, e.g., urbanisation economies, available to the industrial complex in question at different locations. This objection is valid to some extent. Yet on the other hand some of these advantages which are largely a function of the existing configuration of economic and social activity at each possible location, would tend to be reflected in the set of regional prices and costs used in the initial locational comparison.

11. When spatial juxtaposition economies are not so large and pressing as to require full complexes at each possible location, these economies none the less may permit the analyst to consider only a small number of possible split-location patterns, and to some extent to avoid the difficulty mentioned in the text.

7 The Scope and Nature of Regional Science: Channels of Synthesis

I

In this chapter, I indicate in some detail certain of the more recent achievements in regional science method. In pursuit of this objective, I shall rely heavily upon materials written jointly with David Bramhall in a forthcoming book, *Methods of Regional Analysis: An Introduction to Regional Science.*[1]

Consider Figures 7.1 to 7.4. These figures illustrate five channels of synthesis. Each channel aims to fuse certain sets of knowledge and the stronger elements of diverse analytical techniques, some existing and some hopefully forthcoming. Each channel refers to a system of regions, hence an interregional framework. Each channel yields projections of systems of industrial locations, employment, population, resource use patterns, social accounts (Gross Regional Product and regional income, consumption, investment and government outlays), balance of payments and markets. Some also yield systems of central places and urban-metropolitan areas, and intrametropolitan patterns of population, employment, output, transportation, shopping centers, and land use. All the above systems (and their subsystems) in turn are interlaced via interregional systems of interindustry (interactivity) linkages, money flows, commodity flows, population shifts, communications, and diverse sociocultural interactions (inclusive of decision-making processes).

Carefully examine Figure 7.1. This is a first-run diagram of Channel I. Central to this channel is an interregional input-output framework (as indicated in Figure 7.1). This framework is fused with the comparative cost and industrial complex techniques, each in its full dress. The basic procedure for achieving synthesis involves removing from the structural matrix those industries (say aluminum, steel, aircraft) or parts of industries (say, non-local agriculture, and hydrocarbon-based petro-chemicals) whose location patterns are most meaningfully attacked via comparative cost and industrial complex analyses. The results of such analyses specifically indicate how these industries and parts of industries are to be put in place in the final demand (bill of goods) sectors. This synthetic procedure is

Reprinted from *Papers and Proceedings of the Regional Science Association*, vol. VI (1960), pp 9–34.

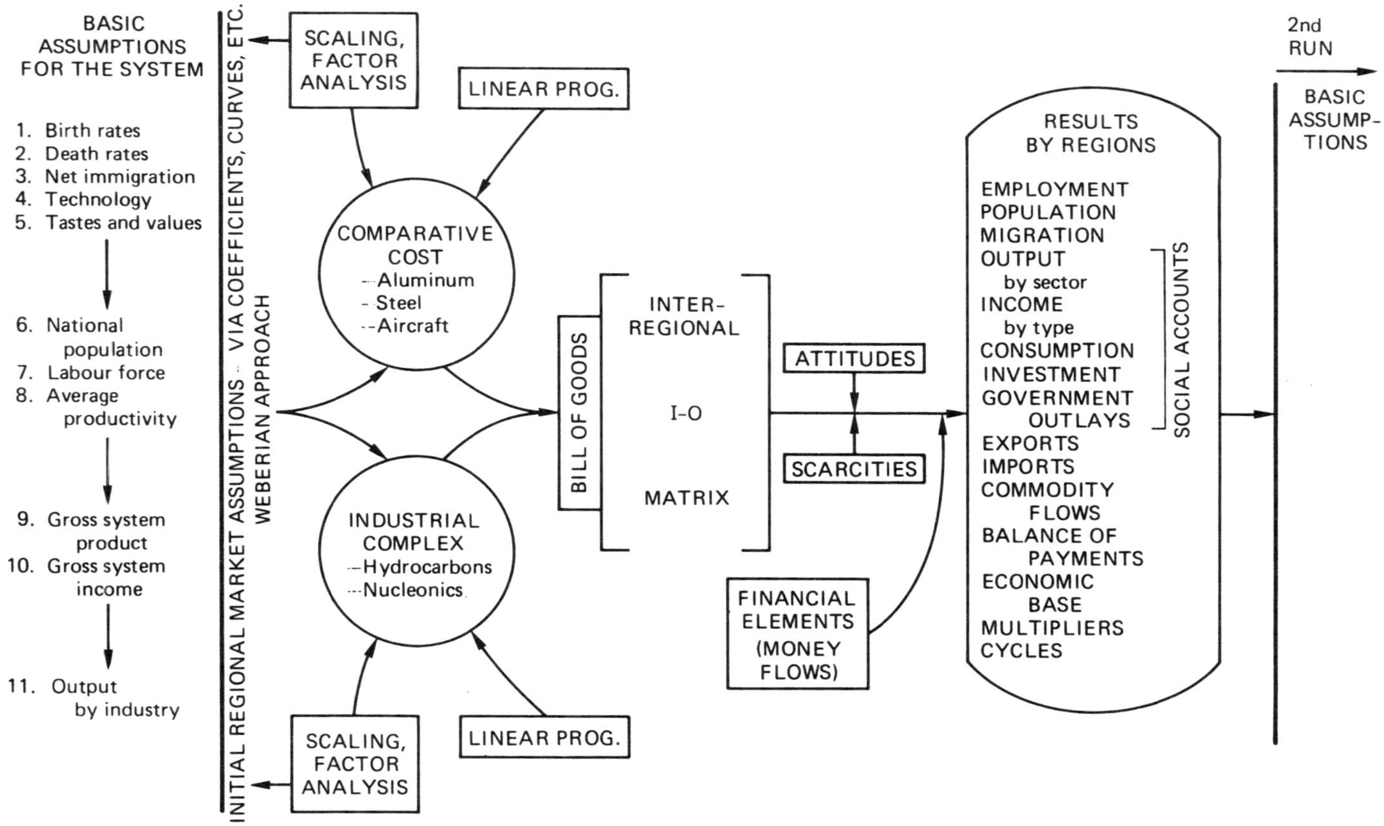

Figure 7.1 First-run diagram: Channel I

suggested in Figure 7.1 by the arrows which course from the Comparative Cost and Industrial Complex circles to the Bill of Goods rectangle.

But to perform meaningful comparative cost and industrial complex analysis à la modern Weberian doctrine requires a set of basic postulates for the system, which in turn can lead to a pattern of initial regional markets. As indicated at the extreme left of Figure 7.1, basic assumptions for the system are made, among others, on birth rates, death rates, net immigration, technology, and tastes and values (items 1–5). From these, system population, its labour force and average productivity (items 6–8) are anticipated for the key future year(s). The social accounts, Gross System Product and Gross System Income (items 9 and 10) are estimated. These are expressed in terms of levels of final demand sectors, such as Government Expenditures and Capital Formation. System outputs by sector or industry (item 11) are derived via input-output or other techniques. Initial regional markets are then established on the basis of current data and via the use of relative growth charts, trend analysis, consumption expenditure studies, and diverse coefficients and curves (such as the coefficient of localisation, location quotient, and localisation curve). These market magnitudes then permit computations via the fused comparative cost – industrial complex – interregional input-output framework to achieve the desired results. However, because of discrepancies, especially between estimates of initial regional markets and the set of regional markets consistent with the results of this computation, it may be necessary to reestimate initial regional markets and rerun the framework. The rerun process is continued until in fact results are obtained which are in harmony with the initial regional market assumptions underlying these results, and with resource scarcities and community attitudes of reality. (See the attitudes and scarcities rectangles of Figure 7.1.)

Time does not permit me even to sketch the detailed tables which are developed in connection with the above procedures. Directly obtainable from these tables are the desired results for each region of the system. These are recorded at the right of Figure 7.1. Sector or industry outputs, sector or industry employment, total employment, population numbers, and net migration estimates are derivable for each region. Gross Product of each region is also obtainable whether measured: (1) at the production stage (in terms of either net output or value added of producing sectors); or (2) at the places where household consumption, government and investment expenditures occur; or (3) at the places where households receive income and government, revenue. It is demonstrated how the detail for the measurement of Gross Regional Product from each of the above three standpoints and for the determination of other basic social accounts is contained in the tables which record the data for the fused framework of Channel I. Hence, regional social accounts are an integral part of the results of Channel I.

Other results pertain to the detail and totals on exports and imports by region, and on interregional commodity flows. These results supplement and to some extent supplant existing interregional commodity flow studies, whose materials can fruitfully be utilised early in the operation of the framework in developing an appropriate classification of sectors, in indicating sectors subject to comparative cost and industrial complex analyses, in establishing initial regional markets, and in estimating interregional input-output coefficients. A closely related result is the magnitude of the Rest of the World Commodity account, a basic account of any balance of payments study.

Ideally, the fused framework of Channel I should embrace the dimension pertaining to financial elements and money flows. (See the appropriate box and arrow in the lower center of Figure 7.1.) Doing so would enable it to yield much more complete balance of payments statements. More important, doing so would tremendously upgrade its validity and usefulness for monetary and fiscal policy. This latter point is appreciated when it is recognised that a complete interregional money flow system embraces or reflects in full the interregional interindustry system and in turn the interregional social accounting system. More specifically, a detailed interregional money flow table contains (1) all the empirical materials (and considerably more) of an interregional interindustry flow table and (2) the magnitudes for all the items listed in a regional social accounts table. Unfortunately, interregional money flow studies are extremely scarce, and relevant data are not readily available. Until such comprehensive interregional money flow studies are developed, the operational use of the fused framework of Channel I must eschew the above conceptual extension, and must be restricted to the use of the limited money flow materials available in secondary ways only.

Still other results of Channel I have reference to economic base, multiplier and cycle analysis. In a real sense, the above framework and associated tables yield a direct and clear-cut, multi-phase picture of the economic base of each region of a system. The use of this framework supersedes the customary economic base-regional multiplier type of study oriented to base-service ratios and other oversimplified concepts. It permits a much superior calculation of diverse types of multipliers (impact effects), encompassing both local revenue variation and interindustry linkage. However, the framework and associated tables can be developed for a relatively small number of regions only, in particular for those regions considered relevant for the one or few specific designs of the framework which it is feasible to implement. Hence, for other regions, especially small areas, the data of the tables of Channel I are to be used as background against which customary economic base-regional multiplier study is to be conducted.

Beyond the identification of structural effects (generally speaking of a long-run nature) based on assumptions of unchanging behavior by consumer, business and government units, the framework of Channel I cannot proceed in an operational manner in the analysis of multiplier and cyclical effects. Even conceptually, the industrial composition variable (if it is judged relevant) or the short-run Keynesian-type interregional trade multiplier can be added to the framework of Channel I in an oversimplified, partial manner only – because of the regional factor at play, the non-additive nature of cyclical and secular effects, etc.

Lastly, a number of refinements of the framework of Channel I are possible. On a single industry or industrial complex basis, interregional linear programming, the comparative cost method and the industrial complex technique may be viewed as three variants of a general analytical approach which aim at the identification of efficient-minimum cost spatial patterns. Consequently, in the framework of Channel I interregional linear programming may substitute for the comparative cost or industrial complex techniques in the limited number of situations when such a step leads to more valid and effective locational analysis for a particular sector or industrial complex. Also where meaningful quantification of attitudes and similar behavioural elements can be achieved by scaling techniques and factor analysis for example in gauging the impact of community attitudes upon industrial location, or in delineating the most relevant set of regions – these techniques too are to be operationally synthesised into the fused framework of Channel I. Thus we have put in Figure 7.1 the boxes designated Scaling, Factor Analysis and Linear Programming.

We now move on to Channel II, another channel which illustrates recent development in regional science method. Channel II like Channel I is designed to achieve a fully operational synthesis of the stronger elements of existing techniques of analysis. Its emphasis however is on urban-metropolitan structure. It substitutes the balanced regional input-output model (incorporating the valid aspects of central place theory) for the interregional input-output scheme. It is illustrated by Figure 7.2 when the rounded major block designated Probability, Gravity, Potential Models is deleted.

As with Channel I, Channel II begins with a set of basic assumptions for the system, proceeds through several steps to initial regional market estimates, and then employs comparative cost and industrial complex techniques supplemented by interregional linear programming and scaling and factor analysis to develop the final demand (bill of goods) sectors by regions. In the process, however, the balanced regional input-output scheme requires the establishment of a meaningful hierarchy of regions of several orders (say a nation (system) which is subdivided into major metropolitan regions, each of which consists of a set of subregions, each of which is composed of a set of local areas, etc.) and a suitable classification of commodities (say national, regional, subregional, local, etc.). As a consequence, the framework of Channel II requires that comparative cost

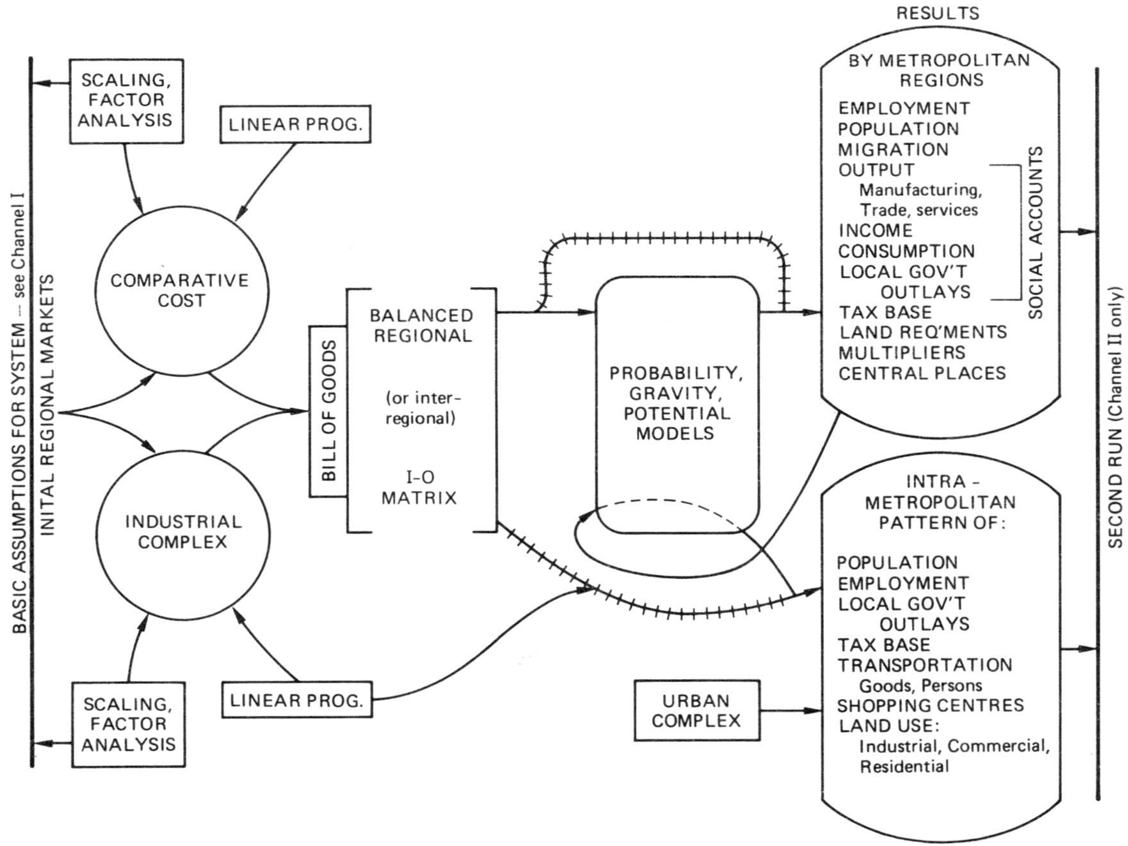

Figure 7.2 First-run diagram: Channels II and III

and industrial complex techniques supplemented by other methods and coefficients, curves, etc. establish constants not only for allocating output of national industries among metropolitan regions, subregions, local areas, etc. but also for distributing regional outputs of regional industries among subregions, local areas, etc., and subregional outputs of subregional industries among local areas, etc.

For metropolitan regions, Channel II yields results, listed at the right of Figure 7.2, which parallel the results yielded by Channel I for its regions. Among these are sector outputs, sector employments, total employment, population, net migration, Gross Product (measured in three ways), income, governmental outlays, net investment, exports, imports, commodity balance of trade, and economic base. Additionally, other types of results such as tax base and commercial, industrial and residential land requirements are explicitly noted because of their strategic importance for the analysis of metropolitan problems.

The framework of Channel II also yields the above results for each subregion of a greater metropolitan region, and for each local area of a subregion, etc. Hence, it basically outlines intrametropolitan structure and pattern (as noted in the lower Results column of Figure 7.2). However, the results on intrametropolitan pattern (and in turn on the internal pattern of each subregion) are not likely to be as firm as those pertaining to the metropolitan region as a whole. They are to be supplemented and reconciled with results on spatial patterns which may emerge from interregional linear programming studies for particular sectors, and from the use of a variety of statistical and graphic techniques which range from simple trend projection of local populations, ratios, etc., through relative growth charts, and finally to multiple regression and covariance methods and factor analysis.

Channel III is motivated by the failure of Channels I and II to achieve an adequate perspective on agglomeration (spatial juxtaposition) forces, and on the spatial behaviour of social masses. It attempts to overcome such deficiency by fusing into the framework of Channels I and II certain gravity, potential and other probability-type concepts. Hence, Channel III is depicted on Figure 7.2 when the rounded major block designated Probability, Gravity, Potential Models and the arrows leading to and from it are considered fully relevant.

One fruitful direction is to introduce into the framework of Channel I a relative income potential concept. First, endogenous producing sectors (those not subject to comparative cost, industrial complex and interregional linear programming analysis) are to be subdivided into (1) non-cost sensitive industries and (2) local service and other market-oriented industries. Second, the time stretch between the base year and the key future year must be considered as a series of successive short time-periods, for each one of which a projection is to be made. Third, the framework of

Channel I is run to project at the end of the first time-period regional levels of those sectors subject to comparative cost, industrial complex and interregional linear programming analysis and of those parts of local service and other market-oriented activities tied to these sectors. (Reruns are eliminated; discrepancies and inconsistencies motivate various units to adjust and change their location, production and consumption levels, etc. in the following time-period.) Fourth, levels of non-cost sensitive sectors and those parts of local service and other market-oriented activities tied to these sectors are projected for each region on an 'as the nation grows' basis, adjusted by a modified relative income potential term reflecting changing interregional position of a region. Fifth, the regional projections for the terminal year of the first period are used as base year materials for regional projections for the terminal year of the second period. And so forth until regional projections are attained for the key future year. As a consequence, a step-by-step discrete path of growth obtains for each major region and the system, involving for the terminal year of each time period estimates by region of output by sector, employment, population, Gross Product and diverse social accounts, tax base and other critical magnitudes. In somewhat similar fashion projections by subregions are obtainable as well when there is introduced into the framework of Channel II a relative income potential concept – a concept which, some may claim, catches more fully than any other concept the effects of the interplay of the host of agglomeration and spatial juxtaposition forces which are so closely associated with markets and accessibility to them.

Another fruitful direction in the use of the probability-type gravity model is in testing the consistency of projections of spatial patterns of several phenomena within metropolitan regions and their subregions. Based on diverse procedures, an industrial location (job) pattern, a population (residential) pattern, a shopping centre pattern, etc. may each be independently projected for the same metropolitan region or subregion. Together, they imply patterns of journey-to-work trips, shopping trips, commodity shipments, etc. Such patterns may be compared with the patterns generated by a probability-type gravity model, when the parameters of such a model are derived from extensive empirical data on trip behaviour and commodity shipments. If the former patterns deviate markedly from gravity model expectations of trip behaviour and commodity shipments, reexamination of the industrial location, population, shopping centre and each of the other independently derived patterns is suggested. Such reexamination may be undertaken in order to effect rearrangement of these patterns without jeopardising the validity of any one and in order to achieve (if necessary, by iteration) greater consistency with gravity model expectations based on current and past transportation data. This direction in the use of the gravity model is particularly of significance because of the relative abundance of transportation data on the one hand, and the limited applicability of other analytical techniques to small area and subregional study on the other hand.

Conceptually speaking (and hopefully operationally in the near future) Channel III can be extended further to embrace urban complex analysis. (Note the rectangle designated Urban Complex at the lower centre of Figure 7.2.) This analysis parallels industrial complex analysis. It purports to attack directly scale and spatial juxtaposition economies as they affect metropolitan structure and functions. It may first involve identification of meaningful complexes of urban activities of different orders, ranging from the central business district type down to the small neighbourhood cluster. Next, reasonable metropolitan patterns (hierarchical sets of urban complexes) are noted. With input-output type data and local price data or estimates, costs and revenues for each urban complex of each order in each metropolitan pattern are then computed. These lead to total cost estimates for each metropolitan pattern, where total cost differences among the metropolitan patterns may be due to differences in cost of transportation (persons and commodities), labour, taxes, land, etc., and in scale economies. (Differences in certain non-quantifiable social costs cannot of course be embraced.) The total cost computations in turn lead to the isolation of the most efficient metropolitan pattern.

In recapitulation, Channel III does operationally attack and at least to some extent does encompass spatial juxtaposition economies on the major region and subregion level, and does bring to bear upon projections of intrametropolitan structure behaviour phenomena of social masses, especially with regard to travel and transportation. Channel III adds the probability-type gravity model to the fused framework of Channels I and II, using it to supplement the techniques already synthesised into these frameworks. Thus Channel III can achieve all of the basic results of Channels I and II and presumably with greater validity. Such validity will be even further enhanced when urban complex analysis can be operationally fused into the framework.

This section has indicated some of the recent developments in regional science methods. There are a number of others, e.g., those associated with Garrison and his students at the University of Washington. These developments clearly demonstrate that if regional science is to be considered as theoretical social science geography, it deals with approaches and techniques which are, by and large, strange to geographers. More significant, it clearly points up the need for graduate training of an entirely different character than is required to attain competence in the well established and proven areas of geography. Such a new training programme can be viewed as justifying the existence of regional science departments.

One more point. It has been claimed by some that regional science is for the most part regional economics. Most of the individual techniques noted in Channels I–III can be traced back to the field of economics. Although this point is true, it must be also recognised that regional science as a field of study is much too young to have developed very many unique approaches of its own; and necessarily must borrow and adapt methods of existing disciplines. It is understandable that regional science borrows heavily from economics, which is generally conceded to be the most advanced of the recognised social science disciplines. Yet I would contend that the particular emphasis on systems of regions, spatial interaction and spatial interdependence, which underlie Channels I–III, the advanced statistical spatial studies of Garrison and his students, and the gravity model formulations of Warntz, Anderson and others are well beyond regional economics. But, whatever one's stand on this issue, I now wish to indicate certain new directions which regional science should and is very likely to probe in the future, and which in my mind establish the fact that regional science purports to be very much more than regional economics.

II

A careful cutting into social processes with a fine-edged analytical scalpel exposes a primary stratum of premises underlying each of the three interregional channels already discussed. These fundamental premises make clear that the three channels are each a special case of a more general interregional system. These premises make clear that the three channels for the most part postulate optimising behaviour of an economic type.[2] Viewed from a broad social welfare standpoint, such channels, therefore, are at best sub-optimising. This point is evident when we review the central techniques employed by each channel.

For example, Channel I begins with a spatial pattern of population and societal behaviour, which is then translated into a set of consumer wants (inclusive of governmental services). These wants set markets, by region – to be later revised to achieve economic consistency. Comparative cost and industrial complex procedures (supplemented by interregional linear programming) are applied. They reflect the businessman attempting to *minimise* cost, given his market – i.e., to *maximise* profits.[3] Interregional input-output is next injected, a technique which essentially establishes requirements, and in the process generates income and employment – but on the premise that inputs are in accord with *best* engineering practice and are supplied from *least-cost* sources. A set of purely descriptive Social Accounts for a system of regions results. Reconciliation is effected to eliminate inconsistencies with a maximum-profit spatial pattern.

Society transcends the economic. Hence, the need for Channel IV which is outlined in Figure 7.3. Central to Channel IV are values, which is noted at the extreme left of Figure 7.3, and which should appear in the figure in even larger and bolder type. I do not wish to deny that a value system

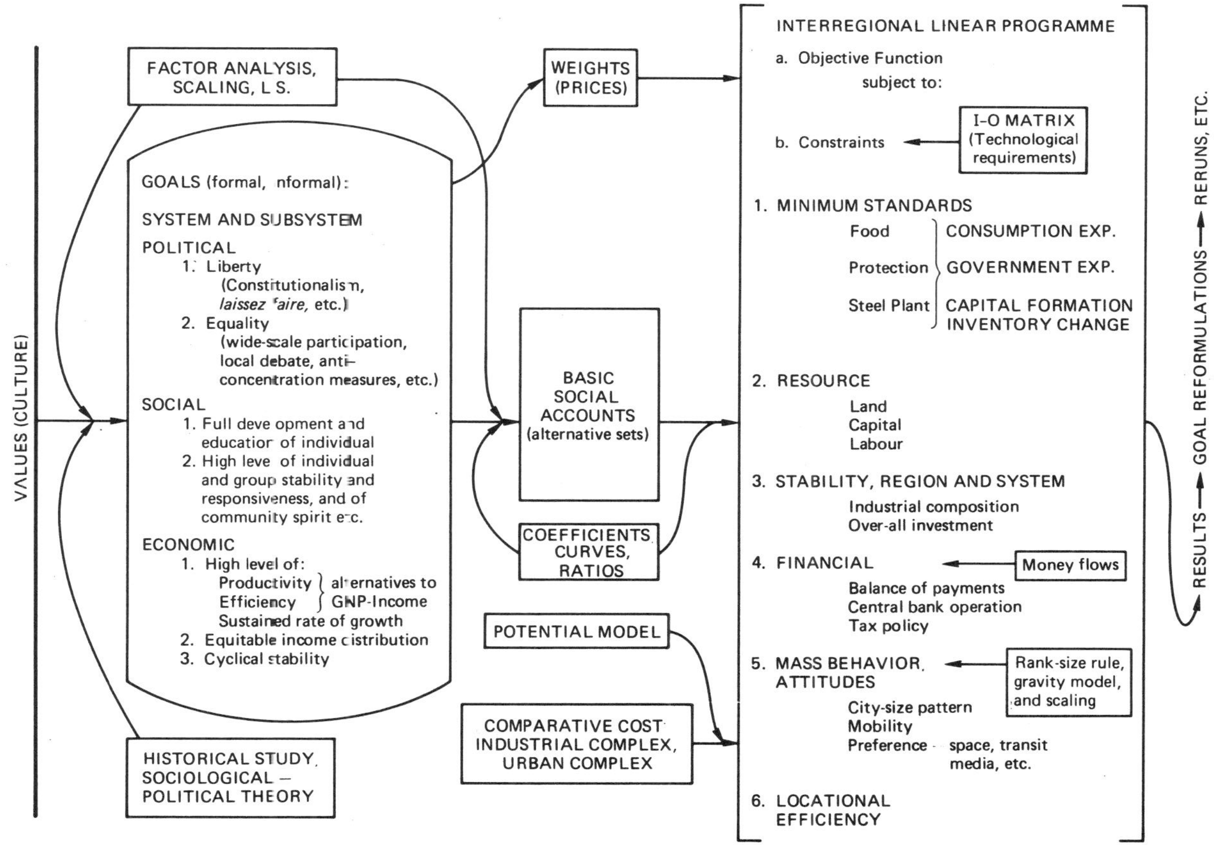

Figure 7.3 Fused framework: Channel IV

underlies Channels I–III. One does. Implicit in the postulate of profit maximising behaviour unconstrained by social institutions and legal restrictions is a value system which places high, if not highest priority, upon *Liberty*, if we may speak in terms of the Great Values of the political scientist.[4] By failing to impose restrictions which assure minimum educational opportunities for all, minimum standards for food and drug products, minimum housing levels, minimum wage payments, equal opportunity to engage in business and enter trades, and so forth, Channels I–III place a relatively low priority on the basic values of *Equality* and *Welfare*. Thus, Channels I–III do rest upon an underlying value system – one which may be said to be in many ways consistent with the system in the United States during the nineteenth century, or at least parts of that century. Yet many other value systems are possible, and do in fact exist. Hence, it is relevant, perhaps even imperative, that we examine the interrelations of value systems with the fused frameworks and their numerous sub-systems relating to a system of regions. In so doing, we shall, in Channel IV, depart to a major extent from the frameworks hitherto developed.

As already indicated, in Figure 17.3 the central theme of Values (Culture) is indicated at the extreme left. Values then lead to System and Subsystem Goals, both formally and informally established. We have, for illustrative purposes, set down in the *Goals* box of Figure 17.3 an abbreviated list of political, economic and social goals – a list which has some relevance for the current culture of the United States (the system). But it is immediately appreciated how accumulated historical study and political, sociological and other social science theory and empirical investigations offer insufficient understanding of this process, that is, how to proceed from values and culture through the decision framework to system and subsystem goals. Thus the quantification of goals rests on a very shaky foundation, and is likely to do so for quite a time despite the promise of new techniques and concepts related to scaling, latent structure and community participation potential. Necessarily, at this time, Channel IV cannot be operational. It must be purely conceptual.

Once a set of specific goals is crudely developed for a system (culture) and its regions (subsystems) such as the one denoted in Figure 17.3, the next step in Channel IV requires the translation of these goals into specific, detailed social accounts. (In Figure 17.3, note the box, Basic Social Accounts.) Here, too, inadequate social science knowledge plagues the investigator. The quantitative implications of certain goals are clear-cut, such as the implications for governmental expenditures of the goal of at least 12 years of schooling for every child. But how can the goal of economic stability (let alone social and political stability) be translated into specific social accounts?

The detailing of social accounts in Channel IV is conditioned significantly by the operational techniques to be subsequently employed. In the version presented here, general interregional linear programming is taken as the central operator. (Observe at the right of Figure 17.3 the major brackets headed Interregional Linear Programme.) This necessitates both (1) the embodiment of goals in an objective function, and (2) the translation of goals into social account magnitudes which serve as limiting constants of constraints.

Many objective functions are possible. A common one is to maximise Gross System Product. Whatever the function, difficulties are typically encountered in the choice of weights, since weights supposedly reflect the values of a society which values, as already noted, too frequently defy quantification. Even market prices are poor indicators of values.

If the objective is to maximise Gross System Product, one group of constraints may relate to minimum standards – constraints which for each region insure the delivery to its households of sufficient food and clothing, the provision of sufficient housing, medical services, education, etc. and the supply of new plant and equipment consistent with system and subsystem goals. (Goals relating to household consumption and govermental services may frequently be governed by levels and patterns of consumption and services already attained.) A second group of constraints may relate to resources and insure that resource requirements do not exceed available supply. At this point, as well as at others, the interregional input-output matrix of coefficients is of considerable utility in calculating requirements. (In Figure 7.3 note the box, I–O Matrix, which feeds into Constraints.)

Other groups of constraints may relate to stability, financial elements (and money flows), and mass behaviour and attitudes. These constraints largely involve relations which can be treated at the conceptual level only. However, industrial composition hypotheses and Keynesian short-run interregional trade multiplier analyses may contribute to the specification of stability constraints. Balance of payments considerations may lead to a precise financial constraint. The rank-size rule, voting records and scaling analysis may contribute to the quantitative statement of a mass behaviour-attitude constraint. Finally, another set of constraints may relate to locational efficiency, whose specification ensues from analysis via comparative cost, industrial complex, and, eventually, urban complex techniques. Through these constraints, the latter techniques become effectively fused into the framework of Channel IV. Also, even a relative income potential model may be fused into the framework of Channel IV at this point via locational pattern constraints on non-cost sensitive industries. (In the lower centre of Figure 7.3, note the boxes, Potential Model, Comparative Cost, Industrial Complex, Urban Complex, which feed into the Locational Efficiency constraints.)

Once the objective function and constraints of the interregional linear programme of the values-social goals Channel IV are set down, a solution may be sought. If no solution is possible the program is infeasible, and the basic set of goals as reflected in the minimum social accounts and diverse constraints must be reset; sights must be lowered. If a solution is possible, and in addition a 'surplus' is yielded, it may be necessary to reset constraints and refashion the programme if the programme's allocation of the surplus among commodities, units and regions is inconsistent with underlying values and goals. When a satisfactory solution is attained, practically all, if not all, the basic social accounts and other magnitudes are obtainable in the same manner as in the previous channels – even when Channel IV is conceptually extended to encompass the growth (capital formation and capacity expansion) process and cyclical features of reality.

This brings to a close my snapshot discussion of Channel IV. In one important way this channel indicates what regional science purports to be – and as I feel at this moment, must be. It clearly demonstrates that regional science is well beyond regional economics, that it should and must have important ties with theory of the diverse social sciences, and that it can contribute in its way to the development of a general theory and empirical framework for the social sciences. At the same time, it is recognised that Channel IV is purely conceptual, and that its operational use lies decades ahead by when, we hope, adequate theoretical and empirical social science knowledge will have accumulated.

This concludes my case. However, to avoid any misleading impressions that regional science is a science of tomorrow, and not of today, I wish to sketch the several variants of Channel V. This sketch is presented as Figure 7.4.

Channel V is an operational channel. It is designed to incorporate significant values and goals which are at least approximately quantifiable into the fused frameworks of Channels I–III. (In Figure 7.4 this incorporation of values and goals is indicated by the top circle at the left.) Thereby Channel V corrects some of the imbalance of Channels I–III.

For example, one variant of Channel V systematically injects in a major way values and social goals into Channel I. (Note that this variant is obtained from Figure 7.4 by deleting the boxes designated Interregional Linear Programme, Weights, Probability, Gravity, Potential Models, Urban Complex, and Intrametropolitan Pattern.) It does so (1) by translating social goals into specific accounts – whether the statement of such goals is relatively simple as in an armaments reduction programme or exceedingly complex as in a political platform, and (2) by putting these accounts in place in the final demand (bill of goods) sectors of the several regions. In general this first variant of Channel V follows the steps of Channel I, but, in the light of values and social goals, constantly requires modification of these steps all the way from the establishment of initial basic assumptions through the reruns.

In like manner, second and third variants of Channel V can be based

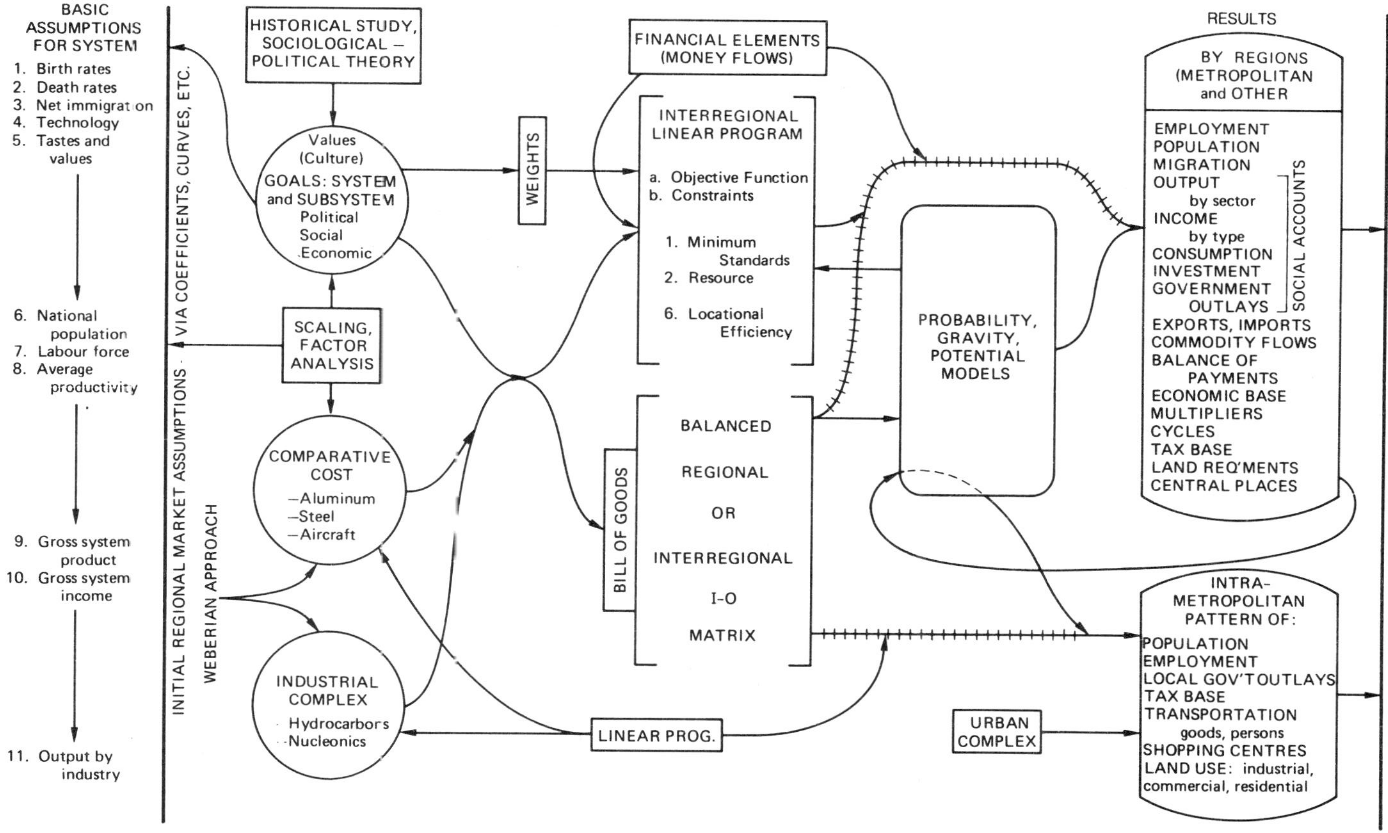

Figure 7.4 Variants of Channel V

upon the introduction of a values and goals box into Channels II and III respectively.

Of all channels, Channel V is the most advanced operationally. Were a far-sighted foundation to provide the necessary funds, the empirical implementation of this Channel could be immediately undertaken and significant results achieved within a reasonable research period. Of all operational channels, Channel V achieves the highest level of synthesis of analytical techniques. Thus, its fused framework is the most effective in attacking the interrelations which pervade an interregional system and its diverse subsystems, both interregional and intraregional. Further, Channel V provides a set of procedures whereby cumulating sociological, anthropological, political and other social science theory and knowledge can be incorporated into regional analysis via an expanding values and goals box in Figure 7.4 – a box which comes to have increasing quantitative content. And as this occurs regional science will come more and more into its own as theoretical social science geography, or as I prefer, a distinct social science discipline with a unique core and combination of subject matter, intent and approach.

Notes

1. By W. Isard *et al.*, 1960 (John Wiley and Sons and MIT Press, New York).
2. This statement is valid for the techniques central to these channels. Exceptions are involved, for example, in the introduction of gravity models for describing patterns of travel desire, and of scaling and related techniques for measuring local attitudes as a qualifying factor.
3. This motivation is assumed to govern the behaviour of businessmen individually and in groups.
4. See Maass, Arthur A. (ed.) 1958, *Area and Power: A Theory of Local Government*, (The Free Press, Glencoe, IL).

8 Gravity, Potential and Spatial Interaction Models

I INTRODUCTION

A system of regions has an intricate structure. Only some of the fine strands which interconnect people, households, firms, social groups, governmental agencies, and a variety of other operating and decision-making units have been isolated and subjected to analysis. In time, with further developments of social science techniques, many more of these strands will be discovered and bared for study under the microscope of the regional scientist. Yet the balanced mind may still be uncomfortable, especially as it observes, perhaps at some distance in space, the geographic pattern in which human beings and their physical structures are massed in metropolises – metropolises which vary widely in size, configuration, and intensity of activity, such intensity tending to diminish in all directions from the core. True, the fine-stranded generalised interdependence schemes as crystallised in interregional input-output and linear programming are powerful analytical tools. Yet have they, even in a small way, been able to cope with spatial juxtaposition (or agglomeration) economies and diseconomies? Even the unsophisticated observer would judge that these economies and diseconomies must underly in a major way the important phenomena of human massing within any system of industrialised regions.

Viewed from another standpoint, is not society more than a matrix of intricately detailed connections among units, à la general equilibrium? Is not the structure of a system of regions more than the sum of the interactions of sets and patterns of units or sectors as conceived by interregional input-output and linear programming? Are there not over-all forces more akin to agglomerative factors which pervade society and confine the multitude of possible interactions among its innumerable units? This consideration motivates the analyst to explore fresh approaches, even though he attributes considerable significance to the more traditionally oriented approaches of input-output and linear programming. He may wish to pursue quite a different approach which has recently been developed. This is the approach associated with gravity, potential, and spatial interaction models.[1]

In the gravity, potential, and spatial interaction models – which for short

---

This chapter was written with David F. Bramhall. Reprinted from W. Isard, D. F. Bramhall, G. A. P. Carrothers, J. H. Cumberland, L. N. Moses, D. O. Price and E. W. Schooler, *Methods of Regional Analysis: An Introduction to Regional Science* (MIT Press, Cambridge, MA, 1960), pp. 493–568.

we shall term gravity models whenever we speak generally of these models – the region is conceived as a mass. The mass is structured according to certain principles. These principles govern in an over-all fashion the range of behaviour of the individual particles, both constraining and initiating their action. Interregional relations may be thought of as interactions among masses. Again general principles may be said to govern the frequency and intensity of such interactions; and by so doing they influence the behaviour of individual units (particles) within each mass. As shall be observed later, this approach may be said to resemble an approach frequently used by physical scientists. For example, Boyle's classic studies of the effects of pressure and temperature on the volume of gases were essentially investigations into the behaviour of masses of molecules; the movement of any individual molecule was not a matter of inquiry.

II THE PROBABILITY POINT OF VIEW

To develop the concept of gravity models it is useful to adopt a rather simple probability point of view.[2] Suppose there is a metropolitan region with population P. Let the region be divided into many subareas. Let there also be known the total number of internal trips taken by the inhabitants of this metropolitan region. We represent this number by the constant T. Further, let there be no significant differences among sub-areas in the tastes, incomes, age distributions, occupational structures, etc., of their populations.

Now, suppose we wish to determine the number of trips which originate in, let us say, subarea i, and terminate in, let us say, subarea j. Assume, for the moment, that no costs and no time are involved in undertaking a trip from one area to another, that is, that the friction of distance is zero. For this hypothetical situation we may expect that for a representative individual in subarea i the per cent of his journeys terminating in subarea j will, *ceteris paribus*, be equal to the ratio P_j/P, which is the population of subarea j divided by the total population of the metropolitan region. That is, if the total population of the metropolitan region is 1,000,000 and that of subarea j 100,000, we may expect the individual to make 10 per cent of the trips to j. Additionally, since a representative individual in subarea i is by our homogeneity assumptions identical with a representative individual in any other subarea, and since his transport time cost is zero, we may estimate the number of trips he undertakes as the average number of trips per capita for the entire metropolitan region. This average is equal to T/P. Designating this average by the letter k, we find that the absolute number of trips which a representative individual in subarea i makes to subarea j is $k(P_j/P)$. That is, if 10 per cent of the total population resides in subarea j, the individual in subarea i will tend to make 10 per cent of his trips to subarea j; if the average number of trips per individual is 20, the individual will make two trips to j.

This reasoning applies to one representative individual in i. But there are P_i individuals residing in subarea i. Therefore, the number of trips to subarea j which these P_i individuals will make will be P_i times the number of trips to subarea j which the representative individual in i makes. That is,

$$T_{ij} = k \frac{P_i P_j}{P} \tag{8.1}$$

where T_{ij} designates the total number of trips taken by individuals in i (i.e., originating in i) which terminate in j. In like manner we can estimate the expected total number of trips for every possible combination of originating subarea and terminating subarea. Thus we obtain for the metropolitan region a set of expected or hypothetical trip volumes (total number of trips) between subareas.

Our next step is to determine the possible effect of the actual distance separating a pair of subareas on the number of trips occurring between them. First, for a typical metropolitan region we obtain actual data on the number of trips between every pair of its subareas. We let I_{ij} represent the *actual* trip volume between any originating subarea i and any terminating subarea j. We divide this actual number by the expected or hypothetical trip volume T_{ij} to derive the ratio of actual to expected trip volume, that is, I_{ij}/T_{ij}. We also note the distance d_{ij} which separates i and j. Finally, we plot on a graph with a logarithmic scale along each axis both the ratio I_{ij}/T_{ij} and distance d_{ij} for this particular pair of subareas. For example, in Figure 8.1 where the vertical axis measures the ratio of actual to expected trips and where the horizontal axis measures distance, we may note point L. Point L refers to a pair of subareas approximately 3.6 miles apart for which the ratio of actual to expected trips is approximately 0.4.

In similar manner, for every other combination of originating subarea and terminating subarea we plot the set of data on the ratio of actual to expected trips and intervening distance. Suppose our data are as indicated in Figure 8.1. They suggest a straight-line relationship between the log of the ratio of actual to expected trip volume on the one hand and distance on the other hand, a simple relationship which is not undesirable for analytical purposes. A straight line may be fitted to the plotted data by least squares or by other methods. Since our variables are the log of the ratio of actual to expected trip volume (the dependent variable) and the log of distance (the independent variable), the equation of the line is

$$\log \frac{I_{ij}}{T_{ij}} = a - b \log d_{ij} \tag{8.2}$$

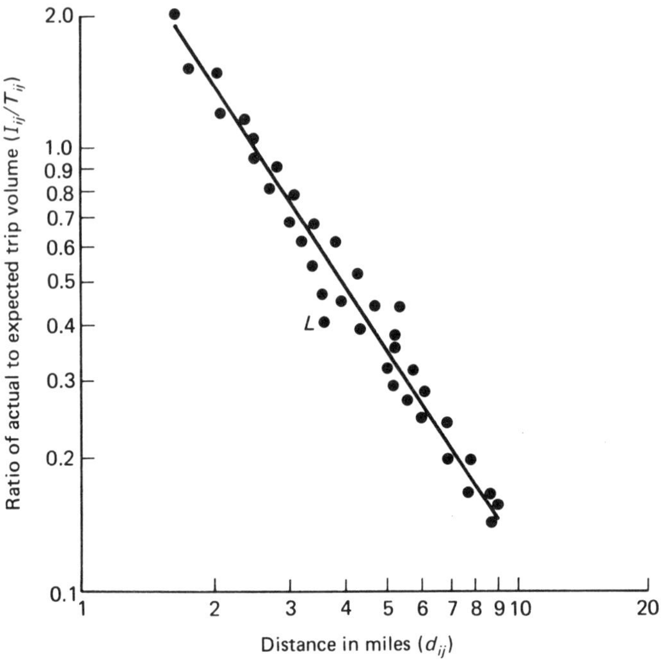

Figure 8.1 Relation between distance and the ratio of actual to expected person trips (hypothetical data)

In this equation a is a constant which is the intercept of the straight line with the Y axis, and b is a constant defined by the slope of the line.[3] Removing logs from equation (8.2) and letting c equal the antilog of a, we have

$$\frac{I_{ij}}{T_{ij}} = \frac{c}{d_{ij}^{b}}$$

or

$$I_{ij} = \frac{cT_{ij}}{d_{ij}^{b}} \tag{8.3}$$

Substituting in equation (8.3) the value of T_{ij} as given in equation (8.1), and letting the constant $G = ck/P$, where c, k, and P are constants as defined earlier, we obtain

$$I_{ij} = G\,\frac{P_i P_j}{d_{ij}^{b}} \tag{8.4}$$

This simple relationship may then be taken to describe roughly the actual pattern of trip volumes within the metropolitan region, *ceteris paribus*. That is, it depicts the interaction of people within the metropolitan region as a function of the populations of subareas and the distance variable when this interaction is reflected in trips.

Suppose we study the relationship of actual to expected magnitude on the one hand, and distance on the other hand, for a number of other phenomena reflecting the interactions of people within the metropolitan mass and among metropolitan masses. We might examine telephone calls, telegraph messages, railway express shipments, money flows, migration, etc. Suppose that for all these phenomena we find, as in Figure 8.1, a close linear association between the log of the ratio of actual to expected volume and the log of distance.[4] We might then conclude that the relationship

$$I_{ij} = G\,\frac{P_iP_j}{d_{ij}^b}$$

reflects a basic principle underlying the structure of metropolitan areas and systems of metropolitan areas. This relationship derived from a probability point of view is essentially the gravity model, although the gravity model did not evolve in this manner.[5]

Additionally, it is to be noted that equation (8.4) can be converted into another useful form. Suppose we are interested in the interaction between a single subarea i and *all* other subareas. We would therefore derive the interaction of i with the first subarea (i.e., I_{i1}) *plus* the interaction of i with the second subarea (i.e., I_{i2}) *plus* the interaction of i with the third subarea (i.e., I_{i3}), *plus* . . ., and finally *plus* the interaction of i with the last or *nth* subarea (i.e., I_{in}). From equation (8.4) we find values for each of the interactions, $I_{i1}, I_{i2}, I_{i3}, \ldots, I_{in}$. By addition we obtain

$$I_{i1} + I_{i2} + I_{i3} + \ldots + I_{in} = G\,\frac{P_iP_1}{d_{i1}^b} + G\,\frac{P_iP_2}{d_{i2}^b} + G\,\frac{P_iP_3}{d_{i3}^b} + \ldots + G\,\frac{P_iP_n}{d_{in}^b}$$

or

$$\sum_{j=1}^{n} I_{ij} = G \sum_{j=1}^{n} \frac{P_iP_j}{d_{ij}^b}$$

Since P_i may be factored from the right-hand side of the equation, we derive, after dividing both sides by P_i,

$$\frac{\sum_{j=1}^{n} I_{ij}}{P_i} = G \sum_{j=1}^{n} \frac{P_j}{d_{ij}^b} \tag{8.5}$$

Note that the numerator of the left-hand side of equation (8.5) is the total interaction of i with all areas including itself,[6] which when divided by the population of i, namely P_i, yields interaction with all areas *on a per capita basis* or more strictly on a *per unit of mass basis*. Interaction on such a basis has been designated *potential at i*, for which we employ the symbol ${}_iV$. By definition, then, ${}_iV = \frac{\sum_{j=1}^{n} I_{ij}}{P_i}$ and, from equation (8.5), we have

$${}_iV = G \sum_{j=1}^{n} \frac{P_j}{d_{ij}^b} \text{ (note 7)} \tag{8.6}$$

Equation (8.6) is the basis of potential models and as developed is a variation of equation (8.4), the basic gravity model.

III THE STEWART–ZIPF HYPOTHESES

Further insight into the significance of the gravity model can be acquired by tracing its historical development. Although the gravity model may have been implied in the writings of Carey,[8] Ravenstein,[9] Young,[10] and Reilly,[11] the first major impetus in its development stems from the work of Stewart[12] and Zipf,[13] who simultaneously worked on this model from independent angles. Stewart formulated his hypothesis in terms of his concept of social physics. As already suggested, in the natural sciences there are laws such as those governing the density, pressure, and temperature of gases that were discovered only because matter was investigated as a mass. For example, if Boyle's investigation had dealt with the rapid and erratic movement of individual molecules as they bounced against each other, it is not likely that his laws would have evolved.

Stewart reasoned that similar relations might underlie the interaction of social units, such as people, which relations could be discovered only by investigating large aggregates of such units. He presented three primary concepts based on Newtonian physics. Following the formula for *gravitational force*, Stewart defines *demographic force* as a constant times the product of two masses divided by the square of the distance separating the masses. Where the population of cities i and j, designated by P_i and P_j respectively, are taken as the relevant masses, demographic force F is

$$F = G \frac{P_i P_j}{d_{ij}^2}$$

where G is a constant corresponding to the gravitational constant.

Stewart develops a second concept corresponding to *gravitational energy*, namely *demographic energy*, where demographic energy E is defined as

$$E = G \frac{P_i P_j}{d_{ij}}$$

It is to be noted that the formula for demographic energy differs from that for demographic force only with respect to the exponent of the distance variable, d_{ij}. More important, both these concepts are derivable from equation (8.4). When the value of b in equation (8.4) is 2, the right-hand side of the equation defines gravitational force (i.e., $I_{ij} = F$). When the value of b in equation (8.4) is unity, the right-hand side of the equation defines gravitational energy (i.e., $I_{ij} = E$).

Stewart's third concept is that of *demographic potential* which corresponds to *gravitational potential*. The demographic potential produced at a point i by a mass at j, which may be designated ${}_iV_j$, is defined as a constant times the mass at j, say P_j, divided by the intervening distance, that is,

$${}_iV_j = G \frac{P_j}{d_{ij}}$$

Thus, the potential created upon the center of Philadelphia by the New York City mass might be defined as a constant times the population of New York City divided by 90 miles. When, as in reality, more than one mass exists, the *total* potential at a point i, designated ${}_iV$, may be taken as the sum of the separate potentials produced by each mass. Thus, if the masses are numbered from 1 to n,

$$\begin{aligned} {}_iV &= G \frac{P_1}{d_{i1}} + G \frac{P_2}{d_{i2}} + \ldots + G \frac{P_n}{d_{in}} \\ &= G \sum_{j=1}^{n} \frac{P_j}{d_{ij}} \quad \text{(note 14)} \end{aligned} \tag{8.7}$$

Since it is possible to compute total potential for every point, or practically speaking for every relevant subarea of a nation or interregional system, we may construct a map of potentials. This Stewart has done with the use of equipotential contour lines. Figure 8.2, which depicts 'weighted'

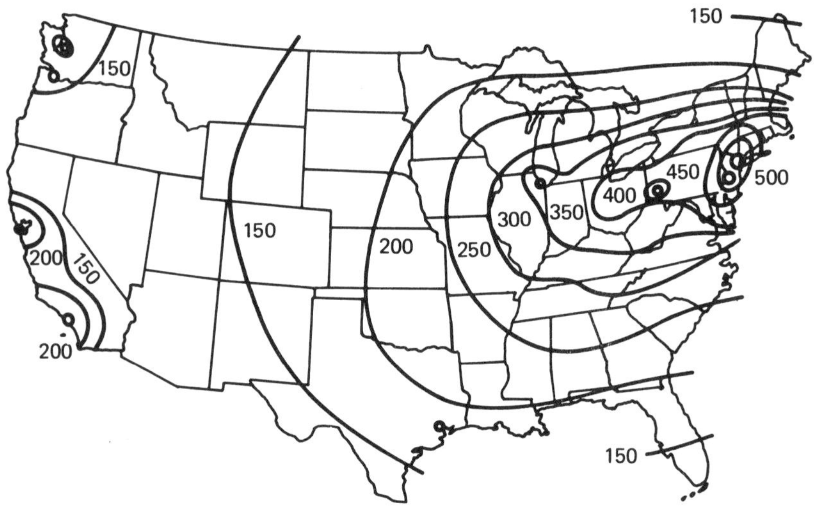

Figure 8.2 Weighted population potentials, United States, 1940. (Potentials are in tens of thousands of persons per mile).
Source: J. Q. Stewart [22], p. 22, as modified in W. Isard [16], p. 67

potentials of population for the United States in 1940[15] is an illustration. In the construction of this map the United States was divided into subareas, the population of each subarea being assumed to be concentrated at a control point. Any one contour line passes through those areas and control points having the same potential. For example, the equipotential contour line 300 (300,000 persons per mile) describes a wide arc which passes through central New England, upper New York, central Michigan, southern Wisconsin, etc., and swings around through Tennessee and North Carolina, all of which areas have approximately the same potential.

In general, the several equipotential maps constructed by Stewart and his associates depict New York City as the peak of demographic potential. East of the Sierras, every general contour line closes around New York City; all other cities are local peaks on the general downhill slope from New York in all directions. (As a consequence, a rural area in New Jersey comes to have a higher potential than the center of Omaha, Nebraska.) West of the Sierras, two separate humps appear. One centres around Seattle, Washington, the other around San Francisco and Los Angeles, the two major peaks of California.

The interpretation to be given to the concept of demographic potential is

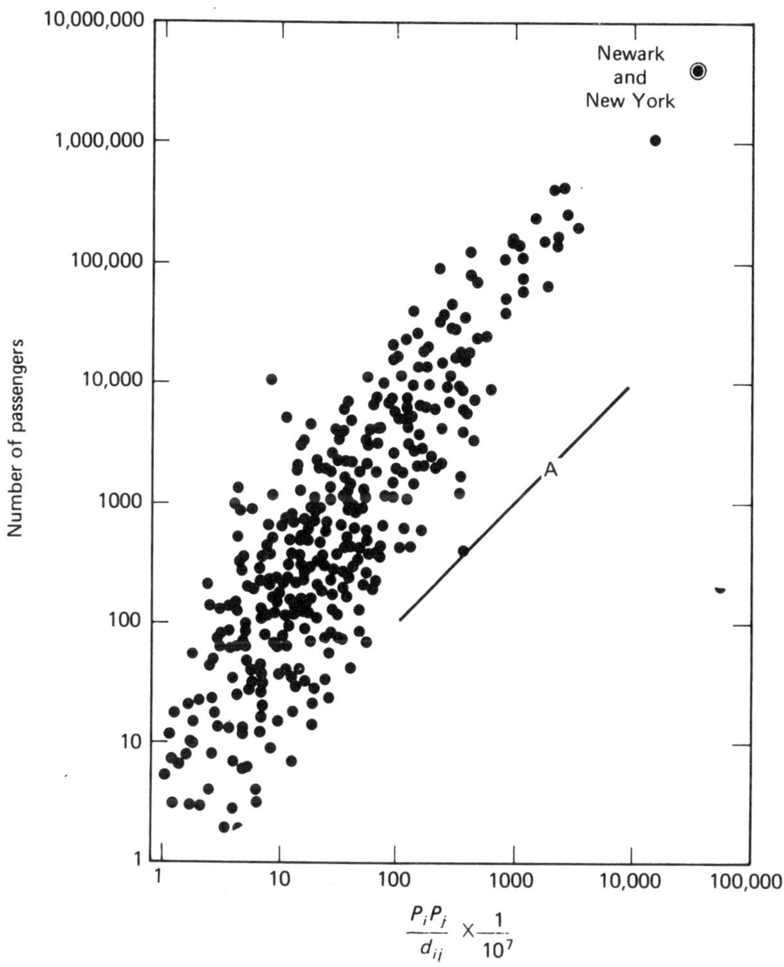

Figure 8.3 The number of passengers traveling by railways between 29 arbitrary cities during 1 month in each quarter of 1933. (Source: G. K. Zipf, [27], p. 396.)

not yet entirely clear. Stewart speaks of population potential of a point as a measure of the proximity of people to that point, as a measure of aggregate accessibility, and more simply as a measure of influence of people at a distance. To point up the significance of this concept and the related concepts of demographic force and energy, Stewart and his associates have conducted a number of empirical studies. He reports high correlation within the United States of the spatial variation of population potential with spatial variation in a wide variety of sociological phenomena. Among others these phenomena include: rural population density, rural nonfarm population density, rural nonfarm rents, farmland values, miles of railroad track per square mile, miles of rural free delivery routes per square mile, density of rural wage earners in manufacturing, and rural death rates.[16]

In the historical development of the gravity model, the works of Zipf have also been influential. Zipf speaks variously of the P/D and P_1P_2/D factors which, using our symbols, are the P_j/d_{ij} and P_iP_j/d_{ij} factors, respectively.[17] These factors closely resemble Stewart's concepts of potential and demographic energy and have been interpreted as identical with these concepts on a number of occasions. However, in empirical analysis, Zipf's use of his relationships or factors differs from Stewart's use of his (Stewart's) demographic concepts. Essentially, Zipf examines for pairs of cities interaction phenomena and the P_iP_j/d_{ij} factor where the entire factor is raised to some power. For example, for each pair of a set of cities for which materials are available he plots on a double-log graph the number of railway passengers moving between them and their P_iP_j/d_{ij} factor. When the data for all the possible pairs of cities are plotted, as in Figure 8.3, Zipf finds a straight-line relationship.

Zipf also finds straight-line relationships between the P_iP_j/d_{ij} factor and other interaction phenomena between pairs of cities, such as bus passenger trips, airline passenger trips, telephone calls, and tonnage of railway express shipments. However, in each case the slope of the straight line refers to the exponent of the entire P_iP_j/d_{ij} factor and not to the exponent of d_{ij} as is true in equation (8.4) which embraces Stewart's concepts. Thus Zipf's findings do not directly test the validity of Stewart's concepts.[18]

IV BASIC ISSUES OF GRAVITY MODELS

Having presented both the probability point of view and some historical background on the gravity model, we must now discuss some general issues which arise particularly in the application of the model. One issue concerns the measurements of the two variables, mass and distance.[19]

In empirical studies mass has been measured in a number of ways. This discussion has used population as a measure of mass. But if intermetropolitan migration is to be studied, it is apparent that employment or income of regions or subareas tends to be a more significant index of mass than population.[20] Or, when the marketing problem for manufactured products is being examined, total dollar volume of retail and wholesale sales tends to be a more significant measure,[21] etc. Clearly, the measure of mass to be employed depends on the problem to be studied, available data, and related considerations. The array of possible measures of mass includes such magnitudes as total investment in facilities, number of families, car registrations, hospital beds, investment in tractors and farm equipment,[22] commodity output,[23] value added in manufacture, Gross Regional Product, economic opportunities, newspaper circulation, church attendance, and school enrollment.

Similarly, distance has been and can be measured in a number of ways. This discussion has measured distance physically along a straight line in terms of miles. However, if a metropolitan traffic study is being conducted, distance in terms of travel time is at least equally important.[24] Or, if industrial location is being analysed, transport cost (or economic) distance is much more significant than physical distance.[25] Thus, as with the measurement of mass, the measurement of distance (many of whose forms are related to the state of transport technology) depends on the problem being attacked, the available data, and related considerations. Among other possible measures of distance are mileage along a specific transport route (waterway, highway, airline, railway, or pipeline), fuel (energy) consumption in transportation,[26] number of gear shifts or stops, and number of intervening opportunities and other forms of 'social distance'.[27]

Another basic issue relates to weights to be applied to the masses. In the specific models of equations (8.4) and (8.6), the weights of the masses are the same and are equal to unity. Should they be the same? According to Stewart[28] and Dodd,[29] different weights are valid. Just as the weights of molecules of different elements are unequal, so should the weights of different kinds of people be different. The average Chinese peasant does not make the same contribution to 'sociological' intensity as the United States urban dweller.[30] But to justify the use of weights in a particular study does not require a physical analogy. Suppose we are studying the volume of first-class or luxury travel. It is reasonable to expect that, *ceteris paribus*, an area with high per capita income will generate a larger volume of such travel than an area of equal population but lower per capita income. One way to correct for such a factor is to multiply the population of each subarea by its average per capita income. Thus equation (8.4) becomes

$$I_{ij} = G \frac{(w_i P_i)\,(w_j P_j)}{d_{ij}^b} \tag{8.8}$$

and equation (8.6) becomes

$$V = G \sum \frac{w_j P_j}{d_{ij}^b} \tag{8.9}$$

where w_i and w_j represent per capita incomes in region i and j, respectively. But since Y_j the total income of region j equals $w_j P_j$, equation (8.9) becomes

$$ {}_iV = G \sum \frac{Y_j}{d_{ij}^b} \tag{8.10} $$

which is the equation for income potential.[31]

In a study of first-class travel it might be desirable to weight population by still another factor, say education level, or even by a third factor, say family size. In such event w_i and w_j would represent composite weights, that is, averages reflecting the relative importance of the several weighting factors.[32] Among other possible weights which might be employed are occupational structure; sex or age composition; racial or ethnic composition; per cent of population receiving income above a certain level, attending church, or owning homes; urban-rural ratio; ratio of mail order to total retail sales; ratio of central business district sales to total metropolitan sales; ratio of capital investment to annual output; average capital investment per employee; etc.[33] Actually, when we weight masses measured in certain ways, we may derive other measures of the same masses. As already noted, when we weight population in subareas by per capita income, we derive total income as the measure of mass in subareas and thereby convert the population potential into an income potential measure. Or, when we weight population in subareas by average number of years of school attendance, we derive total number of man-years of education as the measure of mass for each subarea.

More difficult than the selection of weights or of measures of mass and distance is the choice of exponents for variables in both the potential and demographic energy concepts. From a theoretical standpoint, Stewart maintains that the exponent of the distance variable d_{ij} should be either 1 or 2.[34] However, numerous empirical studies do not support this contention (although the validity of these studies as tests of Stewart's hypothesis is yet to be established). For example, in one study Carroll finds that the appropriate exponent tends to hover around the value of 3.[35] Iklé finds considerable variation in the appropriate exponent of d_{ij}; his values range from 0.689 to 2.6.[36] Carroll and Bevis find that for intrametropolitan travel in the Detroit region the appropriate exponent for total person trips is 1.63.[37] The exponent, however, varies for different types of trips. Finally, Hammer and Iklé in their studies of telephone calls and airline trips find confidence limits of 1.3–1.8 for the exponent of distance and finally state that their data 'fail to justify either the inverse linear or the inverse square "law", which previous investigators had suggested for the distance function'.[38]

Although there is considerable evidence suggesting that the exponent of the d_{ij} variable need not be 1 or 2, depending on the concept employed, there has not been a definitive study of this question; and Stewart and Warntz cogently point to the inconclusiveness of existing studies and question the scientific basis of such studies.[39] For the purposes of the ensuing discussion, the reader may consider b, which is the exponent of d_{ij} in equations (8.4), (8.6), (8.8) and (8.9) as either unity[40] or a variable taking positive values, as a number of investigators have proposed.[41]

A still more difficult issue concerns the exponent to be applied to the measures of mass. (Since this issue and the remaining ones to be discussed in this section are rather nebulous, the reader interested in application may wish to proceed directly to the next section.) In the basic equations (8.8) and (8.9) these exponents are unity. But thoughtful researchers, such as Anderson and Carrothers, have suggested that the power to which mass is raised might be other than unity. Carrothers notes that such factors as agglomeration (deglomeration) economies imply that the exponent to be applied to any mass is a function of the mass.[42] In such a situation the exponents of unequal masses will differ. Thus, equations (8.8) and (8.9) will become, respectively,

$$I_{ij} = G \frac{w_i(P_i)^{\alpha} \cdot w_j(P_j)^{\beta}}{d_{ij}^b} \tag{8.11}$$

$$_iV = G \sum_{j=1}^{n} \frac{w_j(P_j)^{\beta}}{d_{ij}^b} \quad \text{(note 43)} \tag{8.12}$$

As mentioned earlier, Zipf's empirical findings when graphed on double-log scale suggest straight-line relationships of the order of equation 8.11 where $\alpha = \beta = b$, but not necessarily unity, and where $w_i = w_j = 1$. Mylroie's study on intercity travel desire also casts light on the question of the appropriate exponent for mass.[44] Among several cases tested she finds best correlations in equation 8.11 when $\alpha = \beta = \frac{1}{2}$, $b = 2$, and $w_i = w_j = 1$, that is, where

$$I_{ij} = G \sqrt{\frac{P_i P_j}{d_{ij}^2}}$$

Until further sytematic research is pursued on this question, it must remain unresolved.[45] The general equations 8.11 and 8.12 are still relevant, however. They embrace the simple models for which $\alpha = \beta = b = 1$ as well as the more complex models.

A final nebulous issue concerns the general form of the function to describe interaction, in particular the attenuating influence of distance. This issue can be neatly posed by reference to certain empirical materials collected by Carroll and Bevis in their pioneering studies. These materials relate to the effect of distance on the ratio of actual to expected total person trips within the Detroit metropolitan region. Figure 8.4a is a reproduction of a figure which they have presented, except that the straight regression line through the plotted points has been omitted. (Figure 8.4a is constructed in the same manner as Figure 8.1.) For their purposes, it was appropriate to seek a straight-line relationship on double-log scale. Yet, as we observe the set of plotted points, we may be inclined to fit the points with a quadratic (as depicted in Figure 8.4c) rather than a linear function (as depicted in Figure 8.4b). In so doing, we will be abandoning the notion of a gravity model, at least as conceived by Stewart and many others.

Similarly, in many other empirical studies on the effect of distance, it is questionable whether the simple straight-line relationship on double-log scale as given by equations (8.11) and (8.12) is applicable. Although such studies may demonstrate that interaction falls off with increase of intervening distance, many types of function in addition to the ones described by equations (8.11) and (8.12) are consistent with this effect. For example, the attenuating influence of distance (or, as sometimes stated, the friction of distance) is embodied in the simple equations: $I_{ij} = k - bd_{ij}$; or $I_{ij} = aP_iP_j - bd_{ij}$.[46] Similarly, the attenuating influence of distance is implicit in more complex types of functions. One of these is similar to equation (8.11) except that the exponent of d_{ij} (namely b) is itself a function of the size of one or both masses[47]; the appropriateness of this type of function has been suggested by certain statistical tests performed by Anderson.[48] Another of these more complex functions is similar to equation (8.11), except that the exponents of P_i and P_j (namely α and β) are also variables, being different functions of P_i and P_j rather than constants; this type of function has been suggested by Carrothers.[49]

In short, the question of appropriate types of function must be at least theoretically faced. Although various sets of data which have been collected seem to be adequately described by equations such as (8.11) and (8.12), it is still an open question whether or not other types of functions (desirably simple) might not describe such data more satisfactorily, explain their underlying pattern to a greater degree, or both.

V SOME PROBLEMS IN THE USE OF SIMPLE GRAVITY MODELS

Although we have raised some basic questions regarding weights, exponents, and functional forms, we may examine some of the problems associated with the use of the gravity model in the relatively simple forms in which it has been formulated. It is to be borne in mind that gravity models, as input-ouput and other models, can be used for both descriptive

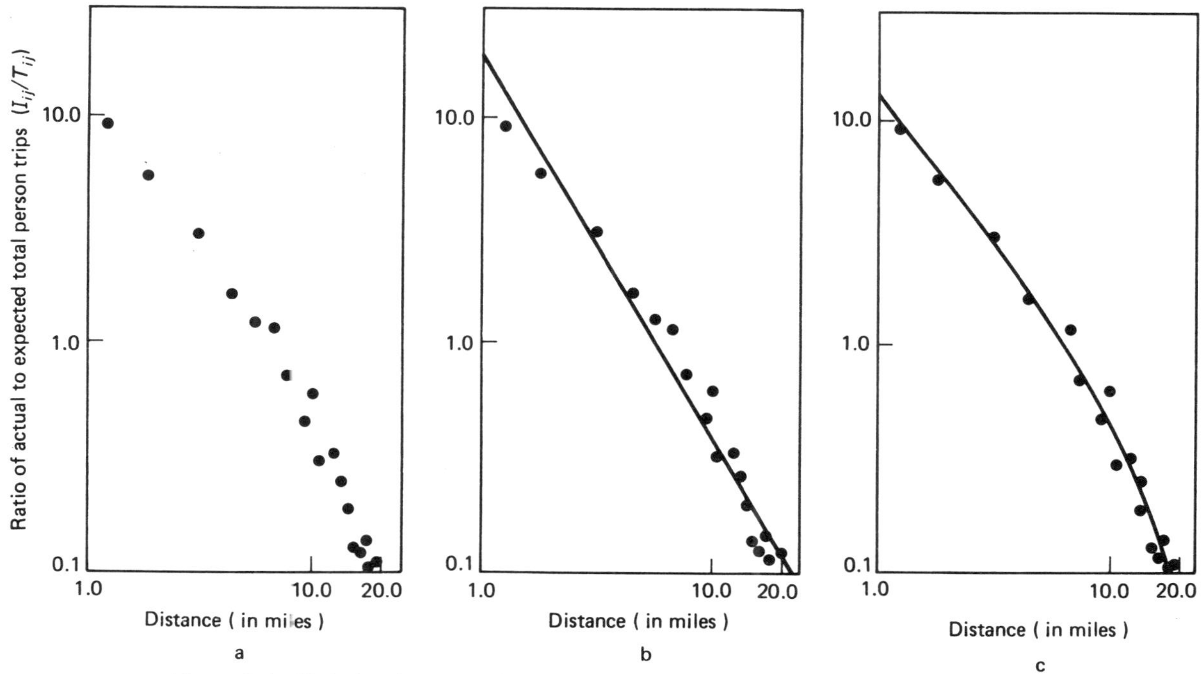

Figure 8.4 Relation between distance and the ratio of actual to expected total person trips, Detroit Metropolitan Region. (Source: J. D. Carroll and H. W. Bevis [5], p. 193, except for curve of c.)

and projective purposes. Hence, the type of problems we encounter is often influenced by the purpose to which the model is put.

One problem common to both types of use concerns the extent to which any whole, integral mass, population, or meaningful aggregate is cut into parts, categorised into sectors, or, in more technical terms, disaggregated. The notion of the gravity model, particularly as developed by Stewart and his associates, pertains to a relatively huge mass composed of a multitude of individual units. Within such a mass it is reasonable to assume that the irregularities, peculiarities, and idiosyncrasies of any individual unit or small subgroup of units are canceled or averaged out. In such a situation the *ceteris paribus* clause is valid to some extent at least. We can then justify a concentration on the two basic variables, distance and mass, and factors which can be encompassed in their weights and exponents, to the exclusion of other variables.

When we seek to employ the gravity model as more than a descriptive tool, problems of a still more difficult character arise. A basic obstacle to its use for projection is the lack of any theory to explain values or functions which we assign to weights and exponents. Currently, the justification for the gravity model is simply that *everything else being equal* the interaction between any two populations can be expected to be directly related to their size; and since distance involves friction, inconvenience, and cost, such interaction can be expected to be inversely related to distance.

VI SOME CONCLUDING REMARKS

In closing, it is indicated that the gravity model constitutes a very promising technique for regional analysis. Yet our review of literature relating to this model demonstrates that much remains to be done in exploring its utility. Solutions to the problems of defining relevant mass and distance, assigning appropriate weights and exponents, establishing a basis of comparison with other research results, and broadening its probability framework, etc., require extensive systematic probing. Only with such research can we extend basic knowledge of the forces underlying gravity models, construct more meaningful and useful descriptive and projective models, and desirably achieve a firm synthesis with other techniques of analysis and bodies of theory.

Notes

1. Development and applications of gravity models may be found in, among others, the references cited at the end of this chapter, especially G. A. P. Carrothers [6, 7]. In order to facilitate the understanding of the mathematical

terms used in this chapter, we have followed traditional notation on gravity models and have refrained from using a strange (and to some extent more complicated) notation which would be consistent with that of the preceding chapters and with any mathematical formulation of the fused frameworks of the chapter to follow.

2. In developing this point of view we draw heavily on J. D. Carroll and H. W. Bevis [4]. Also see S. C. Dodd [9] and F. C. Iklé [15].
3. In figure 8.1, $a = 3.9$ and $b = 1.5$.
4. Studies depicting such close association will be cited at later points in this chapter.
5. Implicit in this model, as in all others to follow, is an appropriate unit of time, say a week, month, or year. Doubling the unit time, *ceteris paribus*, doubles the magnitude of expected interaction. If the reader so cares, he may explicitly introduce a time variable of this sort into the model. Cf. S. C. Dodd [9] for extensive treatment of this factor.
6. For the moment we ignore a discussion of the interaction of subarea i with itself. This point is taken up later in connection with Stewart's concept of demographic potential.
7. It is to be noted that this equation can be derived directly from a simple probability approach as was equation (8.4).
8. As early as the middle of the nineteenth century H. C. Carey [3] observed the presence of gravitational force in social phenomena, stating that the force was in direct ratio to mass and inverse to distance.
9. See ch. 3, pp. 67–9.
10. See ch. 3, p. 69.
11. W. J. Reilly [19] proposed a law of retail gravitation which states that a city attracts retail trade from a customer in its hinterland in proportion to its size (population) and in inverse proportion to the square of the distance separating the customer from the centre of the city. The boundary separating the market areas of two cities i and j competing for customers in a hinterland is thus defined as the locus of points for which $P_i/d_{xi}^2 = P_j/d_{xj}^2$ where d_{xi} and d_{xj} are the distances of cities i and j respectively from any point x on the boundary.
12. See [20–24].
13. See [27, 28].
14. It is to be noted that any mass at i creates potential upon itself. If the distance of this mass from i, that is d_{ii}, is taken as zero, the value of ${}_iV$ becomes infinity. Hence, as Stewart has clearly pointed out, it becomes necessary to conceive of any mass at i or concentrated around i as being located at some finite distance from i. For procedures used to compute this distance, see J. Q. Stewart [20], p. 48; [21], p. 477; J. Q. Stewart and W. Warntz [24], p. 121; G. A. P. Carrothers [7, 8]. An alternative formulation which avoids an infinite value for ${}_iV$ is to add a constant to d_{ij} in the denominator. See T. Anderson [2], p. 178, for a case where $1 + d_{ij}$ is a relevant denominator. Cf. F. C. Iklé [15], pp. 133–5.

   Rigorously (but not practically) speaking, ${}_iV = G \int_s \frac{P}{d_i} ds$ where P is the surface density of the mass over the infinitesimal element of area ds and where d_i is the distance of that element from i.
15. The concept of 'weighted' potentials will be discussed later. For the moment the reader may ignore the term 'weighted'.
16. For full details, see J. Q. Stewart [20]. In a recent article J. Q. Stewart and

W. Warntz [24] compare cities (in the United States and Great Britain) equal in population but located in areas of varying population potential. They find that cities in areas of low potential tend to have larger areas, lower taxes, and a greater excess of births over deaths than cities of equal population in areas of high potential.

17. That is, Zipf's definition of P is the same as our definition of P_j, his of P_1 the same as ours of P_i, his of P_2 the same as ours of P_j, and his of D the same as ours of d_{ij}. See G. K. Zipf [27, 28].
18. In the nontypical case, when the slope of Zipf's straight line is unity, Zipf's use of his so-called P_1P_2/D relationship becomes identical with Stewart's use of demographic energy.
19. We do not discuss in this section problems connected with the use of existing data and the choice of an appropriate set of subareas or regions and control points. Excellent discussion of this latter problem is contained in G. A. P. Carrothers [8]. Carrothers investigates the degree to which a potential calculation actually represents what it purports to represent when different sizes and shapes of regions are delineated and when different spatial distributions of relevant masses are hypothesised. He concludes that the best set of general-purpose regions tends to satisfy as closely as possible the following criteria: (1) absence of concentrations of mass on the peripheries of the region; (2) existence within each region of a definite nodal centre of gravity of mass; (3) coincidence of the centre of gravity of mass with the centre of gravity of the physical area; (4) regular geometric shapes for the physical area of each region; (5) approximately equal areas of adjacent regions when density of mass is relatively uniform; (6) area of region varying in inverse proportion with density of mass. Since a gravity model implies for each region an intraregional distribution of *derived* interaction phenomena or mass proportional to the intraregional distribution of the generating mass (independent variable), these criteria are generally applicable. However, difficulty arises when an initial intraregional distribution of the independent mass, say income, does not conform to the initial intraregional distribution of an existing mass, say population, whose growth or change in interaction is to be derived. Then the initial intraregional distribution of the existing mass does influence the intraregional pattern of its growth or change in interaction; and it then becomes desirable to delineate regions by modifying these criteria to apply to this existing mass as well as the independent mass.
20. Thus W. Isard and G. Freutel [17], pp. 434–9, use income as a measure of mass and develop the concept of income potential to parallel Stewart's concept of population potential. G. A. P. Carrothers [8] uses the income potential concept in his population and migration studies. W. Warntz [26] constructs an income potential map for the United States, 1950. J. Q. Stewart and W. Warntz [23] report for relevant sets of subareas high correlation between income potential and such magnitudes as road density, telephone wire density, average farm size, the size of wholesale dry goods market areas, number of vehicles registered per mile of highway and per dollar of income, number of auto fatalities per registered vehicle, ratio of individual taxes to all income, ratio of number of patents to dollars of income, and ratio of business failures to all businesses.
21. Both C. D. Harris [14] and E. S. Dunn [10] have used retail sales as the relevant measure of mass in their studies.
22. C. D. Harris [14] uses the number of tractors as a measure of mass in analysing the location of agricultural equipment producers.

23. In his study of the spatial variation in the prices of wheat, potatoes, and other agricultural commodities, W. Warntz [26] computes a commodity supply space potential with annual commodity output as mass.
24. For example A. M. Voorhees [25] uses auto-driving time in studying the impact of distance on interaction.
25. In studying the localisation of manufacturing in general, Harris uses a generalised measure of transport cost rather than physical distance in computing his *market potential*. He estimates transport cost by different media according to the following table:

Method of transportation	*Terminal cost per ton*	*Rate per ton-mile*	*Delivery charge at destination*
Truck	\$6.00	4.0¢	
Railroad	\$5.00	2.5¢	\$6.00 (truck delivery)
Ship	\$12.00 (including \$6.00 truck delivery to port)	0.25¢	\$6.00 (truck delivery)

Thus a 400-mile shipment by rail of 1 ton of products costs \$5.00 (terminal cost) plus \$10.00 (line-haul costs over 400 miles) plus \$6.00 (truck delivery charge at destination), or altogether \$21.00. According to the rates of this table, for all distances up to 333 miles it is cheapest to ship by truck. Beyond this distance up to mileages in excess of 3000 it is cheapest to ship by rail.
26. See T. Anderson [2], pp. 177–8.
27. Forms of 'social distance' will be discussed later. See also W. L. Garrison [12], pp. T4–6, which differentiates distance by trip purpose and road type.
28. See [22], pp. 29–30.
29. See [9].
30. Ibid, pp. 247–8. Dodd's discussion of this issue from a sociological standpoint is particularly enlightening.
31. See W. Isard and G. Freutel [17], p. 436.
32. When there are several weighting factors, it is convenient to use double subscripts. The term w_{1i} then represents the subarea i's weight according to the first factor, and the term w_{gi} represents subarea i's weight according to the gth factor. Since it is necessary to derive the composite weight for subarea i, namely w_i, it is also necessary to assign a weight to each of the weighting factors. Thus w_i is the sum of weighted weights, that is,

$$w_i = \sum_{g=1}^{g} c_g \, w_{gi}$$

where c_g is the weight of the gth weighting factor. It also follows that

$$w_j = \sum_{g=1}^{g} c_g \, w_{gj}$$

33. In many of his studies Stewart uses weights of 0.8 for population of the Deep South, 2.0 for population of the Far West, and 1.0 for all other population. These weights were suggested by some of his initial empirical studies.

In studying interaction phenomena as reflected in number of telephone calls and airline trips between pairs of cities, C. Hammer and F. C. Iklé [13] use the gravity model of equation (8.8). They determine weights for different cities which yield the best least-squares fit to actual interaction data. The city weights range from 0.34 for Flint, Michigan, to 2.94 for Los Angeles in the analysis of telephone calls, October 1948. The weights range from 0.14 for Baltimore to 7.61 for Miami in the study of airline trips, March 1950. Hammer and Iklé regard the derived city weights as indicating different 'propensities to interact'. Using rank correlation, they find the weights to be positively associated with (1) the proportion of population who are managers, proprietors, and officials, (2) the average number of transient hotel rooms per capita, and (3) average per capita retail sales. However, the city weights are negatively correlated with median income of families and unrelated persons.

Despite these correlations, the validity of any specific weight derived for a given city is to be questioned, especially for projection purposes. This point will be discussed later.

34. The exponent of 2 in Stewart's demographic force concept is consistent with W. J. Reilly's [19] early study on the law of retail gravitation.
35. See [4].
36. See [15].
37. See [5].
38. See [13], p. 314. Also, W. Isard and M. J. Peck in studying tonnage of Class I railway shipments find such tonnage inversely related to distance raised to approximately the 1.7 power (see [16], pp. 71–3). On the other hand, W. L. Garrison [11] finds little significant relation between trip volumes and distance in analysing trips of rural residents in three counties of Washington State.
39. See [24].
40. Then the right-hand sides of equations 4 and 8 will correspond to Stewart's concept of demographic energy, and those of equations 6 and 9 to his concept of demographic potential.
41. At this point it is instructive to consider the situation where transport cost is used as the measure of distance, and where graduated rate structures (rate increasing less than proportionally with distance) exist, as is true of most modern transport media in industrialised regions. In this situation, if the exponent of 'transport cost' distance is taken to be unity (as C. D. Harris [14]), the exponent of equivalent physical distance (which can always be determined) not only will not be unity but will also become a parameter whose value changes with change in 'transport cost' distance.
42. Other factors such as degree of social integration and political stability may also be interpreted as justifying different exponents for the several masses involved.
43. Here ${}_iV$ would be defined as $\sum_{j=1}^{n} I_{ij} \,/\, w_i(P_i)^a$.
44. See [18]. Data and procedures are extensively presented.
45. For a set of statistical procedures suitable for general use, see G. A. P. Carrothers [8].
46. Another simple function is $I_{ij} = a[(\log P_i + \log P_j)/d_{ij}^b]$.
47. Or b may be a function of d_{ij} itself, or of both mass and distance.
48. See [1].
49. See [7].

References

[1] Anderson, Theodore, 1955. 'Intermetropolitan Migration: A Comparison of the Hypotheses of Zipf and Stouffer', *American Sociological Review*, vol. 20, June.

[2] Anderson, Theodore, 'Potential Models and Spatial Distribution of Population', *Papers and Proceedings of the Regional Science Association*, vol. 2.
[3] Carey, H. C., 1858–1859. *Principles of Social Science* (J. Lippincott, Philadelphia).
[4] Carroll, J. Douglas, 1955. 'Spatial Interaction and the Urban-Metropolitan Description', *Papers and Proceedings of the Regional Science Association*, vol. 1.
[5] Carroll, J. Douglas, and Howard W. Bevis, 1957. 'Predicting Local Travel in Urban Regions', *Papers and Proceedings of the Regional Science Association*, vol. 3.
[6] Carrothers, Gerald A. P., 1959. *Forecasting the Population of Open Areas*, doctoral dissertation (Massachusetts Institute of Technology, Cambridge, MA).
[7] Carrothers, Gerald A. P., 1956. 'An Historical Review of the Gravity and Potential Concepts of Human Interaction', *Journal of the American Institute of Planners*, vol. 22, Spring.
[8] Carrothers, Gerald A. P., 1958. 'Regional Population Projection via Income Potential Models', *Papers and Proceedings of the Regional Science Association*, vol. 4.
[9] Dodd, Stuart C., 1950. 'The Interactance Hypothesis: A Gravity Model Fitting Physical Masses and Human Groups', *American Sociological Review*, vol. 15, April.
[10] Dunn, Edgar S., 1956. 'The Market Potential Concept and the Analysis of Location', *Papers and Proceedings of the Regional Science Association*, vol. 2.
[11] Garrison, William L., 1956. 'Parameters of Spatial Interaction', *Papers and Proceedings of the Regional Science Association*, vol. 2.
[12] Garrison, William L., 1955. 'The Spatial Impact of Transport Media', *Papers and Proceedings of the Regional Science Association*, vol. 1.
[13] Hammer, Carl, and F. C. Iklé, 1957. 'Intercity Telephone and Airline Traffic Related to Distance and the "Propensity to Interact"', *Sociometry*, vol. 20, Dec.
[14] Harris, Chauncy, D., 1954. 'The Market as a Factor in the Localization of Industry in the United States', *Annals of the Association of American Geographers*, vol. 44, Dec.
[15] Iklé, F. C., 1954. 'Sociological Relationship of Traffic to Population and Distance', *Traffic Quarterly*, vol. 8, April.
[16] Isard, Walter, 1956. *Location and Space-Economy*, John Wiley, New York.
[17] Isard, Walter, and Guy Freutel, 1954. 'Regional and National Product Projections and their Interrelations', in *Long-Range Economic Projection*, Studies in Income and Wealth, vol. 16, National Bureau of Economic Research, Princeton University Press, Princeton, New Jersey.
[18] Mylroie, Willa, 1956. 'Evaluation of Intercity Travel Desire', in *Factors Influencing Travel Patterns*, Highway Research Board, Bulletin 119, Washington, DC.
[19] Reilly, William J., 1929. 'Methods for the Study of Retail Relationships', *University of Texas Bulletin*, No. 2944, Nov.
[20] Stewart, John Q., 1948. 'Demographic Gravitation: Evidence and Applications', *Sociometry*, vol. 11, Feb. and May.
[21] Stewart, John Q., 1947. 'Empirical Mathematical Rules Concerning the Distribution and Equilibrium of Population', *Geographical Review*, vol. 37, July.
[22] Stewart, John Q., 1950. 'Potential of Population and Its Relationship to

Marketing', in *Theory in Marketing*, R. Cox and W. Alderson (eds) (Richard D. Irwin, Inc., Homewood, Illinois).
[23] Stewart, John Q., and William Warntz, 1958. 'Macrogeography and Social Science', *Geographical Review*, vol. 48, April.
[24] Stewart, John Q., and William Warntz, 1958. 'Physics of Population Distribution', *Journal of Regional Science*, vol. 1, Summer.
[25] Voorhees, A. M., 1955. 'A General Theory of Traffic Movement', *Proceedings of the Institute of Traffic Engineers.*
[26] Warntz, William, 1959. *Towards a Geography of Price* (University of Pennsylvania Press, Philadelphia).
[27] Zipf, George K., 1949. *Human Behavior and the Principle of Least Effort* (Addison-Wesley Press, Reading, Massachusetts).
[28] Zipf, George, 1946. 'The P_1P_2/D Hypothesis on the Intercity Movement of Persons', *American Sociological Review*, vol. 11, Oct.

9 Impact of Vietnam War Expenditures on the Philadelphia Economy: Some Initial Experiments with the Inverse of the Philadelphia Input-Output Table[1]

This chapter is a report on ongoing research. The research is far from complete and, in fact, will take several more years to finish. There are many questions still to be probed in the Philadelphia Region Input-Output Study. Yet, there are also many interesting results, some of which are worthwhile to report on in a preliminary manner. We anticipate that our thinking and analysis may change significantly as we continue with the basic research.

In one sense, this chapter is a repeat of papers which have been written many times before. We have a regional input-output coefficient table. We have inverted it. Now we have started to play with it via interesting changes in final demand. In these respects, we have little that is new to report.

However, there are three basic aspects of our work which are significant. One pertains to the size and detail of the table itself. A second concerns the disaggregated character of the final demand sectors and particularly of the federal government. The third relates to the question of what should be inverted: (1) the technical input-output coefficients of a region, unadjusted for imports and other factors, or (2) these same coefficients adjusted in any one of many ways, together with parallel adjustments in export allocations in order to obtain more meaningful projections of impacts upon the economy and future magnitudes. We shall take up these aspects in order in the next three sections. Finally, in the last two sections, we shall look at (1) the impact of Vietnam War expenditures upon the Philadelphia economy and (2) offset programs that can absorb the resources which might be released by contraction of the Vietnam War effort.

This chapter was written with Thomas W. Langford, Jr. Reprinted from *Papers of the Regional Science Association*, vol. 23, (1969), pp. 217–65.

I PHILADELPHIA REGION TECHNICAL INPUT-OUTPUT COEFFICIENT TABLE

A preliminary version of the Philadelphia Region Technical Input-Output Coefficient Table was given at the meetings several years ago and published in 1967.[2] Since then, we have revised the table substantially, increasing the nonzero entries by approximately 55 per cent. All told, we currently have within the structural matrix 496 endogenous sectors representing in excess of 26,000 coefficients. In addition, we have developed a final demand subtable comprised of 86 sectors and approximately 7000 coefficients. In short, we have compiled probably the most detailed input-output table ever constructed. It is to be noted that there is no indication that the data collected and the resulting coefficients are any less reliable than the 1958 national input-output study. In fact, it is conceivable that they are significantly more reliable in the sense that they are more accurate and representative of the Philadelphia setting than the national are of the national setting. Full details on the construction of the table are available in the *Working Papers*.[3]

II FEDERAL AGENCY STRUCTURE OF THE PHILADELPHIA REGION

The 86 sectors of the final demand subtable of the Philadelphia study break down as follows: 36 sectors on federal government; 47 on capital formation; 1 on personal consumption (households); 1 on export trade; and 1 on unallocated demand.[4] Although the study has conducted important empirical analyses of the 50 final demand sectors which are not federal agencies, perhaps the most unique part of our analysis concerns the 36 final demand sectors relating to federal agencies. For the first time, we believe, there is presented a highly detailed and comprehensive portrayal of federal activities within a major metropolitan region.[5] Increasingly in future regional input-output work it will be necessary to develop such detail as the involvement of the federal government in metropolitan and regional life continues to mount.

In our initial classification of federal government agencies, we separated military (Department of Defense) agencies from civilian agencies. We further classified military agencies within the region with respect to their primary activity, namely: (1) procurement, (2) manufacturing and research and development, and (3) military base operation. We devised this specific sectoring pattern in order to be able to identify those agencies which have unique expenditure patterns, while being able to aggregate those agencies having similar expenditure patterns or missions. Table 9.1 presents the sectoring scheme as it was used in this study.

Table 9.1 Federal government agencies, Philadelphia SMSA

Sector	*Title*
9101	US Department of Agriculture
9102	US Post Office Department
9103	Social Security Administration
9104	Internal Revenue Service
9105	Veterans Administration, Medical Institutions
9106	Veterans Administration, Administrative
9107	US Coast Guard
9108	US Mint
9109	Bureau of Customs
9110	US Treasury
9111	US Department of Housing and Urban Development
9112	National Park Service
9113	General Services Administration
9114	Civilian Federal Government Agencies, n.e.c.*
9116 9117	Defense Personnel Supply Center
9118 9119	Defense Industrial Supply Center
9120 9121	US Army Electronics Command
9122 9123	US Marine Corps Supply Center
9124 9125	US Naval Aviation Supply Office
9126	Defense Contract Administration Services
9128	US Naval Supervisor of Shipbuilding
9129	US Naval Facilities Engineering Command
9130	Frankford Arsenal
9131	US Army Corps of Engineers
9132	Naval Air Engineering Center
9133	Naval Air Development Center
9134	US Naval Publications and Forms Center
9135	Military Hospitals
9136	US Naval Air Station
9137	US Naval Shipyard

Note: *Sector 9114 was included in the initial study for 1959 but was deleted from this chapter due to lack of adequate data at the time of preparation.

Government agencies are not significantly different from private organisations insofar as input-output analysis is concerned. Like private organisations, government agencies are engaged in producing goods or services for outside consumption, and in producing these outputs they require specific amounts of various resources.

The fact that governmental agencies are engaged in providing services does not make them unique. However, an important distinction is that government services are not sold on the open market and thus not subject to a direct market valuation. Moreover, many governmental services are not intended for or cannot be defined as benefiting individual consumers but rather apply to society as a whole.

The nonexistence of market prices poses some distinct problems. Typically, a workable method is to evaluate governmental services by setting the value of output as equal to the resource costs or the dollar expenditures of the agency. Such a procedure was used in this study. When transfer payments were involved, they were excluded from the expenditures of the originating agency and included in those of the recipient agency or government. HUD grants, for example, appear as materials and service expenditures by the local governmental agencies receiving the funds.

To obtain the data on expenditures by federal agencies, special questionnaires were developed which conformed to the accounting terminology and classification used by the agencies. For fiscal year 1960, only bench mark data relating to employment, payroll, and total expenditures were usually available – so that the detailed expenditure information generally had to be extrapolated from a more recent period for which they were available – but, for the fiscal year 1968, data were obtained directly from the questionnaire.

Full details relating to the development of the data are available in chapter 15, vol. III, *Working Papers*.[6]

It should be noted that the Philadelphia region is a major procurement center for defense needs. Consequently there are two distinct purchasing patterns which should be borne in mind. One pattern represents actual expenditures for goods and services procured for usage by the military services. The other expenditure pattern represents purchases of goods and services consumed by the agency in the achievement of its mission.

As examples of the above two patterns, consider the activities of the Defense Industrial Supply Center (DISC), sectors 9118 and 9119. This agency catalogs and is responsible for the procurement of approximately 690,000 industrial supply items for the armed services. The procurement categories include: hardware, abrasives, metal bars, sheets, shapes, blocks, tackles, rigging, fibre rope, cordage, twine, bearings, chain, wire rope, cable fittings, electrical wire, and cable. The procurement level of this agency for the fiscal year 1960 was estimated to be approximately $120, 300,000. The bulk of these purchases are made on a nationwide basis, with little emphasis given to the local or regional suppliers.

Table 9.2 shows the allocation of the total procurement activity to the specific RIS sectors, and the amount which is estimated to be procured within the Philadelphia region for sector 9119. Thus, out of the $120.3 million in procurement by the agency located within the region, only $9,066,753 or 7.5 per cent of the total represents purchases within the region.

To procure these goods, DISC required the services of 783 employees.

Table 9.2 Procurement activities, fiscal year 1960. Defense Industrial Supply Center

RIS sector		*Total procurement* $	*Regional procurement* $
2298	Cordage, twine	3,430,956	147,188
2499	Wood products, n.e.c.	435,366	11,145
2821	Plastic, mtls, resins, etc.	734,311	111,909
2851	Paints, varnishes, enamels	2,203,054	212,154
2852	Putty, calking compounds	1,468,743	105,456
2891	Glue, gelatin	710,973	120,297
3079	Misc. plastic products	2,263,685	764,889
3291	Abrasive products	4,349,687	606,781
3293	Steam packing, pipe cover	4,349,687	729,443
3312	Blast furnace, steel mill prod.	8,211,558	515,686
3315	Steel wire, nails, spikes	4,105,719	219,245
3356	Roll, draw, extrude: nonfe.	5,800,265	59,163
3357	Draw, insulate nonfe. wire	9,660,210	82,112
3429	Hardware, n.e.c.	7,159,775	234,125
3441	Fabricated struc. steel	7,843,680	458,071
3452	Bolts, nuts, screws, rivets	31,154,332	2,931,623
3481	Misc. fabricated wire prods.	7,629,546	434,121
3497	Metal foil, leaf	2,185,731	33,442
3499	Fabricated metal prods. n.e.c.	3,088,943	246,807
3562	Ball, roller bearings	4,579,460	626,470
3566	Mech. power transmission eq.	3,052,973	292,170
3981	Brooms, brushes	2,222,422	124,456
9842	Transport. serv. (allocator)	3,658,924	—
Total		$120,300,000	$9,066,753

These employees received wages and salaries of approximately $18,678,374. Other expenditures of $2,006,538 were incurred for the operations of the activity. Table 9.3 presents the disaggregated pattern of these operating expenditures. Regional distributions are not available; however, DISC personnel estimated at least 95 per cent were purchased within the Philadelphia region.

The value of keeping the procurement expenditure pattern separate and distinct from that of the operating expenditure pattern is clear when it is noted that in the Defense Industrial Supply Center the procurement increased 221 per cent in the period 1960–66. During this period, operating expenditures increased by only 21 per cent. It has been estimated that the two patterns of expenditures by commodity remained essentially the same from 1960–66.

Table 9.3 Operating expenditures, fiscal year 1960. Defense Industrial Supply Center

RIS sector		*Operating expenditures*
2522	Metal office furniture	$ 25,977
2542	Metal office, store fixture	15,586
2621	Paper mills	37,348
2641	Paper coating, glazing	20,748
2645	Die cut paper, cardboard	20,748
2649	Converted paper prods, n.e.c.	20,748
2655	Fibre cans, tubes, drums, etc.	16,599
2841	Soap, other detergents	17,064
2911	Petroleum refining	46,849
3069	Fabricated rubber products, n.e.c.	17,196
3315	Steel wire, nails, spikes	16,948
3421	Cutlery	12,935
3481	Misc. fabricated wire prods.	17,246
3555	Printing trades machinery	17,416
3571	Computing, acct. machines	955,047
3572	Typewriters	15,647
3579	Office machines, n.e.c.	32,094
3641	Electric lamps	34,526
3651	Radio, TV receivers	13,037
3861	Photographic equip., sup.	36,073
3951	Pens, mech. pencils, parts	30,152
3952	Lead pencils, arts mtls.	30,152
3953	Marking devices	17,230
3955	Carbon paper, ink ribbons	25,845
4811	Telephone communications	284,886
4890	Communication services, n.e.c.	113,822
4911	Electric company, systems	50,081
4941	Water supply	25,040
9842	Transport. serv. (allocator)	39,498
9888	Households (wages, salaries)	18,678,374
Total		$20,684,912

III METHODOLOGY AND THE CHOICE OF AN INVERSE

Once the data on expenditures by federal agency were developed, it became necessary to decide upon a methodology whereby the impact of federal expenditures might *first* be explored. It was clear that an inverse should be developed; yet a decision had to be made on the exact characteristics of the matrix to be inverted.

On first thought, it would seem logical that if the impact of changes in federal expenditures upon the Philadelphia economy is to be examined, one should not base the inverse on the technical input-output coefficient table. Rather one should employ these coefficients after they have been adjusted (reduced) so as to reflect not the full technical requirements but only that fraction of the technical requirements locally supplied. That is, that part of any technical requirement provided by imports should be eliminated. However, although for our base year 1959 we had estimated the percentages of intermediate local demand for every commodity which were furnished by local suppliers and by importers, it was questionable whether all the coefficients along any row should be reduced by the same import percentage factor. It seemed unreasonable to assume that all purchasing industries buy a particular input from local suppliers and importers in the same ratio. Yet no other simple criterion was available for immediate use.

Moreover, it is questionable whether the 'average' percentage of a commodity which was supplied locally, as estimated for the base year 1959, and as reflecting the total operation of an economy, should be applied to *changes* in final demand sectors. It is clear that if we were to assume greater elasticity in the supply of a commodity from importers than local suppliers, such averages might overstate the impact on the local economy. Yet, again there was no simple criterion which we might have employed to adjust our import and locally supplied percentages, upward and downward respectively.

Furthermore, there was a still more basic factor to consider. Once technical input-output coefficients are adjusted (based on knowledge of the local economy's import patterns, competitive cost position, etc.) to reflect only the percentage of each input which is supplied locally, and once these adjusted coefficients are inverted, the resulting inverse has no value for use in studies of other economies. For example, if we were to invert a table of adjusted Philadelphia coefficients, the resulting matrix would reflect much too much the peculiar import and competitive cost position of Philadelphia to have potential for the study of other metropolitan regions.

Finally, when one probes the question at its deepest level, one recognises that the exact type of inverse to be used depends on the nature of the economic and social problem being examined. If one were interested in the impact of a new integrated iron and steel works or the relocation of a major facility to another far distant region with which Philadelphia has little trade, the argument is perhaps strongest for using an inverse of technical coefficients adjusted downward to reflect imports. However, if one is interested in the impact of the Vietnam War expenditures on the Philadelphia economy, and the resources that might be released for social programmes and economic growth, then it becomes necessary not only to adjust technical coefficients of the Philadelphia region downward to reflect imports from other regions but also to adjust the amounts of Philadelphia exports (that is the Philadelphia-produced commodities that must be delivered to other importing regions) which are required to support the

Vietnam War production assigned to these other regions. Not to do so for Philadelphia as well as other regions would then yield inconsistent Vietnam War impact analysis, region by region – inconsistent in the sense that for the nation as a whole its Vietnam War programmes were being produced with inputs smaller than those technically required. But to identify appropriately our allocation of Philadelphia's exports for the Vietnam War effort in effect requires an interregional input-output table. Such a table does not exist. Moreover, even to make this identification crudely at this time requires considerably more resources than federal agencies have been willing to devote to regional and multiregional input-output analysis. Hence, one might well argue that it is better *not to adjust technical input-output coefficients downward to reflect imports than to do so and not develop simultaneously appropriate 'Vietnam War' export allocations.*[7]

In the light of all these considerations, we decided to use the unadjusted (unreduced) Philadelphia technical input-output coefficients and to examine impacts with an inverse of these unadjusted coefficients. We note that such a use is consistent with the assumption that Philadelphia is a closed economy, which it is not. Therefore we must adjust the results (from the use of unadjusted coefficients) to account for the changed levels and composition of both Philadelphia imports and exports to be anticipated. Although it is again questionable whether we can effect a valid adjustment of such results, it is also clear that if we were able to do so crudely, we would have a generally useful tool. For we would be able to make projections of other economies on the basis of an application of the inverse of unadjusted Philadelphia coefficients where the magnitudes yielded by this inverse would need to be adjusted to take into account the particular import-export patterns and competitive cost positions of these other economies.

Finally, although the magnitudes of our initial exploration of impact are in terms of the unadjusted coefficients – as if Philadelphia were a closed economy – later explorations of impact will be based on inverses constructed from adjusted coefficients and export allocations. By comparing two or more sets of projected magnitudes, we hope to gain much insight into ways and means by which particular import-export patterns and competitive cost positions can be incorporated into 'unadjusted' regional input-output projections.

IV IMPACT OF VIETNAM WAR EXPENDITURES

Having sketched the data and the basic structure of the federal government sectors, we may now discuss a few possible applications. One critical question on the minds of many social scientists and policy makers these days is the impact of the Vietnam War on national and regional economies.

With reference to the Philadelphia economy, there are several ways in which this question can be attacked. Having identified the actual employment and expenditures for each of the 35 aggregate governmental sectors for the fiscal year 1968, as shown in Table 9.4, we then chose to estimate the extent and nature of the changes in these expenditures which can be attributed to the Vietnam War.[8] We took this step by projecting what the expenditures for each agency for fiscal year 1968 would have been had there been no war. Such projections were based upon the trends of agency expenditures for a period of years, discussions with appropriate government and private officials, and staff judgement. To illustrate our procedures, we briefly describe our treatment of four sectors.

Sector 9103, Social Security Administration

The nature of this service agency suggests that its level of operations is not directly related to the level of defence activities. Examination of the annual operating expenditures of the regional and district offices suggests that there has not been a significant reduction in the growth rate of the agency during the past years as a result of increasing military expenditures.

Sector 9111, Department of Housing and Urban Development

Although this agency is not directly related to defence activities, the expenditures of the agency and its employment appear to have been significantly restricted during the period of the Vietnam conflict. The exact changes are not readily determined due to changes in scope of mission and programmes administered by the agency. It is however a widely held opinion that had there been no war, significant additional monies would have been made available through this agency to aid housing, planning, and development. We estimate that there would have been approximately 150 more persons employed, representing a 20 per cent increase in employment. We also estimate that the dollar amount of grants and loans processed by the various constituent agencies would have been approximately 100 per cent greater.

Sectors 9118 and 9119, Defense Industrial Supply Center

This agency provides procurement activity for general industrial products. From a 1960 level of 783 employees, the agency expanded to a level of 2687 in 1964. It has witnessed a relatively slow decline in personnel in recent years because of increase in productivity of personnel in handling procurement actions.

Review of the annual data suggests that the procurement levels of the agency would have stabilised at approximately $125 million per year in the

Table 9.4 Actual 1968 federal government activity, by agency

		Employment	*Payroll* $	*Output* $	*Local purchases* $	*Total impact* $
US Agriculture Dept.	9101	576	6,500,667	10,007,001	8,934,575	41,716,666
US Post Office	9102	20,478	135,885,336	170,052,953	161,698,002	745,340,533
Social Sec. Admin.	9103	2,480	19,058,659	21,645,659	20,738,171	93,862,663
Int. Rev. Service	9104	4,013	29,658,579	31,762,717	30,850,872	141,315,016
Vet. Admin: Med. Inst.	9105	1,943	16,577,299	21,794,463	19,862,893	87,694,717
Vet. Admin: Admin.	9106	1,690	12,165,129	12,951,073	12,548,995	57,473,204
US Coast Guard	9107	332	749,063	996,978	933,122	3,980,847
US Mint: Philadelphia	9108	537	4,748,499	31,337,276	7,851,611	34,276,721
Customs Bureau	9109	218	2,594,448	2,671,894	2,617,214	11,970,079
Bureau of Accts: Disburs.	9110	142	920,259	1,570,760	1,313,160	5,983,877
US Dept. Hous., Urban Dev.	9111	725	5,250,944	225,614,152	54,863,240	293,636,690
National Park Service	9112	337	3,558,384	4,231,797	4,031,238	18,576,824
General Services Admin.	9113	446	3,476,250	21,049,918	13,945,865	69,035,708
DPSC Oper.	9116	5,252	41,279,440	48,100,681	44,574,632	203,127,380
DPSC Proc.	9117			2,051,908,094	103,498,924	509,926,899
DISC Oper.	9118	2,608	20,485,300	25,782,778	22,969,766	104,102,531

DISC Proc.	9119			283,600,000	23,919,927	118,647,679
Army Electron., Oper.	9120	3,020	30,465,974	47,866,707	45,477,900	217,311,119
Army Electron., Proc.	9121			956,052,222	163,246,299	880,023,238
USMC Oper.	9122	2,232	18,167,122	21,561,377	19,975,235	91,541,207
USMC Proc.	9123			181,099,090	14,745,616	73,988,354
ASO Oper.	9124	2,320	20,029,425	26,069,436	19,773,489	91,613,208
ASO Proc.	9125			605,013,172	66,077,529	325,063,958
DCASR	9126	1,446	12,835,355	14,660,589	14,198,111	65,157,715
USN Sup. Shipbldg	9128	111	1,210,000	13,975,000	12,622,742	62,025,845
USN Facils. Eng. Comd.	9129	180	1,833,425	23,630,774	13,915,661	77,465,087
Frankford Arsenal	9130	5,809	56,741,000	490,182,000	91,121,098	407,055,193
Corps. Engineers	9131	613	4,980,668	31,274,366	19,213,087	94,020,305
NAEC	9132	2,423	19,909,063	123,009,063	65,364,648	305,785,848
NADC	9133	2,967	30,306,985	92,945,807	45,912,139	213,470,410
Nav. Pub. Form Ctr.	9134	159	13,050,934	65,867,084	56,000,760	260,300,150
Military Hosps.	9135	3,369	23,062,088	45,107,585	37,395,260	162,243,319
Navl Air Stn: WG	9136	1,033	8,636,801	12,295,963	9,568,176	44,167,642
Nav. Shipyard: Phila.	9138	12,194	113,293,000	295,293,000	206,059,055	1,009,058,222
Total		79,653	657,430,096	6,010,981,429	1,435,819,012	6,920,958,854

absence of the war. Employment would have been reduced by approximately 108 persons, and operating expenditures by approximately $1850 thousand.

The generalised nature of the procurement items suggests that cessation of the war would have had little, if any, effect upon the spatial pattern of the procurement by four-digit SIC.

Sectors 9120 and 1921, Army Electronics Command: Philadelphia

This agency would have undergone significant changes had there been no Vietnam War. In April, 1965, it was publicly announced that the operations of the agency were to be transferred to Fort Monmouth, New Jersey. This decision was changed within three months, since such a relocation would have drastically reduced the efficiency of the agency at a time when the build up for Vietnam was beginning. In fact, 400 additional personnel were authorised for the agency in August, 1965.

It is therefore reasonable to assume that had there been no Vietnam War the agency would have been relocated. This would have resulted in a 100 per cent decrease in the operations in Philadelphia. However, it is estimated that the effect of the relocation upon the procurement purchases from the relative concentration of supplier firms in the Philadelphia region would have been negligible. A no Vietnam War situation would have, however, decreased the total procurements to the prewar level of approximately $500 million.

Once employment, payroll, and total and local expenditure estimates by agency for a no Vietnam War situation were derived for fiscal 1968, these estimates were subtracted from actual 1968 employment, payroll, and total expenditures. The differences were converted to percentages of actual 1968 magnitudes. These percentage changes are listed in Table 9.5.

To obtain some notion of the magnitudes involved turn to Table 9.4. For fiscal 1968, total employees summed over all federal sectors in the Philadelphia region were 79,653; total payroll was $657.4 million; and total expenditures were $6.010 billion, of which approximately 24 per cent was goods and services produced locally. Had there been no Vietnam War we estimate that this employment would have been 71,509, a net reduction of 10.2 per cent; the associated total payroll would have been $586.4 million, a net reduction of 10.8 per cent; and the total expenditures would have been $3.466 billion, a net reduction of 42.3 per cent. However 33 per cent rather than 24 per cent of these expenditures would have been local.

Having crudely identified these differences between what is and what would have been with respect to federal government operations, we ask the next question: what would be the differences in total impacts upon the Philadelphia economy? To help answer this question we multiplied the

Table 9.5 A no Vietnam War situation, 1968: estimated changes from 1968 actual

Sector	*Employment (per cent)*	*Local purchases (per cent)*	*Total impact (dollars)*
9101	+ 64.76	+ 64.88	+$ 24,690,663
9102	+ 5.00	+ 4.76	+ 32,375,754
9103	0.0	0.0	0
9104	+ 4.98	+ 4.98	+ 6,424,369
9105	+ 14.98	+ 15.00	+ 11,999,002
9106	− 7.34	− 7.34	− 3,849,844
9107	0.0	0.0	0
9108	0.0	0.0	0
9109	0.0	0.0	0
9110	0.0	0.0	0
9111	+ 20.69	+ 92.24	+ 250,264,277
9112	+ 13.35	+ 13.35	+ 2,262,756
9113	+ 28.25	+ 28.27	+ 17,800,343
9116	− 33.36	− 33.30	− 61,699,644
9117	—	− 65.75	− 307,069,950
9118	− 4.14	− 4.14	− 3,932,432
9119	—	− 55.92	− 60,525,423
9120	−100.00	−100.00	− 217,311,119
9121	—	− 46.85	− 375,304,650
9122	− 40.64	− 37.56	− 31,369,832
9123	—	− 83.66	− 56,430,039
9124	− 13.79	− 15.75	− 13,151,614
9125	—	− 21.75	− 64,492,068
9126	− 10.10	− 10.10	− 6,001,120
9128	0.0	0.0	0
9129	− 36.67	− 61.58	− 40,658,028
9130	− 5.32	− 23.39	− 85,656,542
9131	0.0	0.0	0
9132	0.0	− 16.50	− 46,788,023
9133	0.0	0.0	0
9134	0.0	0.0	0
9135	− 34.70	− 34.70	− 51,352,610
9136	− 54.02	− 47.22	− 15,988,914
9138	− 13.89	− 19.54	− 182,074,370
Total	− 10.22	− 19.71	− 1,277,839,058

inverse by each of the 35 local expenditure vectors to determine each government sector's impact on each of the 496 endogenous sectors. These impacts were then summed to obtain the total federal government impact. Thus for fiscal year 1968, we list in column 4 of Table 9.4 the set of local expenditures (total dollars of inputs required) by each government sector from the Philadelphia region, and, in the last column, the total impact in terms of total dollars of inputs required directly and indirectly of the

Philadelphia region. For example, the local expenditures of the Naval Shipyard in Philadelphia (sector 9138) were $206 million; total impact was $1.009 billion.

To determine what would have been, had there been no Vietnam War, we list in column 1 of Table 9.5 the percentage change of employment, in column 2 the percentage change in local expenditures (dollar value of inputs required from the Philadelphia region), and in the last column the total net impact in terms of dollar value of inputs directly and indirectly required of the Philadelphia region by each government sector. Thus, employment in sector 9101 would have increased by 64.76 per cent while that in sector 9116 would have decreased by 33.36 per cent. Total dollar change in inputs directly and indirectly required would have corresponded to a $24.7 million increase for sector 9101 and a $61.7 million decrease for sector 9116. The grand totals at the bottom of the columns indicate that (1) federal employment would have been 10.22 per cent smaller, (2) local expenditures of federal agencies would have been 19.71 per cent smaller, and (3) direct and indirect expenditures on locally supplied inputs would have been smaller by an amount of $1.278 billion.

Comparing the last columns of Tables 9.4 and 9.5, we observe that the total inputs directly and indirectly required of the Philadelphia region by actual 1968 federal government operations are approximately $6.9 billion; without operations associated with the Vietnam War, the same requirements would have been $5.6 billion or approximately 18.4 per cent smaller.

Thus far, the total impacts and change in impact has been presented by each agency, summed over all 496 sectors. It is equally, if not more, important to examine the impact of all federal programmes as a whole upon each of the 496 endogenous sectors of the Philadelphia economy. Accordingly, we present in columns 1, 2, and 3 in Table A1 of the Appendix of the original version of this chapter direct, indirect, and total requirements by all federal agencies as a whole for each of these sectors for fiscal year 1968. These data are summarized by major economic sectors in Table 9.6. For example, from columns 1, 2, and 3 of this table we note that the agricultural, forest, and fisheries sector furnishes $1.4 million of output directly to federal agencies in the region, another $269.9 million indirectly, for a total of $271.3 million. Aside from households, the electrical machinery sector furnishes *directly* the largest dollar amount of product, namely, $118.7 million; but, directly and indirectly, the services sector furnishes the most, namely, $789.1 million.

Next, we wish to examine on net the estimated Vietnam War impact. The relevant data are recorded by each of the 396 sectors for fiscal year 1968 in columns 4, 5, and 6 of Table A1 in the Appendix of the original version of this chapter. They record respectively the changes in the amounts of direct federal requirements, of indirectly generated requirements of federal activity, and of the total of these two. Where no sign is indicated , the change represents an increase as a result of Vietnam War operations. Where a negative sign appears, the change represents a decrease as a result of the Vietnam War operations. Again, we present in Table 9.6, in columns 4, 5, and 6, the relevant data by major economic sector.

For example, as a result of the Vietnam War operations, direct expenditures for construction activities are estimated to be $5.4 million greater than if there had been no war. The indirectly generated expenditures are estimated to be $8.5 million and the total impact is therefore approximately $13.9 million. The total of all direct federal expenditures as a result of Vietnam War operations is estimated at $284 million; the indirectly generated expenditures are estimated at $999 million; and the total of direct and indirect at $1283 million.

Thus far, we have examined in Tables 9.4 and 9.5 the over-all impacts of each federal agency on the Philadelphia economy. In Table 9.6 and Appendix Table A1 of the original version of this chapter, we have examined the impact upon each sector of the Philadelphia economy of over-all federal activity. It is now useful to focus on the impact of specific selected federal activities upon each sector of the Philadelphia economy. For this purpose, we choose to concentrate on the impacts of the Army Electronics Command: Philadelphia Office, and of the Naval Shipyard in Philadelphia.

We first consider the Army Electronics Command. Its impact, as discussed above, is via (1) expenditures associated with its staff employed at the Philadelphia office and (2) the procurement orders given to Philadelphia enterprises. In Table 9.7 and Appendix Table A2 of the original version of this chapter, columns 1, 2, and 3 refer to operations; columns 4, 5, and 6 refer to procurement activity. Table 9.7 indicates impacts by major economic sectors. Appendix Table A2 indicates impacts by selected four-digit sectors. From Table 9.7, it is seen that the direct impacts of its operation in fiscal year 1968 are largely in households (wage and salary payments), and the printing and publishing industry. Its indirect impact (largely reflecting the expenditure associated with employee earnings) is spread over all sectors. The direct impact of its procurement activity is largely on the three major economic sectors: electrical machinery, equipment and supplies (SIC 36), transportation services, and general services, (the SIC definitions for the latter two categories may be found in Appendix Table A1 of the original version of this chapter). From the details of Appendix Table A2 of the original version of this chapter, we see that the direct impact on services represents entirely the need for procuring business services, excluding advertising.

As already indicated, had there been no Vietnam War, the Army Electronics Command would have in all probability relocated to Fort Monmouth, New Jersey. Hence, the impact of the Vietnam War on the Philadelphia economy via the Electronics Command: operations can be

Table 9.6 Direct, indirect, and total impacts of regional federal agency expenditures by major economic sector

Major sectors	*Actual Federal impact Fiscal year 1968*			*Estimated Vietnam War impacts (net) (actual 1968 Federal impacts less estimated Federal impact in a no Vietnam War Situation)*		
	Direct (1) $	*Indirect (2)* $	*Total (3)* $	*Direct (4)* $	*Indirect (5)* $	*Total (6)* $
Agric., forest, fish.	1,389,224	269,906,578	271,295,802	922,167	28,735,823	29,657,990
Extractive	221,596	5,187,139	5,408,735	54,279	962,593	1,016,872
Construction	32,658,045	103,201,984	135,860,029	5,414,474	8,498,171	13,912,645
Ordnance	1,374,659	18,778	1,393,437	1,022,862	–585	1,022,277
Food	36,628,042	411,959,988	448,588,030	19,014,734	80,714,907	99,729,641
Tobacco	12,525	19,259,134	19,271,659	970	3,277,155	3,278,125
Textiles	6,808,304	83,051,591	89,859,895	4,247,500	18,078,680	22,326,180
Apparel	27,820,580	63,167,172	90,987,752	17,417,150	11,517,702	28,934,852
Lumber, wood	2,583,222	20,623,123	23,206,345	481,564	2,838,970	3,320,534
Furn., fixtures	2,529,032	17,298,968	19,827,999	311,145	3,407,563	3,718,708
Paper	3,371,538	115,830,215	119,201,753	536,828	24,218,704	24,755,532
Print., publish.	37,427,648	110,771,256	148,198,904	11,406,933	23,930,044	35,336,977
Chemicals	18,222,983	102,419,377	120,642,360	7,548,664	18,869,772	26,418,436

Petroleum ref.	11,538,946	96,608,685	108,147,631	1,632,339	17,609,995	19,242,334
Rubber, plastics	1,171,056	17,167,089	18,338,145	505,661	3,429,218	3,934,879
Leather	887,242	17,192,131	18,079,373	600,575	3,049,526	3,650,101
Stone, clay, glass	6,845,638	34,645,751	41,491,389	2,428,343	6,880,848	9,309,191
Primary metals	23,093,353	93,823,337	116,916,690	9,809,125	22,055,974	31,865,099
Fab. metal prods.	17,119,676	43,524,566	60,644,242	7,356,927	8,660,453	16,017,380
Machinery	47,264,981	37,692,922	84,957,903	13,680,061	8,427,665	22,107,726
Elec. machinery	118,762,675	67,252,257	186,014,932	46,060,622	20,725,461	66,786,083
Transp. equip.	54,330,738	90,231,603	144,562,341	9,802,189	16,740,157	26,542,346
Scient., cont. inst.	11,407,057	17,289,033	28,696,090	3,482,990	4,440,767	7,923,757
Misc. mfgr.	1,115,469	29,508,851	30,624,320	439,993	5,199,265	5,639,258
Transp. services	85,454,784	150,225,901	235,680,685	41,765,461	31,271,210	73,036,671
Utilities, comm.	40,124,148	213,089,307	253,213,455	5,384,987	39,160,998	44,545,985
Wholesale trade	1,581,382	167,794,207	169,375,589	219,686	31,874,141	32,093,827
Retail trade	502,161	429,845,479	430,347,640	55,964	73,288,686	73,344,650
Fin., ins., R.E.	3,076,882	301,437,920	304,514,802	47,733	51,950,831	51,998,564
Services	129,872,973	659,236,755	789,109,728	50,997,000	96,725,569	147,722,569
Local government	49,830,320	127,985,145	177,815,465	–49,682,355	24,347,182	–25,335,173
Households	660,792,133	1,567,893,601	2,228,685,734	70,961,818	308,664,578	379,626,396
Total	1,435,819,012	5,485,139,842	6,920,958,854	283,928,389	999,552,023	1,283,480,412

Table 9.7 Direct, indirect, and total impacts of US army electronics command, Philadelphia Office, fiscal year 1968, by major economic sector

Major sectors	*From operations*			*From procurement activity*			
	Direct (1) $	*Indirect* (2) $	*Subtotal* (3) $	*Direct* (4) $	*Indirect* (5) $	*Subtotal* (6) $	*Total* (7) $
Agric., forest, fish.	—	8,396,514	8,396,514	—	11,529,033	11,529,033	19,925,547
Extractive	—	122,095	122,095	—	245,406	245,406	367,501
Construction	—	2,375,422	2,375,422	—	5,598,562	5,598,562	7,973,984
Ordnance	—	511	511	—	740	740	1,251
Food	—	13,745,463	13,745,463	—	18,040,123	18,040,123	31,785,586
Tobacco	—	663,754	663,754	—	870,000	870,000	1,533,754
Textiles	—	2,304,388	2,304,388	1,716	3,142,336	3,144,052	5,448,440
Apparel	—	2,088,278	2,088,278	22	2,747,454	2,747,476	4,835,754
Lumber, wood	—	608,914	608,914	22,332	968,113	990,445	1,599,359
Furn., fixtures	12,503	511,624	524,127	2,480	1,359,718	1,362,198	1,886,325
Paper	318,219	6,096,634	6,414,853	68,700	5,387,167	5,455,867	11,870,720
Print., publish.	11,016,320	3,879,778	14,896,098	70,398	5,956,609	6,027,007	20,923,105
Chemicals	—	3,349,667	3,349,667	2,795	5,491,228	5,494,023	8,843,690

Petroleum ref.	53,764	3,062,305	3,116,069	1,373	4,429,979	4,431,352	7,547,421
Rubber, plastics	—	555,228	555,228	7,055	939,394	946,449	1,501,677
Leather	—	585,057	585,057	—	766,357	766,357	1,351,414
Stone, clay, glass	—	789,393	789,393	4,987	1,781,174	1,786,161	2,575,554
Primary metals	—	1,934,318	1,934,318	1,819	5,885,989	5,887,808	7,822,126
Fab. metal prods.	—	1,014,305	1,014,305	1,267,054	2,608,085	3,875,139	4,889,444
Machinery	151,292	809,073	960,365	2,087,825	3,027,907	5,115,732	6,076,097
Elec. machinery	—	1,055,972	1,055,972	26,785,878	10,768,792	37,554,670	38,610,642
Transp. equip.	—	3,174,309	3,174,309	74,460	4,239,969	4,314,429	7,488,738
Scient., cont. inst.	—	526,689	526,689	342,716	2,330,884	2,673,600	3,200,289
Misc. mfgr.	49,366	933,914	983,280	18,983	1,313,622	1,332,605	2,315,885
Transp. services	328,031	4,625,227	4,953,258	8,096,752	7,857,275	15,954,027	20,907,285
Utilities, comm.	2,014,753	6,967,157	8,981,910	—	10,689,007	10,689,007	19,670,917
Wholesale trade	—	5,314,866	5,314,866	—	8,595,216	8,595,216	13,910,082
Retail trade	—	14,784,600	14,784,600	—	19,381,735	19,381,735	34,166,335
Fin. ins., R.E.	417,792	9,909,474	10,327,266	—	14,416,685	14,416,685	24,743,951
Services	649,885	21,153,598	21,803,484	37,619,247	31,322,418	68,941,665	90,745,149
Local government	—	4,125,879	4,125,879	—	6,433,757	6,433,757	10,559,636
Households	30,465,974	46,368,813	76,834,787	—	100,703,324	100,703,324	177,538,111
Total	45,477,900	171,833,219	217,311,119	76,476,592	298,828,058	375,304,650	592,615,769

roughly measured by the numbers in columns 1, 2, and 3 of Table 9.7 and Appendix Table A2 of the original version of this chapter.[9] Both direct and indirect requirements of output from sectors of the Philadelphia economy would have decreased by the amounts listed in these columns.

Had there been no Vietnam War, we estimate, as already noted, that the procurement activities of the Electronics Command would have been at the prewar level of $500 million. Procurement from sources in the Philadelphia region, however, would not have been adversely affected from a relative standpoint by the *relocation*. Thus, total procurement from the Philadelphia region would have been at the estimated prewar level of $86.8 million rather than at the actual fiscal year 1968 level of $163.2 million, a reduction of 46.8 per cent. Accordingly, the impact of the Vietnam War via the procurement activities of the Electronics Command may be taken to be roughly 46.8 per cent of the figures listed in columns 4, 5, and 6 of Table 9.7 and Appendix Table A2 of the original version of this chapter.[10]

Next consider the impacts of the Naval Shipyard in Philadelphia. Data on the expenditure pattern of the Naval Shipyard were available for both fiscal year 1960 and 1968. On the basis of these data and annual summary data for the intervening years, the changes due to the Vietnam War, for example the intensified shipbuilding, are clearly discernible. We are then able to approximate expenditure patterns had there been no Vietnam War. We estimate that the level of Naval Shipyard expenditures in the Philadelphia region would have been 19.5 per cent less, that is, it would have been at a level of $165.8 million instead of the actual fiscal year 1968 level of $206 million. Its employment would have been 13.89 per cent less than the actual fiscal year 1968 employment of 12,194. Its direct and indirect impact would have been reduced by $182 million.

The breakdown of these over-all magnitudes are given in Table 9.8 for major economic sectors and in Appendix Table A3 of the original version of this chapter for selected four-digit sectors. Among the manufacturing industries, primary metals and machinery are most affected. For example, the direct and indirect impacts on the four-digit sector 3312, blast furnaces, steel works and rolling mills, are $4.39 million and $4.38 million, respectively. Because of the shipyard's large employment and payroll, its indirect impacts are diffused over many sectors, as is clear from column 2 of Table 9.8.

Since impacts in terms of employment as well as dollar output are also significant, we present in Table 9.9 the employment impacts upon the Philadelphia region of the Electronics Command, in both its operations and procurement activity, and the Naval Shipyard, as a result of their increased levels engendered by the Vietnam War. Among the major economic sectors most affected are: printing and publishing; electrical machinery, equipment, and supplies; transportation services; retail trades; and general services.

Table 9.8 Direct and indirect impacts US Naval Shipyard, Philadelphia, Fiscal Year 1968, increased operations due to Vietnam War, by major economic sector (actual 1968 impact less estimated impact in a no Vietnam War situation)

Major sectors	*Direct (1)*	*Indirect (2)*	*Total (3)*
Agric., forest, fish.	$383	$7,350,065	$7,350,448
Extractive	1,331	130,187	131,518
Construction	39,238	2,126,689	2,165,927
Ordnance	—	399	399
Food	1,603,373	10,391,568	11,994,941
Tobacco	970	481,369	482,339
Textiles	9,630	1,755,477	1,765,107
Apparel	8,878	1,526,219	1,535,097
Lumber, wood	27,917	527,386	555,303
Furn., fixtures	216,623	382,925	599,548
Paper	75,927	2,797,906	2,873,833
Print., publish.	106,770	2,551,548	2,658,318
Chemicals	580,864	2,880,074	3,460,938
Petroleum ref.	1,243,203	2,646,704	3,889,907
Rubber, plastics	31,666	414,152	446,418
Leather	380	422,784	423,164
Stone, clay, glass	886,153	921,639	1,807,792
Primary metals	4,390,437	4,380,667	8,771,104
Fab. metal prods.	1,088,471	1,369,515	2,457,986
Machinery	6,422,813	933,906	7,356,719
Elec. machinery	2,440,441	1,528,520	3,968,961
Transp. equip.	24,834	2,377,691	2,402,525
Scient., cont., inst.	258,610	305,220	563,830
Misc. mfgr.	32,923	690,156	723,079
Transp. services	984,310	4,222,764	5,207,074
Utilities, comm.	2,879,729	5,691,543	8,571,272
Wholesale trade	219,686	4,596,184	4,815,870
Retail trade	63,127	10,721,447	10,784,574
Fin., ins., R.E.	5,640	7,884,501	7,890,141
Services	879,671	16,410,995	17,290,666
Local government	—	3,415,660	3,415,660
Households	15,738,664	39,975,248	55,713,912
Total	40,262,662	141,811,708	182,074,370

V OFFSET PROGRAMMES AND RELEASED RESOURCES

It is not necessary to dwell upon the point that had there not been Vietnam War expenditures in the Philadelphia SMSA, there would have been more of other governmental and private enterprise activity within the region. It is now well-recognised that the employment which was channelled into Vietnam War activities could and would have been profitably occupied in

Table 9.9 US Army electronics command and US Naval Shipyard employment direct and indirect impacts Vietnam War, by major economic sector (actual 1968 employment impacts less estimated employment impacts in a no Vietnam War situation)

	US Army Electronics command		
	Operations	*Procurement*	*US Naval Shipyard*
Direct	3,020	0	1,694
Indirect	10,749	22,056	9,188
Agric. forest, fish.	1,330	1,847	1,166
Extractive	6	13	7
Construction	133	311	121
Ordnance	0	0	0
Food	438	581	400
Tobacco	15	19	11
Textiles	164	223	125
Apparel	223	292	163
Lumber, wood	53	85	48
Furn., fixtures	41	96	44
Paper	238	222	118
Print., publish.	1,022	389	176
Chemicals	105	168	108
Petroleum ref.	49	69	60
Rubber, plastics	27	50	22
Leather	43	73	41
Stone, clay, glass	46	107	86
Primary metals	81	261	399
Fab. metal prods.	54	244	143
Machinery	76	511	427
Elec. machinery	79	2,161	256
Transp. equip.	113	156	87
Scient., cont., inst.	43	230	46
Misc. mfgr.	72	99	54
Transp. services	463	1,498	488
Utilities, comm.	587	622	456
Wholesale trade	308	491	342
Retail trade	1,783	2,332	1,302
Fin., ins., R.E.	258	391	217
Services	2,489	7,876	1,936
Local government	410	639	339
Total	13,769	22,056	10,882

other economic pursuits. Hence, when the Vietnam War effort is reduced, hopefully by 100 per cent, the amount of the inputs of any of the 496 sectors directly and indirectly required by the Vietnam War effort should not be viewed as indicative of the amount by which the output of that sector is likely to be contracted. Rather, that amount should be viewed as resources (and associated productive capacity) released for the support of other government programmes and private undertakings, or as output that may be foregone in order for the economy to turn out products of higher priority.

This view can be developed in some detail. For example, one might estimate, or speculate, what levels of educational programmes and low-income housing construction could have been accomodated in the Philadelphia region had there been no Vietnam War. This question can be examined from the standpoint of the monies that might have been available or critical inputs in short supply which would have been released. In terms of monies, the difference, $284 million, between actual fiscal year 1968 federal agency expenditures of $1.443 billion and our estimate of federal agency expenditures of $1.159 billion had there been no Vietnam War is a reasonable indication of the monies that might have gone into social programs in the Philadelphia region and also of the monies that might become available when and if the Vietnam War comes to a halt. For such a sum of money, namely $284 million, one can develop many mixes of reasonable programmes which could have been or could be developed as an offset policy. One mix to meet some of our urban needs, for example, might involve the following expenditures:

Elementary and secondary education	$170,400,000
Institutions of higher education	56,800,000
Low income housing construction	56,800,000
	$284,000,000

The detail of expenditures for each of these programmes at these suggested levels are presented in columns 1, 3, and 5 respectively of Table 9.10 (by major economic sectors) and Appendix Table A4 of the original version of this chapter (for selected four-digit sectors). As can be expected, payments to households dominate the programmes in support of elementary and secondary education and institutions of higher learning. In the former, for example, $120 million of the $170 million expenditure level is for such payments. In the case of the programme for low income housing construction, however, payments to households represent less than 50 per cent of total expenditures, and purchases of stone, clay, and glass products, and fabricated metals constitute roughly 10 per cent and 9 per cent, respectively, of total expenditures.

The support of these programmes at the specified levels, just as the Vietnam War programmes, require, indirectly, various other inputs which are to be identified. These indirect input requirements are listed in columns 2, 4, and 6 respectively of Table 9.10 and Appendix Table A4 of the original version of this chapter.

The over-all total of the indirect requirements by programmes are as follows:

Table 9.10 Direct and indirect impacts of hypothetical offset programmes, fiscal year 1968 by major economic sector

Major sectors	*Elementary, secondary education*		*Institutions of higher education*		*Low income housing construction*	
	Direct (1) $	*Indirect (2) $*	*Direct (3) $*	*Indirect (4) $*	*Direct (5) $*	*Indirect (6) $*
Agric., forest, fish.	566,920	30,654,481	137,115	9,827,058	57,015	9,470,829
Extractive	—	468,320	—	186,703	277,204	963,651
Construction	624,005	7,935,730	944,527	3,130,718	14,892	3,104,983
Ordnance	—	1,726	—	564	—	645
Food	2,235,139	48,200,656	797,870	15,343,760	—	15,302,863
Tobacco	—	2,293,130	—	726,130	—	733,045
Textiles	—	7,901,886	18,005	2,582,668	1,914	2,621,226
Apparel	—	7,211,139	23,402	2,292,275	319	2,317,315
Lumber, wood	335,347	2,302,078	11,132	774,640	1,596,638	1,420,346
Furn., fixtures	433,498	1,794,107	752,430	588,310	536,219	575,449
Paper	668,018	11,400,502	333,636	4,086,441	166,153	4,766,564
Print., publish.	3,411,924	10,521,273	1,356,808	3,901,021	15,849	4,399,568
Chemicals	407,587	10,419,337	278,650	3,392,485	446,868	4,101,422

Petroleum ref.	413,220	10,395,660	236,345	3,430,526	987,555	4,055,427
Rubber, plastics	18,418	1,737,405	7,624	572,029	177,003	646,474
Leather	—	2,018,955	—	641,552	—	643,901
Stone, clay, glass	397,601	2,793,899	30,318	1,109,927	5,536,962	2,113,467
Primary metals	55,472	6,344,321	3,122	2,415,544	1,241,145	4,368,368
Fab. metal prods.	302,929	3,623,526	59,933	1,359,556	4,860,972	2,021,158
Machinery	661,691	2,495,557	424,696	922,914	103,392	1,157,986
Elec. machinery	81,110	3,764,314	328,588	1,299,016	717,690	1,382,857
Transp. equip.	114,168	10,925,494	13,689	3,497,274	1,488	3,630,253
Scient., cont., inst.	548,742	1,353,281	358,364	429,623	78,077	534,860
Misc. Mfgr.	734,515	3,204,571	262,333	1,020,922	23,721	1,088,649
Transp. services	551,074	15,041,891	302,971	5,311,350	1,858,312	6,421,584
Utilities, comm.	4,610,854	22,794,684	2,271,603	8,071,559	554,834	8,892,467
Wholesale trade	—	18,050,941	—	6,846,110	7,493,774	6,689,623
Retail trade	—	51,075,497	147,737	16,174,810	—	16,328,065
Fin., ins., R.E.	19,016,299	24,803,187	2,118,357	11,396,141	773,429	11,861,582
Services	8,541,982	71,654,162	584,018	24,325,844	3,014,996	25,215,458
Local government		13,667,845	53,108	4,532,662	861,397	4,877,639
Households	119,542,925	145,908,667	34,766,764	49,288,112	25,371,334	59,477,293
Other, n.e.c.	6,126,562	—	10,176,855	—	30,848	—
	170,400,000	552,760,222	56,800,000	189,478,244	56,800,000	211,185,017

Table 9.11 Comparison of impacts (direct plus indirect) of Vietnam War and offset programmes, fiscal year 1968

Major sectors	*Vietnam War impacts (net)* (1) $	*Impacts of offset programmes*: *Elementary, secondary education* (2) $	*Institutions of higher education* (3) $	*Low income housing const.* (4) $	*Total impacts of offset programmes* (5) $	*Offset—War* (6) $
Agric., forest, fish.	29,657,990	31,223,401	9,964,173	9,527,844	50,715,418	21,057,428
Extractive	1,016,872	468,320	186,703	1,240,855	1,895,878	879,006
Construction	13,912,645	8,559,735	4,075,245	3,119,875	15,754,855	1,842,210
Ordnance	1,020,181	1,726	564	645	2,935	–1,017,246
Food	99,729,641	50,435,795	16,141,630	15,302,863	81,880,288	–17,849,353
Tobacco	3,278,125	2,293,130	726,130	733,045	3,752,305	474,180
Textiles	22,326,180	7,901,886	2,600,673	2,623,140	13,125,699	–9,200,481
Apparel	28,934,852	7,211,139	2,315,677	2,317,634	11,844,450	–17,090,402
Lumber, wood	3,320,534	2,637,425	785,772	3,016,984	6,440,181	3,119,647
Furn., fixtures	3,718,708	2,227,605	1,340,740	1,111,668	4,680,013	961,305
Paper	24,755,532	12,068,520	4,420,077	4,932,717	21,421,314	–3,334,218
Print., publish.	35,336,977	13,933,197	5,257,829	4,415,417	23,606,443	–11,730,534
Chemicals	26,418,436	10,826,924	3,671,135	4,548,290	19,046,349	7,372,087

Petroleum Ref.	19,242,334	10,808,880	3,666,871	5,042,982	19,518,733	276,399
Rubber, plastics	3,934,879	1,755,823	579,653	823,477	3,158,953	775,926
Leather	3,650,101	2,018,955	641,552	643,901	3,304,408	–345,693
Stone, clay, glass	9,309,191	3,191,500	1,140,245	7,650,429	11,982,174	2,672,983
Primary metals	31,865,099	6,399,793	2,418,666	5,609,513	14,427,972	–17,437,127
Fab. metal prods.	16,017,380	3,926,455	1,419,489	6,882,130	12,228,074	–3,789,306
Machinery	22,107,726	3,157,248	1,347,610	1,261,378	5,766,236	–16,341,490
Elec. machinery	66,786,083	3,845,424	1,627,604	2,100,547	7,573,575	–59,212,508
Transp. equip.	26,542,346	11,039,662	3,510,963	3,631,741	18,182,366	–8,359,980
Scient., cont., inst.	7,923,757	1,092,023	787,987	612,937	3,302,947	–4,620,810
Misc. mfgr.	5,639,258	3,939,086	1,283,255	1,112,370	6,334,711	695,453
Transp. services	73,036,671	15,592,965	5,614,321	8,279,896	29,487,182	–43,549,489
Utilities, comm.	44,545,985	27,405,538	10,343,162	9,447,301	47,196,001	2,650,016
Wholesale trade	32,093,827	18,050,941	6,846,110	14,183,397	39,080,448	6,986,621
Retail trade	73,344,650	51,075,497	16,322,547	16,328,065	83,726,109	10,381,459
Fin., Ins., R. E.	51,998,564	43,819,486	13,514,498	12,635,011	69,968,995	17,970,431
Services	125,566,965	80,196,144	24,909,862	28,230,454	133,336,460	7,769,495
Local government	25,335,173	13,667,845	4,585,770	5,739,036	23,992,651	49,327,824
Households	379,626,396	265,451,592	84,054,876	84,848,627	434,355,095	54,728,699
Other	—	6,126,562	10,176,855	30,848	—	—
Total	1,261,322,712	723,160,222	246,278,244	267,985,017	1,221,089,218	40,233,494

Elementary and secondary education	$552,760,222
Institutions of higher education	189,478,244
Low income housing construction	211,185,017
	$953,423,483

The total of direct plus indirect impacts, by major economic sectors, for each of these three programmes is recorded in columns 2, 3, and 4 of Table 9.11 and their combined totals in column 5. For example, their combined requirements (direct plus indirect) of labour services of households is $434 million. The breakdown of the combined totals for selected four-digit sectors is presented in column 2 of Appendix Table A5 of the original version of this chapter. For example, their direct plus indirect requirements for the output of sector 3312, blast furnaces, steel works, and rolling mills is $7.8 million.

It is appropriate to ask two basic questions at this point. One relates to the question: in what ways would the Philadelphia economy have been different today had there been no Vietnam War and had the three programmes listed above taken place at the respective levels? A second question is: to what extent could these three programmes, at the respective levels listed, have effectively constituted an offset policy in fiscal year 1968 had the Vietnam War come to a halt in that year?

A preliminary answer to both these questions is obtained from the data of Table 9.11 and Appendix Table A5 of the original version of this chapter. In Table 9.11, we list in column 1 our estimates of the Vietnam War impacts upon the Philadelphia economy by major economic sectors. In column 5, as already indicated, we list the combined total impact of the three programmes at the specified level. Subtraction of column 1 from column 5 yields the figures of column 6. These figures list the amounts by which each major economic sector would have been favourably affected when there is no negative sign in front of it, and adversely affected when there is a negative sign in front of it. For example the first sector, agriculture, forestry, and fishery products would have been favourably affected (the demand for its output would have increased) to the amount of $21 million. In contrast, the electrical machinery sector would have been adversely affected (demand for the output decreased) by the amount of $59 million. Perusal of this column indicates certain major effects, in particular the adverse effects in the electrical machinery sector and the transportation services sector and the favourable effects in the local government and household sectors. From the standpoint of social welfare, such changes cannot be said to be, in general, undesirable. The availability of more income in the local government and households sectors could be viewed as desirable, particularly as such income is more likely to accrue to the relative benefit of low income segments of our population. The decline in output in the electrical machinery and transportation services sectors is likely to be a burden, relatively speaking, on the middle and higher income groups. But these groups may be judged to be well able to forego some of the improvement in their income position experienced during the Vietnam War years.

The over-all total of column 6 is a negative figure of $40 million. This constitutes 3.1 per cent of the total of column 3, Table 9.5 which represents the total impact of the war. This percentage decline is of minor significance particularly in light of the various errors which unavoidably exist in the data.

On the question of offset policy, interest goes well beyond the magnitudes that relate to the highly aggregated major economic sectors. An investigator wishes to have the data in as fine detail as possible so as to be able to pinpoint any critical areas of contraction. Accordingly, we present such detail for selected four-digit sectors in Appendix Table A5 of the original version of this chapter. Note that the major contractions are $48 million in sector 7400 (research and development), $10 million in sector 3312 (blast furnaces, steel works, and rolling mills) and $9 million in sector 2311 (men's, youths', and boys' suits, coats, and overcoats). Of these three contractions, only that in sector 2311 could be of real concern since both the research and development and the steel industry need not have expanded as much as they did during the Vietnam War. (Our economy would have been just as well off, and possibly much better off, if some of the $10 million of basic steel produced in Philadelphia steel works had been embodied as structural steel in new schools and hospitals instead of being lost to productive use by society as bulkheads, etc. in aircraft carriers.)

VI CONCLUSIONS

We wish to reiterate that our research is far from finished. We report upon it at this time to make some of it available to interested scholars. Also, certain of its results may already be useful for policy formulation.

Although the work on the technical input-output coefficients for the Philadelphia region is complete, and our materials on the federal agency structure of the Philadelphia region are also in satisfactory shape, we still must engage in further explorations on appropriate methodology. Such methodology would pertain to the specific adjustments of the technical coefficients, on account of import-export patterns and competitive cost position, which are required or desirable before a matrix is inverted. We then will be in a much better position to design a methodology to attack different kinds of social problems whether they relate to the closing down of wartime operations and conversion programmes or to diverse development programmes. Pending the attainment of such improved methodology, we still assert that it is desirable to use the methodology that we have, in particular that employed in this paper, to attack basic social problems – especially if we combine this methodology with intuition and good judgement. It is in this context that we have examined the impacts of the Vietnam War on the Philadelphia region and suggested one approach toward the development and consideration of offset programmes.

Notes

1. The research on the development of the Philadelphia Region Input-Output Table has been conducted under grant no. NsG–497–39–010, National Aeronautics and Space Administration, and grant no. WP.00938, Federal Water Pollution Control Administration. The authors are indebted to Edward T. Horn and William Knapp for their invaluable assistance and to the Philadelphia Federal Executive Board and its chairman, Warren Phelan, for replies to unending questions.
2. W. Isard, E. Romanoff, and T. W. Langford, Jr., *Working Papers*, Philadelphia Region Input-Output Study, I, II, III, and IV (Philadelphia: Regional Science Research Institute, 1967).
3. Op. cit.
4. For purposes of inversion, however, the households sector has been incorporated within the structural matrix.
5. A tentative listing of approximately 102 federal agencies with offices located in the Philadelphia SMSA was developed with the assistance and cooperation of the Philadelphia Federal Executive Board. The board is an interagency group comprised of principal executives of each of the major agencies within the region.
6. Op. cit.
7. We are not aware at this time of any studies designed to identify such 'export allocations'.
8. Sector 9114, civilian agencies, n.e.c., was not included because of lack of data.
9. Strictly speaking, these numbers should be adjusted for change in exports to the Fort Monmouth area were the Electronics Command located there; these adjustments, however, would be minor.
10. In the absence of information to the contrary, the agency procurement expenditures for the Vietnam War were assumed to have been similar in *pattern* to the prewar situation.

10 Some Notes on the Linkage of the Ecologic and Economic Systems[1]

I INTRODUCTION

This morning I want to describe in a *non-technical* manner some of our recent research, in particular some of the ways in which a new conceptual framework is being forged to link both the economic and ecologic systems. I also want to illustrate, again without supporting *technical* discussion, how we put into this framework some numbers that we judge to be valid for the quantification of the relationships involved. In this paper, I shall use very extensively materials already published[2] and those contained in a forthcoming book[3] which fully develops the technical materials underlying this research. While the empirical materials and basic concepts were developed by a team of research analysts, the shortcomings of the specific views to be presented in this paper are solely my own responsibility.

Before presenting the framework and empirical materials, let me first provide some perspective and background on our research. Let me quote from our forthcoming book on *Ecologic-Economic Analysis for Regional Development*. We state that

> the combination of the two words, *ecologic* and *economic*, while an unusual one, is set forth as a true combination, in the sense of a synthesis of analyses of two systems within the world of actuality. Throughout this book, we stress the need for this synthesis. Having constantly in mind the mounting air, water, and sonic pollution problems of society, its solid waste problem, its need for open spaces, its diverse impingements on the environment, we assert that no longer can regional development and regional planning be treated in their traditionally narrow garb. Emphasis on the strict economics of such development and planning, with only passing consideration of physical environment and design, let alone social, political and other cultural factors, can no longer be tolerated. Whether we look at the problems of planning and development within the New England states, or metropolitan Budapest, or the environs of Lake Baikal, or in innumerable other regions of the world, we are confronted with the reality that control of the disturbance to the ecological and physical processes, and design of the environment are key

Reprinted from *Papers of the Regional Science Association*, vol. 22, (1968), pp. 85–96.

elements of economic development and planning work. Even conscious redesign of the environment may be.

The above view represents a new dimension which must be brought into our current-day analysis. It is around this new dimension that I wish to centre my remarks this morning.

II THE CONCEPTUAL FRAMEWORK

The conceptual framework which we have developed is in essence a framework for a linear system. It approaches the problem from a traditional input-output or, more strictly, a linear programming standpoint. In fact, my first step in developing this framework is to set down as the upper left major block of cells in Table 10.1 a traditional coefficient table, with columns representing sectors (industries and activities) and rows representing commodities associated with these sectors as outputs and resources. Thus we list by column some typical economic sectors (Agriculture, Textiles, . . ., Petroleum Refining, etc.), and by row some typical economic commodities (Wheat, Cloth, . . ., Crude Oil). Since those who are assembled here this morning are thoroughly aware of the nature of the input-output and linear programming techniques, and the virtues and limitations of a coefficient matrix for both description and projection, I need not spend any time on this first step.[4]

My second step involves the hypothesis that one can fruitfully view the ecologic system in input-output or activity analysis format. Thereby, the ecologic system is considered as a very large set of interdependent activities, involving as inputs and outputs many commodities, only a few of which correspond to those of the economic system.[5] Not only do these activities feed upon each other, that is provide each other with inputs, but also to the exogenous economic system they furnish final outputs or deliveries (in the sense of exports to the economic system). In turn these ecologic activities are dependent on the economic system (that is currently require or receive from it) various commodities (in the sense of imports) to be used as inputs. These imports from the economic system are at the same time exports or deliveries of the economic system to the world exogenous to it, while the exports from the ecologic system are at the same time imports into the economic system, again from the world exogenous to it. To the extent that one system's imports and exports are the other system's exports and imports respectively, to that extent the ecologic and economic systems can be effectively linked. And correspondingly the typical input-output or activity analysis coefficient table for the economic system can be many times enlarged to encompass a comprehensive accounting framework for a fused ecologic-economic system.

In effect my second step is to add three major blocks of cells to Table

Table 10.1 An extended input-output coefficient table

		Economic activities					*Ecologic processes*			
		Agriculture	Textile	. . .	Petroleum refining	Sport Fishing	. . .	Plankton production	Herring production	Cod production
Economic commodities	Wheat									
	Cloth									
	. . .			Economic system: intersector coefficients			Ecologic processes: their input and output coefficients re: economic commodities			
	Crude oil									
Ecologic commodities	Water intake									
	Alkalinity									
	. . .			Economic sectors: their input and output coefficients re: ecologic commodities			Ecologic system: interprocess coefficients			
	Detritus									
	Plankton							+	−	
	Herring								+	−
	Cod					−				+

10.1 to complement the major block in its upper left. The first of these major blocks, at the lower right of Table 10.1, relates to the *ecologic system* and what might be designated its *interprocess coefficients* (which parallel the *interindustry coefficients of the economic system*, in the upper left major block). The second new major block is at the lower left of Table 10.1. Here are listed in coefficient form the inputs and outputs of ecologic commodities (imports from and exports to the ecologic system), by economic sector. Finally, the third new major block is at the upper right of Table 10.1. Here are listed in coefficient form the inputs and outputs of economic commodities, by ecologic process.

III ILLUSTRATIVE COEFFICIENTS

To point up the significance of the conceptual framework as briefly sketched in Table 10.1, as well as to cast light upon the problems of empirical implementation, we now present some representative types of coefficients.

Let us begin by considering the petroleum refining industry. In the upper part of Table 10.2 are listed its typical inputs and outputs of economic commodities in coefficient form. For example, the first coefficient, −.612006, is the dollar's worth of input of crude petroleum and natural gas per dollar output of the industry, and the next to last coefficient, −.089378, is wages and salaries per dollar output of the industry.

But, as is well known, the petroleum refining industry has inputs and outputs other than economic commodities. One set of these is concerned with water intake from, and pollutants discharged via water to the environment. To consider this set, we first need a workable classification of non-economic commodities. Such a classification with reference to water use and pollution has been developed and is presented in Table 10.3. For example, the fourth item in Table 10.3 is water intake for cooling (in 1000 gallons per dollar intake) while the last item but two is toxic materials. If we now list the set of noneconomic commodities of Table 10.3 along the left-hand tab in the lower half of Table 10.1, we are able to set down all the coefficients listed in the bottom half of Table 10.2 for petroleum refining. (For example we explicitly list in Table 10.1 water [intake] and alkalinity.) Thus we have taken a first step in linking the economic and ecologic systems empirically. Needless to say, petroleum refining involves other noneconomic commodities, for example air, and air pollutants; and other economic activities involve still other noneconomic commodities. Accordingly we need to extend the classification system of noneconomic commodities still further, and conduct further laborious work at obtaining the data for the development of coefficients.

Proceeding on, we may ask: from where do the inputs from the ecologic system to the economic system come? How are they produced? To begin to answer this question, let me shift away from the petroleum refining industry to one on which we have been conducting research recently, namely, the recreation industry. Let me also center my attention on the subject of recreational planning and development which has concerned us in the last year.

In the process of our research we have developed a rather fine classification system of recreational activities, and accordingly have developed the coefficients for a meaningful complex of these activities. One of these activities is sport fishing from boats with outboard motors. We therefore list this economic type activity at the head of one of the columns of left-hand part of Table 10.1 and set down its coefficients. One of these

Table 10.2 Input-output coefficients: RIS sector 2911, petroleum refining

Sector			*Coefficient*
RIS 1311	Crude petroleum and natural gas		−.612006
1509	Construction, maintenance, and repair		−.001410
2652	Set-up paperboard boxes		−.008054
2655	Fibre cans, tubes, drums, and similar products		−.001557
2812	Alkalies and chlorine		−.000935
2818	Industrial organic chemicals, n.e.c.		−.027180
2819	Industrial inorganic chemicals, n.e.c.		−.009190
2911	Petroleum refining		+.954950
2992	Lubricating oils and greases		−.014954
3411	Metal cans		−.005294
4811	Telephone communications		−.001210
4890	Telegraph and other communications		−.000100
4911	Electric utilities		−.007629
4920	Gas companies and systems		−.012336
4941	Water supply		−.001082
4990	Sanitary and other systems		−.000142
6020	Interest		−.005088
6301	Insurance, nonlife		−.001242
6510	Real estate services		−.007680
7301	Business services excluding advertising		−.017228
7310	Advertising		−.002312
7400	Research and development		−.000390
7500	Automotive repair		−.001040
9000	Local and state taxes		−.008049
9100	Federal income tax		−.014755
9826	Office supplies		−.000280
9842	Transportation costs		−.040831
9888	Wages and salaries		−.089378
9899	Residual		−.063589
WPC1001	Water intake, sanitary use,	1,000 gal./$ output	−.000455
1002	" production	"	−.022827
1003	" cooling	"	−.114861
1011	Water discharged, sanitary use	"	+.000323
1012	" production	"	+.070219
1013	" cooling	"	+.050100
1031	Biochemical oxygen demand BOD, 5-Day,	1,000 lb./$ output "	+.000065
1032	Ultimate oxygen demand, UOD	"	+.000076
1033	Chemical oxygen demand, COD	"	+.000169
1041	Suspended solids	"	+.000084
1042	Settleable solids	"	+.000124
1047	Turbidity	"	+.000365
1051	Alkalinity	"	+.000051*
1052	Acidity	"	+ n.a.
1061	Oils and greases	"	+.000012
10951	Phenols	"	+.000003

Note: *The pH factor for this industry is in the range of 4.5–10.7, with 7.0–8.0 most likely.

Source: Isard *et al.* [2; p. 94].

Table 10.3 Proposed water pollution classification (WPC) code

WPC Code	*Water related item*	*Units*
1000	Water intake, total[a]	1,000 gal./$
1001	Sanitary use	"
1002	Production use	"
1003	Cooling	"
1004	Boiler feed	"
1008	Irrigation	"
1009	Other, n.e.c.	"
1010	Water discharge, total	"
1011	Sanitary use	"
1012	Production	"
1013	Cooling	"
1019	Other, n.e.c.	"
1020	Water consumed	"
1031	Biochemical oxygen demand, BOD 5-Day	1,000 lb./$
1032	Ultimate oxygen demand, UOD	"
1033	Chemical oxygen demand, COD	"
1040	Solids, total[b]	"
1041	Suspended solids	"
1042	Settleable solids	"
1047	Turbidity	"
1048	Colour	"[c]
1051	Alkalinity	"
1052	Acidity	"
1061	Oils and greases	"
1062	Surfactants	"
1070	Pathogenic (disease causing) organisms	–[d]
1080	Temperature	–[d]
1090	Other pollutants	"
1095	Toxic material[e]	"
1096	Radioactive waste	–[d]
1099	Not classified	–

Note: [a] Cost of water intake is given by SIC code 4941–water supply.
[b] 1040 = 1041 + 1042.
[c] In addition, colour should be specified by kind by its wave length.
[d] No one satisfactory measure was decided upon at this time, although thermal pollution may be specified most satisfactorily in terms of millions of BTU's per dollar output.
[e] Phenols, which fall in this category, may be identified by a five-digit code such as WPC 10951.

Source: Isard *et al.* [2; p. 90].

coefficients is of course fish; and in the context of the particular area of New England which we have studied, namely Plymouth Bay, a chief fish is cod. We have estimated that per 1000 man days of sport fishing in outboards, 6200 pounds of cod are required (in the sense of providing a satisfactory recreational outcome). Therefore we can set down that number in the cell corresponding to the cod row and sport fishing (outboard motor) column of Table 10.1.

Clearly, for cod to be an input in an activity of the economic system, it must be yielded as an output of a process of the ecologic system. We may designate this process the cod production activity. We have studied it, and we find that in the context of Plymouth Bay it requires, among others, the set of inputs listed in the last column of Table 10.4. That is the output of cod requires, as food inputs, herring, invertebrate carnivores and small fish. In turn, herring consumes (requires as inputs) plankton; and invertebrate carnivores and small fish consume herbivorous invertibrates which live on detritus. All these activities are part of a basic food chain in the ecologic system, and their input-output data are to be recorded in the lower right major block of Table 10.1 on *ecologic subsystem: interprocess coefficients*. Each of the production processes listed at the head of Table 10.4 is to be listed as a column in the right-hand part of Table 10.1 (which relates to ecologic processes) and every one of the commodities listed at the left-hand tab of Table 10.4 is listed among the ecologic commodities in the lower part of the left-hand tab of Table 10.1. A few of these are explicitly recorded in Table 10.1.

The food chain, however, represents only one basic set of linkages in the ecologic system. The proper study of the ecologic system – and its relation to the economic – must cover numerous other basic linkages. Biogeochemical cycles, such as the phosphorous cycle, and photosynthesis are obviously two sets of basic relationships which cannot be ignored. Consider, for example, photosynthesis. In Table 10.4 the -10 entry at the top of the herring production column indicates that 10 pounds of plankton are required per pound of herring. Now, in the Plymouth Bay setting, one gram of plankton may be considered as a mixture consisting of 0.86 grams of phytoplankton, and 0.14 grams of zooplankton. The zooplankton largely grazes upon phytoplankton, and consumes phytoplankton as an input, as indicated in the last column of Table 10.5. The phytoplankton itself is produced via photosynthesis as indicated in the first column of Table 10.5. Note the inputs of light in the form of incident solar radiation at the ocean surface, measured in gram calories per square centimetre per minute, and the input of dissolved phosphate in sea water, etc. Lastly, in columns 2–5 of Table 10.5 we take into account the fact all the carbon fixed by phytoplankton photosynthesis is not available for consumption by zooplankton; some is used up by the phytoplankton's own respiration processes.

Table 10.4 Food chain requirements for cod production

Classification number	*Process* / *Commodity (lbs)*	*MX 61–10* *Plankton production*	*MX 60* All marine plant production	*MX 6001* Detritus production	*MX 718* Production of herbivorous invertebrates	*MX 719* Herring production	*MX 722* Small fish production	*MX 721* Carnivorous invertebrates (large gastropods) prod.	*MX 725* Cod production
MX 61–10	Plankton	+1				−10			
MX 60	All marine plants		+1	−1					
MX 6001	Detritus			+1	−10				
MX 718	Herbivorous invertebrates (miscellaneous)				+1		−10	−10	
MX 719	Herring								−1.167
MX 722	Small fish					+1	+1		−1.167
MX 721	Carnivorous invertebrates (large gastropods)							+1	−8.333
MX 725	Cod								+1

The above empirical materials illustrate the kinds of data which can be developed to fill in the cells in three of the major blocks of Table 10.1. We now turn to the question of data to fill in the cells of the fourth major block at the upper right of Table 10.1. Here, the seemingly difficult problem is resolved once the need for an appropriate classification of commodities is recognised.

Reconsider the commodity, cod. In Table 10.1 it appears as an ecologic (non-economic) commodity in the lower half of the commodity classification system – a commodity which is yielded as an output (designated by a '+' sign) by an ecologic process, and used as an input (designated by a '−' sign) by an economic process. It could, however, be also classified directly

Table 10.5 Plankton and related production activities

Classification	*Commodity*	*Units*	*PMX 61–14* *Gross phytoplankton production by photosynthesis*	*P1MX 61–15* *Utilisation of phytoplankton matter by respiration @ 2°C*	*Utilisation of phytoplankton matter by respiration @ 7°C*	*P3MX 61–15* *Utilisation of phytoplankton matter by respiration @ 12°C*	*P4MX 61–15* *Utilisation of phytoplankton matter by respiration @ 17°C*	*PMX 61–11* *Phytoplankton in pounds (dummy unit conversion process)*	*PMX 61–10* *Plankton formation*	*PMX 117* *Zooplankton production*
		Process								
MX 11	*Solar radiation*	G. cal per CM^2 per min.	−1.79							
MẊ 420	Below 0.55 mg. at./M^3 Phosphorous as phosphate	Mg. at. per M^2	−0.06							
MX 42	Above 0.55 mg. at./M^3	Mg. at. per M^2	−0.00							
MX 44	Water transparency: $t = i/k$		−68.68						−0.86	
MX 61–11	Phytoplankton @ 3.98 g.c./M^2	Pounds						+1.00		−1.7
MX 61–12	Phytoplankton stock	Grams carbon per M^2	±35.54							
MX 61–13	Phytoplankton	Grams carbon per M^2		−.0201	−.0281	−.0400	−.0565			
MX 61–14	Phytoplankton-gross production by photosynthesis	Grams carbon carbon per M^2 per day	1.00					-277		
MX 61–15	Loss of phytoplankton matter due to *respiration*	Grams carbon per M^2 per day		+1.00	+1.00	+1.00	+1.00			
MX 61–10	*Plankton*	Pounds							+1.00	
MX 711	*Zooplankton* @ .68 g.c./M^2	Pounds							+0.14	+1.00

as an economic commodity or resource, along side of land and labour; and the amount of cod which is provided by the natural environment, just as the amount of land and labour, represents the stock of a resource, which behaves as a constraint on the levels of economic activities which use these resources. Likewise the timber, or grass (for grazing purposes), or even the phosphate (basic to agricultural production) which natural production processes yield may be listed among the economic commodities in the upper half of the left-hand tab, and viewed as scarce economic resources whose use must not exceed current stocks. Given this viewpoint, then, the positive entries in the cells in the upper right block of Table 10.1 would represent outputs of ecologic processes which are provided as resources to the economic system; and negative entries would represent the use of these economic resources in natural production processes themselves, as the consumption of herring (an economic commodity) by cod, or the consumption of corn by wildlife, or the use of phosphorous in noneconomic processes in the phosphorous cycle.

However, this discussion should make clear the point that often a classification of commodities involves somewhat arbitrary decisions. Is land an economic commodity to be recorded in the upper half of the list of Table 10.1, or an ecologic commodity to be recorded in the lower half. Likewise with water, air, fish, livestock, etc. But while arbitrary elements do enter any classification system, for the purposes of our conceptual framework it makes no difference how we classify commodities, provided there is full accounting of all inputs and outputs of all commodities relevant for the problem being examined. Similarly, while a classification of processes as economic and ecologic may facilitate our comprehension of a problem, basically such classification is not essential for analysis, provided all processes relevant for a problem are identified and accurately described in an empirical manner.

IV REGIONAL CLASSIFICATION

There is one more general comment I wish to make. It is a well-known axiom in Regional Science that in any applied study the definition of relevant regions depends in large measure on the problems being attacked. Now, when we attack problems such as air pollution, water pollution, solid waste disposal, and sonic pollution (observe that these problems are of quite a different nature than those which have plagued regional analysts over the centuries), we clearly must introduce the physical environment properly into our conceptual framework. To do so requires in my mind a drastically different set of regions than those to which we have been accustomed. I envisage at least three new major types of regions, namely, LAND, AIR and MARINE. Each of these three major types of region is composed of sub-systems; each has an economic sub-system, an ecologic

Table 10.6 A further extended input-output coefficient table

			Land		*Air*		*Marine*	
			Economic (*Phila*)	*Ecologic* (*Phila*)	*Economic*	*Ecologic*	*Economic*	*Ecologic*
Land	*Economic*	*Phila*						
	Ecologic	*Phila*						
Air	*Economic*							
	Ecologic							
Marine	*Economic*							
	Ecologic							

sub-system, and other sub-systems. I have depicted in Table 10.6 these three major regions with their economic and ecologic systems. Note that this system of regional classification does not contain traditional regions such as the Philadelphia region or New England. However, such traditional regions can easily be incorporated. Under both the economic and ecologic sub-systems of the major region LAND, I have a sub-system region designated Philadelphia. A typical column could be the petroleum refining activity; and in that column would be listed all its input and outputs, some of which would be economic and noneconomic commodities that may be classified as land-type (such as crude oil, gasoline, and solid wastes); others would be commodities of marine-type (such as clean water from a river system, and water pollutants delivered to the marine region); and still others would be commodities of an air-type (such as unpolluted air itself, and particulates).

Similarly, I could list other subsystem regions (with their activities) relevant for the major environmental region LAND. I can also do so for the major environmental regions AIR and MARINE; but here the relevant subsystem regions may differ. The Philadelphia region need not be a meaningful subsystem region for the AIR and MARINE major regions.

In short, I have reached a point in my thinking where we must consider rather revolutionary changes in our conception of regions appropriate for planning purposes. In this matter the nature of our problems are compelling us to traverse quite unexpected ground in our effort to attack them effectively.

V CONCLUSION

I now wish to conclude this set of notes with a few additional comments.

(a) Our conception of regional planning has become greatly expanded. No longer can it be concerned with the economy and its interrelated sectors, nor with society and its interrelated behaving units. It must proceed further and encompass the interrelationships of economic and other social behaving units with the natural processes which are the essence of the ecosystem.

(b) While the input-output and linear programming techniques are useful in depicting the interrelations between the economic and the ecologic systems, these techniques are only two of several that may prove to be efficient in depicting these interrelationships and in attacking problems which stem from such interrelationships. It is easily envisaged how non-linear programming techniques, activity complex techniques, and comparative cost analysis may also be extended for use in ecologic analysis and in synthesised ecologic-economic analysis.

(c) It is only by the development of such techniques that we can hope to reach viable or workable solutions to the pressing environmental problems which loom on the horizon.

Notes

1. The author is associated with the Regional Science and Landscape Analysis Project, Harvard University and the Regional Science Research Institute, Philadelphia, USA. He is indebted to the Environmental Sciences Services Administration (US Department of Commerce), the National Science Foundation, the Federal Water Pollution Control Administration and Resources For the Future, Inc., for their financial support of this research.

2. See Isard *et al.* [2].
3. See Isard *et al.* [3].
4. The reader who is unfamiliar with these techniques may refer to Isard *et al.* [1; chs 8 and 10] and the references contained therein.
5. For a standard discussion of ecologic systems, see Odum [4].

References

[1] Isard, W. *et al.*, 1960. *Methods of Regional Analysis: an Introduction to Regional Science*, (The Technology Press of MIT and John Wiley and Sons, New York).

[2] Isard, W., K. Bassett, C. Choguill, J. Furtado, R. Izumita, J. Kissin, E. Romanoff, R. Seyfarth and R. Tatlock, 1967. 'On the Linkage of Socio-Economic and Ecologic Systems', *Papers, Regional Science Association*, vol. 21, pp. 79–99.

[3] Isard, W., C. Choguill, J. Kissin, R. Seyfarth, R. Tatlock, K. Bassett, J. Furtado and R. Izumita, 1971. *Ecologic-Economic Analysis for Regional Development*. (Free Press, New York).

[4] Odum, E. P., 1959. *Fundamentals of Ecology*. Philadelphia: Saunders Co.

11 Potentials and Problems of Economic-Ecologic Models for Management of Multiregion Systems

I INTRODUCTION

In recent decades, we have experienced rapid advances in both spatial theory and operational models governing regional and interregional relations. For example, central place and related urban theory has developed tremendously since the first statement by Christaller while, at the same time, the extremely simple operational basic-service ratio has been replaced by multiregional linear and non-linear mathematical programming models. In this paper we aim to couple an extension of the framework of a system of central places or nodal regions, inclusive of decision-making functions, with operational models to be able to cope more effectively with the problems of depicting and understanding the world's ecological system and managing the global environment.

Important first steps have already been taken which link the ecologic system to the economic system. In their simplest form, they can be stated in terms of linear systems. It is well-recognised that at least certain parts (sectors) of the ecologic system can be fused with a Leontief-type interregional input-output model. Other parts can be captured in the framework of Kolosovsky's 'areal complexes of productive forces' or 'industrial complex analysis' to use the terminology current in the United States.[1] Still other parts are susceptible to fruitful depiction and control in terms of linear and recursive programming models. In this paper, we purport to outline the more significant problems and challenges in applying these state-of-the-art methodologies to regional and multiregional development inclusive of the impact on the ecologic system and environmental quality.

II IDENTIFICATION OF REGIONS

As is almost obvious, the modeling and management of economic-ecologic systems requires the identification of appropriate level or levels of spatial

This chapter was written with Roger Van Zele and Phyllis Kaniss. Reprinted from *Geoforum*, vol. 7, (1976), pp. 203–13.

aggregation or disaggregation. As is also obvious, the diverse activities of each nation of the world impinge not only upon the immediate environment of that nation but also, through spatial diffusion and related processes, upon the world environment and thus upon the immediate environment of all other nations. Thus we are compelled to consider the global system as such and conduct analysis of the global system of relationships.[2]

While this level of analysis is necessary, it is not sufficient. At least one more level is required – say that which corresponds to continental masses on a physical basis, or to the boundaries of nation-states on a political basis. Political realities dictate that any operational model employ national boundaries as political regions as a second level of analysis.

Within the highly industrialised nations with complex social systems, many major economic-ecologic problems are multiregional in nature; thus, for these nations at least, further disaggregation by their major regions is required, and the methods employed must capture the basic interrelations of forces within these regions as they interplay over time in the multiregional system of the nation. For example, the United States must be broken down into, say, the New England region, the Great Lakes region, the Middle Atlantic region, and so forth. Likewise, the USSR must be disaggregated into a meaningful set of regions.

Finally, we may be able to pursue still better operational analysis when we disaggregate major regions within a nation into urbanised areas, SMSA's, or river valleys. For example, the Middle Atlantic states may be fruitfully partitioned into the New York SMSA, the Philadelphia SMSA, and so forth.

While it is easy to conceptualise a resonable hierarchy of spatial units, the question of the proper delineation of space into regions and subregions for an operational economic-ecologic model is extremely difficult. This subject has been discussed extensively in the literature (e.g. Isard and Langford, [11]; Miernyk *et al.*, [16]). The concensus is that the answer to this central question very much depends on the characteristics of the problem and the situation being studied, the financial resources available, and the predilections of the investigator.

III ACTIVITIES AND COMMODITIES

Having systematically defined a regional (spatial) framework for analysis, we must next address in a systematic and comprehensive fashion (a) the range of activities of interest, and (b) the set of commodities. To be complete, the framework fashioned must be broad enough to include activities and commodities of interest at all levels of the spatial hierarchy described above; that is, from the sub-regional to the global level. To the set of regular economic commodities and activities must be added all relevant production and consumption residuals (pollutants), treatment (depollution) activities and commodities, and all basic ecologic processes, e.g., the food chain or the nitrogen-oxygen cycle in the atmosphere.

Toward this end, one useful and efficient way to organise the data available is in an activity analysis input-output format, with (industrial) plant-by-plant data organised on worksheets at least at the four-digit SIC level (Isard *et al.*, [12]). Using this format, the data is stored in the most disaggregated form, and whenever an analysis requires greater aggregation, such aggregation can be effected quickly and easily, perhaps through the use of low-cost computer programmes developed for both storage and aggregation purposes. For example, one general interrelations table with reference to general environmental regions – land, water, air – is depicted in Figure 11.1, in which activities and commodities are represented by columns and rows, respectively (Isard *et al.*, [9]). The zones shown in the sketch would correspond to the parts of the subregions of interest. A portion of a detailed classification system which could be employed is suggested in Tables 11.1–11.4.[3]

Now with the available data neatly stored in the activity analysis format, what types of models emphasising the interdependence of the economic and ecologic subsystems of the real world can be exploited, and what are the advantages and significant limitations of each? Three existing types of models for economic-ecologic analysis which are among the most useful for capturing the basic interrelationships between economic and environmental systems and which address the problems of regional development and the location of economic activities are as follows: (1) input-output analysis, (2) a hybrid of input-output and comparative cost analysis, industrial complex analysis,[4] and (3) mathematical programming.

Input-Output Analysis

Input-output (I-O) was developed as a means to depict and investigate the interrelationships between the elements comprising the economic system of a region or multiregion system. As originally developed, I-O presents a detailed picture of the production and distribution characteristics of individual industries (sectors) of different regions and the nature of the interrelationships among them. The strengths of the I-O model lie in its abilities to provide a comprehensive and systematic snapshot of a regional or multi-regional economy at a particular point in time and to reflect in a concise and internally consistent manner a large amount of information about commodities and processes and their interrelationships, in and among regions. In addition, it imposes a desirable statistical discipline on further data collection and empirical investigations and can be used in a limited way for projections.

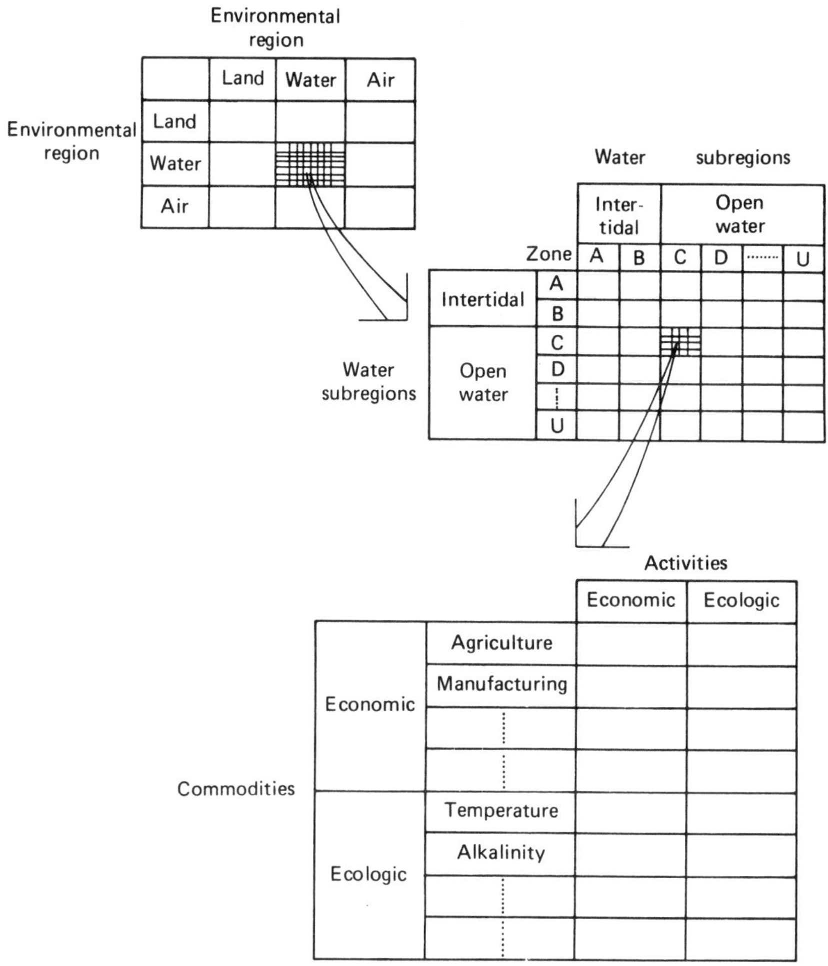

Figure 11.1 A general interrelations table

While the I-O model in its original formulation, like most other economic models of past decades, took no cognisance of what have come to be called 'environmental externalities' or production and consumption 'residuals', the situation now can be remedied. Because of its systematic nature, the I-O model, perhaps more so than many other economic models, has lent itself to extensions whereby ecologic and economic-ecologic linkages can be studied.[5]

Table 11.1 Selected air pollution commodity codes

No.	*Commodity*
AP 1000	Gases
1100	Sulfur compounds
1110	Hydrogen sulfide
1120	Sulfur oxides
1121	Sulfur dioxide
. . .	. . .
1200	Carbon compounds
. . .	. . .
1300	Nitrogen compounds
. . .	. . .
1400	Halogens
. . .	. . .
AP 2000	Liquids, entrained droplets, and saturated vapors
2100	Acid mists
2110	Hydrochloric acid mist
. . .	. . .
2200	Water vapour
. . .	. . .
2300	Smoke and oil vapours
. . .	. . .
AP 3000	Particulates
. . .	. . .

Industrial Complex Analysis

Input-output, however, fails to adequately treat economics of scale, localisation economies, urbanisation economies and regional price variations resulting in or associated with the use of different factor proportions. Accordingly, a technique called industrial complex analysis[4] (Isard, [5])

Table 11.2 Selected water pollution classification codes

No.	*Item*	*Units*
1000	Water intake, total	1,000 gal./$
1001	Sanitary use	"
1002	Production use	"
.	.	
.	.	
1010	Water discharge, total	"
1011	Sanitary use	"
1012	Production use	"
.	.	
.	.	
1020	Water consumed	"
1031	Biochemical oxygen demand, BOD 5-day	1,000 lb./$
.	.	
.	.	
1040	Solids, total	"
.	.	
.	.	
1051	Alkalinity	"
.	.	
.	.	
1070	Pathogenic organisms	*
1080	Temperature	*
1090	Other pollutants	*
.	.	
.	.	
.	.	

Note: *No one satisfactory measure yet decided upon.

Table 11.3 Solid waste commodity codes

No.	*Commodity*	*No.*	*Commodity*
SW 1	Paper	SW 8	Ferrous metals
SW 2	Wool	SW 9	Non-ferrous metals
SW 3	Textiles	SW 10	Petroleum
SW 4	Plastics	SW 11	Chemical
SW 5	Rubber	SW 12	Soil, sand, ash
SW 6	Glass	SW 13	Other
SW 7	Food		

Table 11.4 Selected air pollution control systems codes

No.	*Control system*
E 0000	No control
0100	Closed system
E 1000	Mechanical separation
1100	Cyclones
1110	High-efficiency cyclones
1120	Low-efficiency cyclones
⋮	⋮
1200	Floating roof tanks
⋮	⋮
1300	Vapour manifold
⋮	⋮
E 2000	Electrostatic precipitation
⋮	⋮
E 3000	Filters
⋮	⋮
E 4000	Wet scrubbers
4100	Venturi scrubbers
⋮	⋮
4200	Spray towers
⋮	⋮
E 5000	Incineration
5100	Afterburners
⋮	⋮
5200	Flares
⋮	⋮

Table 11.5 Classification of resources, commodities, processes and qualities

$y_1, \ldots, y_h$	inputs of resources $h' = 1, \ldots h$; $y_h \leqslant 0\ y_f$	$y_1, \ldots, y_f$	inputs of economic resources	$f = 1, \ldots, f$; $y_f \leqslant 0$
		$y_{f+1}, \ldots, y_h$	inputs of ecologic resources	$h = f + 1, \ldots, h$; $y_h \leqslant 0$
$y_{h+1}, \ldots, y_m$	inputs or outputs of intermediate commodities $m = h + 1, \ldots, m$ $y_m \lesseqgtr 0$			
$y_{m+1}, \ldots, y_p$	outputs of commodities $p = m + 1, \ldots, p$ $y_p \geqslant 0$	$y_{m+1}, \ldots, y_n$	outputs of economic end-products	$n = m + 1, \ldots, n$
		$y_{n+1}, \ldots, y_p$	outputs of ecologic commodities (inclusive of pollutants)	$p = n + 1, \ldots, p$

$x_1, \ldots, x_r$	processes $x_r \geqslant 0$ $r = 1, \ldots, r$	$x_1, \ldots, x_j$	economic processes	$j = 1, \ldots, j$ $x_j \geqslant 0$
		$x_{j+1}, \ldots, x_k$	depollution (pollution control) processes	$k = j + 1, \ldots, k$ $x_k \geqslant 0$
		$x_{k+1}, \ldots, x_r$	ecologic processes	$r = k + 1, \ldots, r$ $x_r \geqslant 0$

$s_{p+1}, \ldots, s_q$	environmental qualities (public end-products; standards) $q = p + 1, \ldots, q$ $s_q \geqslant 0$

has been developed further (Isard, [7]) – specifically to analyse the problem of environmental degradation associated with complexes of activities, e.g., a major industrial complex including the transportation, residential and other subsidiary economic, political, social and cultural activities associated with its operations. This technique circumvents the restrictiveness of the I-O assumption of a single process and a single product per sector and thus allows a more comprehensive description of the economic-ecologic system.[6]

To extend this model for environmental application, one useful scheme for the classification of activities and commodities is as presented in Table 11.5. This system of classification is by no means the only one possible, but at this time the disaggregation of the system into five classes of commodities, three classes of activities and one class of qualities (standards), seems most relevant. Further disaggregation would not improve the analysis sufficiently to justify the increased data problems, and a less detailed classification scheme might result in the loss of some important possibilities for complex analysis.

Proceeding from this classification scheme, complexes can be constructed in many ways. Consider the following upon which a complex might be focused: (1) critical economic resources, e.g., an energy-fuels complex; (2) critical ecological resources, e.g., air, land and water in their 'natural' state with regard to an urban complex or new town; (3) critical intermediates, e.g., a petrochemicals raw material complex; (4) critical economic end-products, e.g., housing and food for subsistence; (5) key economic processes, e.g., heating, power production, and transportation as energy-using processes; and (6) critical ecological processes, e.g., aeration of water systems and photosynthesis. In addition to these, complexes can focus on combinations of commodities, of processes or of commodities and processes from the above listing. There may be many other ways in which complexes can be developed, and it should be noted that the suggested classification scheme is useful only insofar as it identifies the problem of selecting and/or managing a meaningful complex.

The interest underlying the development of complex analysis is that of improving society's welfare through the rearrangement of the spatial pattern of behaving units within society and through the control of aggregate spatial behavior. In using this analysis, one typically thinks of new types of industries or commercial developments which might be promoted in a given region to improve welfare there without detracting too much from the welfare of another region in ways which are socially efficient. To extend the conceptual framework to encompass environmental impacts, the costs and benefits of the pollution of the ecologic system, together with the consumption of ecologic resources and the disruption of ecologic processes associated with such new projects, must be introduced into the analysis. This extension should not be limited, however, to the consideration of only the costs and benefits related to new or expanded activities in the region, but also should consider the possibility of the contraction or cessation of activities in a region, even including the preclusion of growth. The emphasis must be on change rather than on promotion and growth. It is possible that the consideration of potential irreversible environmental degradation may dictate the preclusion of growth as defined from the standpoint of economic efficiency.

Mathematical Programming

A third type of model which can be employed in the context of regional development and which can be extended to reflect economic-ecologic linkages is the family of mathematical programming models. These optimising models generally take the form as follows:

Maximise: πX

Subject to: $-AX \leqslant R$

Where X is an $(n \times 1)$ column vector of activity levels, $x_g(g = 1, \ldots, n)$ being the level of the gth activity

π is a $(1 \times n)$ row vector of unit contributions to the objective function, π_g being the contribution of one unit of activity g to the objective function

A is an $(m \times n)$ matrix of coefficients relating to the m commodities on each of which a constraint is imposed. When negative, the element a_{fg} represents an input and, when positive, an output of commodity or resource $f(f = 1, \ldots, m)$ per unit level of activity g

R is an $(m \times 1)$ column vector of constraint magnitudes, e.g., R_f specifies the supply of resource f available for use.

To extend the programming format to include pollutants as outputs of economic activities and treatment activities as depollution processes to eliminate pollution, the pollutants must be added to the formulation as commodities $m + 1, \ldots, p$ and the depollution processes as activities $n + 1, \ldots, q$. Accordingly, X becomes a$(q \times 1)$ column vector, πa$(1 \times q)$ row vector with non-positive values for $\pi_{n+1}, \ldots, \pi_q$, A a $(p \times q)$ matrix, and R a $(p \times 1)$ vector with $R_{m+1}, \ldots, R_p$ representing permissible levels of pollution generation.

Like input-output and industrial complex analysis, the programming models emphasise general interdependence via the coefficient matrix although, unlike I-O, mathematical programming is an optimising technique. The strong points of mathematical programming for use in regional development studies generally include its comprehensive nature, its usefulness for pointing out existing inefficiencies in patterns of location, transport flows and resource use, for attacking problems of resource scarcity and for the introduction of certain normative elements into the model via the constraint set. Through the use of mathematical programming, the trade-offs which are being made within the economic-ecologic system are made explicit. Thus this tool is a useful aid to the decision-maker, and further, when used recursively with intervening periods of bargaining, serves as a tool for resolving goal conflicts (Thoss, [25]; Isard, [6]). These applications and their implications will be discussed further in Section IV.

These models have certain limitations and the reader is referred to the extensive literature on this subject.[7] The models cannot by themselves or in conjunction resolve most of the problems encountered in the management of economic-ecologic systems. Informed qualitative judgement cannot be replaced with exclusively quantitative methods. These models, however, are useful as tools of description and prediction to supplement judgement – particularly in making explicit certain costs and benefits for the evaluation of alternative policies.[8]

An application

Before closing this section, we might refer to a basic study which has confronted the problem of identifying a meaningful region, of selecting appropriate activities and commodities, especially with regard to depollution activities, of designing a relevant set of techniques, and of effecting pertinent investment and production cost analysis relevant for policy determination. This is a model developed by Rose for use by the New York State Department of Environmental Conservation for environmental management of urbanised areas.[9]

The model has two basic components, i.e., an economic submodel and an environmental submodel. The economic component is represented by a partially dynamic input-output system in which the structure of the economy is related to the level of environmental regulation, as well as to time. Pollution abatement is conceived of as an integral part of each industry's production process and is treated as an intermediate process. For many purposes, such as assessing the costs of pollution control and the repercussions on the rest of the economy, the economic component will suffice, but additional valuable insights can be achieved by analysis of the intermediate depollution process; hence an environmental submodel in which the nature of pollution abatement is treated also has been developed.

The environmental submodel is comprised of various activities (technologies) used to treat pollution in each of the industries of an economy. Pollution generated within the economy is considered as the input to the different depollution activities and cleaner air or a reduced amount of pollution is treated as output. Using an activity analysis format, as suggested above, a high level of disaggregation is achieved which allows examination of the various technological alternatives which might be used to meet a given air quality standard. A linear programming solution allows the determination of the least-cost response of industry to changes in environmental regulations.

The model has been used empirically to determine the annual cost of an optimal strategy to meet uniform emission reduction requirements for particulates in a region of New York state (Isard *et al.*, [8]; Rose, [19]). It was found, for example, that the total annual direct costs (covering both fixed charges on investment and annual operating costs) of achieving at least 99.5 per cent abatement of particulate emissions in manufacturing sources in the Syracuse SMSA in 1975 would total $11.6 million, an increase of $4.8 million over the cost of base year (1972) levels. This empirical application was effected by applying a specially designed cost-estimating methodology to certain engineering parameters of pollution control operations compiled by the New York State Department of Environmental Conservation to arrive at structural and capital coefficients of the input-output components and the abatement (cost-effectiveness) coefficients of the abatement submodel. These abatement (depollution) coefficients, which represent the dollar value of different inputs to the abatement process required to eliminate one unit of pollution, were calculated for various types of control equipment, each at various cleaning efficiencies. For example, it was found that for each one-ton reduction of process particulate emissions achieved through the use of a cyclone operating at 90.0 per cent average efficiency in plants classified as SIC 2011, $6.39 of services of control equipment and $5.93 of power, labour and other inputs would be required annually.

IV PROBLEMS OF MULTI-LEVEL MANAGEMENT OF THE ENVIRONMENT

A basic shortcoming of the various models thus far developed is their inability to pinpoint conflicts that certainly will arise among the regions of a nation and among the nations of the world in managing the environment. To help attack this shortcoming, we look at the world as a multi-level system involving planning and environmental management at all levels, and attempt to determine the optimal distribution of decision-making functions at these levels. What follows is some conceptual thinking which we hope, in time, can be embedded in the operational framework already described.

We raise the question: 'To what degree should decisions on environmental quality be allocated among the different regions and nations of the world, i.e. centralised or decentralised?' Within a given nation, the problem involves allocating responsibility for different types of decisions to the national, regional, and local levels of a planning network. At the international level, the question is one of determining which aspects of environmental management cannot be handled adequately by the individual action of nations, but rather must be assigned to an international decision-making body. In each case, the problem requires a determination of the advantages and disadvantages associated with centralisation versus decentralisation of different types of decision-making.

To begin, we recognise that any spatial configuration of decision-making – 100 per cent centralisation, 100 per cent decentralisation, or anything in-between – involves certain costs and benefits. These costs and benefits are influenced by, among others, three major factors which are as follows:

1. *Information costs* – the costs involved in collecting, processing, and transmitting information on local conditions. Some contend that the collection and transmission costs rise with the degree of centralisation, whereas processing costs fall, due to scale economies.
2. *The existence of externalities* – interdependent decision situations in which the actions of one decision-making unit on a given level affect the outcomes of other decision-making units on that level. These externalities – whether they involve technological or social interdependence – most often lead to suboptimal outcomes for all units and therefore encourage centralisation up to the level at which the externalities may be 'internalised'.
3. *The existence of interlevel conflict* – which leads to resistance of lower levels to the yielding of autonomy to higher-level constraints. This factor favours decentralisation of decision-making in order that individual sovereignty may be retained. The current resistance to the formation of an international environmental management organisation seems to stem from the fact that most nations do not consider the costs of international externalities to be high enough to justify any sacrifice in national sovereignty – but inevitably such costs will come to be perceived to be high enough.

Thus, for any given type of policy issue, the choice of centralisation versus decentralisation involves a tradeoff between the costs and benefits that arise as a result of these three factors, as well as others. In some cases, the costs from externalities are extremely high and far outweigh the information costs and reduction in autonomy associated with centralisation. In other cases, the situation may be just the opposite. By carrying through a comparison of the aggregate costs of different spatial configurations of decision-making for different kinds of decisions, we obtain an overall picture of an optimal distribution of decision-making functions. In general, we may say that when the costs of externalities – however those externalities are defined – outweigh the costs of information collection and transmission and the 'costs' of sacrificed local autonomy, the related decisions should be centralised up to the level at which the externalities are internalised. On the other hand, decentralisation should be encouraged in those areas in which there are no significant externalities. These general rules yield a division of decision-making labour among levels of a hierarchy which is responsible for the very efficiency of hierarchical organisation. Each of the lower levels, though constrained in some areas by higher level dicta, is in other areas autonomous and self-reliant and able to handle routine contingencies without asking higher authority for instructions. This keeps communication channels from becoming overloaded, prevents higher level planners from becoming occupied with petty detail, and keeps attention from being diverted from more important matters. A hierarchical planning scheme thus allows each level to focus on the degree of detail for which it is suited; the lower levels handle localised, detailed planning, and the higher levels concern themselves with broader issues and more aggregated forms of data.

Once the allocation of different kinds of decision-making to the different levels is made, a hierarchical planning network functions as a stratified decision system. When we go from higher to lower levels, we move from broader to narrower concerns and from higher to lower goals and objectives. Each level solves its own decision problem – the solution to any level's decision problem representing a constraint on the next lower level's decision-making. To illustrate, consider a three-level environmental planning system consisting of world, national and regional levels. Taking into account the various costs and benefits sketched out earlier, imagine that through the political process, a concensus is reached as to which levels of the planning system will be given responsibility for managing which types of environmental problems. For example, the world unit might be empowered to set limits on the amount of mercury to be discharged to the world's waters, to regulate the handling of plutonium and other nuclear wastes, or to set limits on chemical discharges that threaten the ozone layer. However, decisions on the level of emissions of nitrous oxides or carbon monoxide or on the quality of streams and lakes within any given nation might be made at the national, rather than the world, level – if there are no significant externalities between nations. In turn, specific plans to meet national emissions and effluent standards – including land-use and transportation planning – might be carried out at the regional level.

Let us sketch one possible general decision-making model for a multilevel system. Assume that each level, in making its decisions, operates on the basis of (1) a *model*, representing the structure of its level as well as the structure of the significant interactions or interdependencies of the next lower level, and (2) a *utility* or *objective function* which takes into account how the lower-level activities contribute to the higher level objectives or goals. The decision-making unit then proceeds to maximise the value of its objective function, subject to the constraints of its model and (if it is not the top level) subject to any higher level constraints. In this way, it chooses a course of action which places constraints on some of the next lower level's activities – namely, in those areas in which there are perceived externalities between units. The next lower level then, in turn, constructs a model of *its* structure inclusive of interactions of its lower-level constituents and in accord with the maximisation of some objective relevant to *that* level (e.g. *national* goals and priorities) – and respecting its constraints from above – chooses constraints to be placed on some of the activities of its lower level.

In terms of the general programming model discussed earlier, the model might specify at the world level

$$\max GWP = \pi X$$
$$\text{s.t.} - AX \leqslant R$$

where GWP = Gross World Product, and $R_{m+1}, \ldots, R_p$ represent permissible levels of pollution generation.

Assume the world unit is empowered to set limits on the permissible levels of pollutants of variety $m + 1$, $m + 2$, and $m + 3$ for each nation, J, $J = A, \ldots, U$. Then the world unit's decision problem would be to select the constraints on the emission of these three pollutants in each nation, or to set the levels of $\bar{R}^J_{\bar{p}}$ where

$$\sum_{g=1}^{q} a^J_{\bar{p}q} \, x^J_g \leqslant \bar{R}^J_{\bar{p}}$$

for $\bar{p} = m + 1, m + 2, m + 3$ and $J = A, \ldots, U$.

Next, at the national level, the problem becomes one of setting constraints on the levels of activities of the various regions in accord with the maximisation of national goals. Thus we might have for nation J

$$\max GNP^J = \pi^J X^J$$
$$\text{s.t.} - A^J X^J \leqslant R^J \leqslant \bar{R}^J$$

where

$$\bar{R}^J_h = R^J_h \text{ for } h \neq \bar{p}$$

$$\bar{R}^J_h \geqslant R^J_h \text{ for } h = \bar{p}$$

Note that each nation may choose to emit less of the pollutants than the level dictated by the world unit, but not more. The national level thus places its constraints on pollutants, say, $m + 1, \ldots, m + 8$ where the levels of $m + 1$, $m + 2$, $m + 3$ must not exceed the world-set level. Then, for each region j of nation J, the national unit sets the following constraints:

$$\sum_{g=1}^{q} a^{j}_{\bar{p}g}\, x^{j}_{\bar{p}g} \leqslant R^{j}_{\bar{p}} \quad \text{for } \bar{p} = m + 1, \ldots ,$$

$$m + 8 \text{ and } j = \text{a}, \ldots , \text{u}$$

In a similar fashion, the regions, constrained by national limits, proceed to choose local constraints on pollutants, $m + 1, \ldots , p$, which maximise regional objectives.

A hierarchical scheme of environmental management gives rise to at least two kinds of conflict, *inter*level and *intra*level. Intralevel conflict – conflict between decision-making units on the same level – can be observed where, say, the economic activities of one nation create adverse ecological repercussions on another. To what degree should the activities in the first nation be cut back in order to improve the environment in the second? Obviously the two will differ sharply in their answers. The resolution of this kind of conflict must be carried out on the next higher level, i.e. the world level, in which arena the levels of constraints are arrived at through the interplay of political processes. A somewhat different kind of conflict exists *between* levels; that is, in cases where higher level objectives clash with lower-level objectives. For example, a world organisation entrusted to ensure the ultimate survival of the world may seek to place restrictions on the use of nuclear reactors to generate energy. However, at the national level, many nations concerned with a more circumscribed planning horizon may place their national objectives of lower-cost energy above considerations of world survival. This conflict must also be resolved in an international arena in which compromises are made between the values of the different nations.

Notes

1. See Saushkin [22] and Isard [5]. Also called activity complex analysis (Isard *et al.*, [9]) and now sometimes called complex source analysis by the US Environmental Protection Agency.
2. Strictly speaking, the purist might insist that intergalactic analysis is essential, but at present we have no prospect for a commensurate operational model; hence we shall limit our discussion to the global environment.
3. The codes shown here make up part of a classification system used to organise data accumulated over the years for selected urban regions of New York State (USA).
4. Also called activity complex analysis, complex source analysis, and areal complex of productive forces.

5. For examples of different types of proposed extensions to the I-O model to account for economic-ecologic interdependencies, see Leontief [14], Rose [19], and the collection of papers on I-O analysis in environmental modelling presented by the IEEE [4].
6. In doing this, however, the degree to which the model is operational can be reduced.
7. For discussions of limitations of the types of models considered here, see, for example, Isard, [5], Miernyk [15], Norton [18] and Georgescu-Roegen [1], [2].
8. For some examples of how quantitative models of the type described here can be combined into operational channels of synthesis for use in the management of economic-ecologic systems, see Russell and Spofford [20], Russell *et al.*, [21], Spofford *et al.*, [23]; [24], Heady *et al.*, [3], Nichol and Heady [17], and Isard and Van Zele [13].
9. Another study of the counties of the Appalachian region of New York State involves an urbanised area suitably disaggregated by a one square kilometre grid to permit use of an air pollution dispersion model for policy purposes. This study involves the identification of key air and water pollutants and the specification of relevant (optimum mix) treatment activities (determined partly by linear programming methods) to identify costs to be incurred by municipalities and specific industrial firms to meet different possible regulations. The analysis is based on projection of the growth of industrial sectors using comparative cost and input-output techniques and allows for feedback and adjustment of industrial and population growth in view of resource limitation (such as assimilative capacity of a river system and quality standards imposed such as those on SO_x concentrations in the atmosphere). The results will permit certain evaluations in monetary terms of different environmental policy options permitted by federal and state law, and of existing federal and state legislation, as well as help determine the impact on unemployment in these areas of relatively slow growth. For example, in the Elmira–Corning area, any growth of employment can be achieved only by incurring the exceedingly high costs that would be involved in forcing all industry and municipalities to go from primary or secondary treatment to tertiary treatment of their water wastes. See Isard and Cesario [10].

References

[1] Georgescu-Roegen, N., 1971. *The Entropy Law and the Economic Process*. Harvard University Press, Cambridge.

[2] Georgescu-Roegen, N., 1975. Energy and economic myths; *Southern Econ. J.* vol. 41, pp. 347–381.

[3] Heady, E. O., H. C. Madsen, K. J. Nichol and S. Hargrove, 1972. *Agricultural and water policies and the environment*. CARD Report 40T, Center for Agricultural and Rural Development (Iowa State University, Ames, Iowa).

[4] Ieee, 1973. *Systems, Man and Cybernetics*, SMC-3.

[5] Isard, W., 1960. *Methods of Regional Analysis: an Introduction to Regional Science* (The MIT Press, Cambridge, MA).

[6] Isard, W., 1974a. World environmental conflicts: some relevant analytical frameworks; *Papers, Peace Science Society International*, vol. 23, pp. 139–153.

[7] Isard, W., 1917b. Activity-industrial complex analysis for environmental management, in Isard and Cesario, *Working papers on the Use of Regional*

Science Techniques in Environmental Management, Ithaca, NY.

[8] Isard, W., F. Cesario and T. Reiner, 1974. *Marginal pollution analysis for long-range forecasts, with particular reference to urban regions of the United States*. Center for Urban Development Research, Cornell University, Ithaca, NY.

[9] Isard, W., K. E. Bassett, C. L. Choguill, J. G. Furtado, R. M. Izumita, J. Kissen, R. H. Seyfarth, and R. Tatlock, 1971. *Economic-ecologic analysis for Regional Planning* (The Free Press, New York).

[10] Isard, W., F. J. Cesario (eds), 1974. *Working papers on the use of regional science techniques in environmental management*. Center for Urban-Development Research (Cornell University, Ithaca, NY).

[11] Isard, W., T. Langford, 1971. *Regional Input-Output Study* (The MIT Press, Cambridge, MA).

[12] Isard, W., T. Langford and E. Romanoff, 1966–1968. *Philadelphia Region, Input-Output Study, Working Papers*, vols 1–4. Regional Science Research Institute, Philadelphia, Pa.

[13] Isard, E., R. Van Zele, 1975. Practical regional science analysis for environmental management, *Int. Reg. Sci. Rev.* vol. 1.

[14] Leontief, W., 1970. Environmental repercussions and the economic structure: an input-output approach; *Rev. Econ. Statist.* vol. 52, pp. 262–271.

[15] Miernyk, W. H., 1965. *The Elements of Input-Output Analysis* (Random House, New York).

[16] Miernyk, W. H. *et al.* 1967. *Impact of the Space Program on a Local Economy: an Input-Output Analysis* (West Virginia University, Morgantown).

[17] Nichol, K. H., E. O. Heady, 1974. The national assessment model; Center for Agricultural and Rural Development, (Iowa State University, Ames, Iowa).

[18] Norton, G. A., J. W. Parlour, 1972. The economic philosophy of pollution: a critique; *Environ. Plann.* vol. 4, pp. 3–11.

[19] Rose, A., 1974. *A dynamic interindustry model for the economic analysis of pollution abatement in New York State*. Unpublished PhD dissertation (Cornell University, Ithaca, New York).

[20] Russell, C. S., W. O. Spofford, Jr., 1972. A quantitative framework for residuals management, in A. V. Kneese and B. T. Bower (eds), *Environmental Quality Analysis* (The Johns Hopkins Press, Baltimore).

[21] Russell, C. S., W. O. Spofford, Jr., and E. T. Haefele, 1974. The management of the quality of the environment, in: J. G. Rothenberg and I. G. Heggie (eds): *The Management of Water Quality and the Environment*. Macmillan Press, London.

[22] Saushkin, J., 1962. Large areal complexes of productive forces of the Soviet Union; *Papers, Reg. Sci. Assoc.*, vol. 8, pp. 93–104.

[23] Spofford, W. O., Jr., C. S. Russell and R. A. Kelley, 1972. Operational problems in large-scale residual management models; Conference on Economics of the Environment, Universities-National Bureau Committee for Economic Research and Resources for the Future.

[24] Spofford, W. O., 1974. The Delaware Valley integrated residuals management model: a summary; Conference on Regional Residuals-Environmental Quality Management Models, World Health Organization and Resources for the Future.

[25] Thoss, R., 1975. Resolving goal conflicts in regional policy by recursive linear programming; *Papers, Reg. Sci. Assoc.*, vol. 33, pp. 59–76.

12 Matching Conflict Situations and Conflict Management Procedures

I INTRODUCTION

Over the course of history, man has confronted innumerable conflicts and almost as many times has tried out different procedures to resolve (manage) these conflicts. Although he has been able to survive, we cannot say that he has been successful in developing a set of effective procedures to handle his conflicts. His survival must be attributed much more to flexibility in adapting to diverse environmental situations under crises than to an ability to resolve internal conflict. Today man again confronts conflicts of overwhelming magnitude and still has only a hodge-podge of procedures available to manage them.

It is the purpose of this paper to continue our comprehensive and systematic survey and analysis of conflict management procedures to stimulate their development and make them more commensurate with the size of the problem. We begin to develop a methodology of general use for identifying and even constructing conflict management procedures applicable to each of a significant number of the many conflict situations that have come to exist and are possible.

In section II we recognise that in one or more respects every complex conflict situation is unique; yet, there are many important characteristics common to a number of these situations. Hence, we try to set down systematically characteristics that may be important to one or more possible complex conflict situations. Thus when seeking to identify relevant conflict management procedures for a given conflict situation, we are in a position to characterise as fully as possible the important properties of that situation, as well as its total scope.

Having in mind the need to match the unique as well as non-unique elements of any given conflict situation with one or more useful conflict management procedures, we discuss in Section III key properties of conflict management procedures. We are then in a position to consider in Section IV the process of matching a specific conflict situation (a target) with an 'ideal' conflict management procedure. For the moment we posit that the target is a non-changing one, although in reality this is not the

This chapter was written with Christine Smith. Reprinted from *Conflict Management and Peace Science*, vol. 5, 1, (1980), pp. 1–26.

case. Moreover, the properties of a conflict management procedure may also change; but such change is generally smaller and taken to be of negligible magnitude.

The above identification of an 'ideal' conflict management procedure for a specific conflict situation implies that there exists complete information. In reality, complete information is far from attainable. We typically must get along with highly inadequate information. In practice, then, we cannot claim to be able to match a given conflict situation with an ideal conflict management procedure. Rather, we can only aim to match a conflict situation with one or more admittedly imperfect procedures, but ones which are adequate in certain important respects and are among the best available for that conflict situation.

To make possible this practical matching process, we develop in section V the concept of an index of inadequacy for evaluating actual conflict management procedures. This index can be computed for each actual CMP for any specific conflict situation.

In Section VI we present the method in a more rigorous mathematical manner. The non-technical reader need only glance through this section. Finally, we discuss in Section VII the overwhelming number of shortcomings of the matching and evaluation processes. However, we suggest that these shortcomings are no more severe than those for all social science operational models except those that attack the most simple problems.

II KEY CHARACTERISTICS OF CONFLICT SITUATIONS

Although each conflict situation is unique, there are certain important general characteristics whose presence or absence may be noted in many, if not every situation. The purpose of this section is to identify such general characteristics so that any important conflict situation that does arise can be easily classified with respect to them.

Our classification of general characteristics is presented in Tables 12.1–12.3. Table 12.1 pertains to *structural properties of the underlying systems* from which the participants come. These properties are distinct from the nonsystemic structural properties of a particular conflict situation which are listed in Table 12.2, and from the characteristics of participants (behavioral units) which we take to be non-structural properties and which are listed in Table 12.3.

In Table 12.1 we have constructed a tentative list of properties under several headings: cultural, social psychology, social groups and social organisations, economic system, political system, and system information. For example, under cultural we have:

1. need to have (exercise) power,

2. need to have identity,
3. need to achieve,
   etc.

We consider neither these subclasses nor the properties listed under each as final. They simply represent what we consider for the moment to be a reasonable set of categories and properties, to be subsequently modified. Obviously, system properties are important in considering the use of any conflict management procedure. For example, a procedure which involves a step at variance with the legal system, religious principles, or underlying driving force of a participant's system (nation) is not likely to be a relevant procedure.

In Table 12.2 we list the *non-systemic structural characteristics* of a conflict situation by the following categories: nature of disagreement (issues and stakes), participants involved, coalition-related properties, nature of reference point, and so forth. Obviously, we need to know the nature of disagreement, the number of participants involved, and various other characteristics listed in Table 12.2 before we can proceed to identify a possible relevant conflict management procedure.

In Table 12.3 we list the relevant *participants' characteristics*, which are psychological, level of education (sophistication), control of resources, and so forth.

III KEY PROPERTIES OF CONFLICT MANAGEMENT PROCEDURES

To effect a matching, we need to know the properties of relevant conflict management procedures that can meet the requirements of a conflict situation. Again, based on our existing knowledge of the literature on conflict management and resolution, we set down in Table 12.4 a listing of these properties in a format that at the time of writing seems most useful for the matching process. Our set of categories and listing of properties is only tentative and is subject to significant change. The categories are structural properties, time-related properties, motivating properties, psychological properties, solution properties, and implementability of solutions.

IV STEPS IN THE MATCHING PROCESS

To discuss effectively the steps in the matching process it is best to begin with a particular illustration involving a specific conflict situation (CS) and a specific conflict management procedure (CMP). In this illustration we

Table 12.1 Systemic structural characteristics

Cultural

1. Need to have (exercise) power
2. Need to have identity
3. Need to achieve
4. Type of legal system (e.g., Roman, Common, Islamic, Old Testament)
5. Language and communication capabilities

Social psychology

10. Conservative (e.g., a Have nation)
11. Bold and Daring (e.g., a Have-not nation)
12. Intermediate between 10 and 11
13. Solutions sought via group action (or bureaucratic means)
14. Solutions sought via individualistic action
15. Calculating (methodical) approach
16. Abstract (logical) approach
17. Intuitive (religio-mystic) approach
18. Experimental approach
19. Precedent (tradition) oriented

Social groups and social Organisations

20. Division within society with regard to public purpose
21. Division within society as reflected in socio-economic and political inequalities
22. Division within society as reflected in the history and time period (duration of protracted social conflict)
23. Division within society as reflected in the probability of imminent revolution
24. Division within society as reflected in absence of major conflict in recent past
25. Significant experience with conflict management

Economic system

30. Developed
31. Developing
32. Significant dependence on foreign aid, trade and/or investment
33. Largely small individual entrepreneurship
34. Dominated by large domestic corporations
35. Significant presence of foreign multinationals
36. In take-off period
37. High rate of growth
38. Low to modest rate of growth
39. No growth
40. Recently declining
41. Long-run declining and stagnation

Political system

50. Military dictatorship (authoritarian)
51. Socialistic-communistic, independent
52. Socialistic-communistic, dependent
53. 'Western' democracy, 2 party
54. 'Western' democracy, many party

55. }
56. } other types
57. }

58. 'Big Power' nation
59. Stability (electoral standing) of incumbent government
60. Nature of hierarchical decision-making structure
61. Functional centralisation (decentralisation)

System information

70. Low (e.g., some 4th World nations)
71. Medium
72. High

Table 12.2 Non-systemic structural characteristics

I *Nature of disagreement (issues and stakes)*

100. Qualitative – major
101. Qualitative – minor
102. Quantitative – major
103. Quantitative – minor

II *Participants involved*

110. Two
111. Several
112. Representing themselves
113. Representing larger constituency
114. Technical support team available
115. Technical support team not available
116. Mediator present
117. Mediator not present

III *Coalition related properties*

120. Presence of coalitions by type
121. Stability of coalitions by type
122. Potential for new coalitions by type

IV *Nature of reference point*

130. Current standpoint
131. Nature of current demand
132. Nature of current offers
133. Threat points possible
134. Ideal points possible

V *Time related properties*

150. Historical factors to be considered
151. Static or slowly changing
152. Dynamic
153. Highly dynamic (explosive)
154. Multiple applications of procedure envisaged
155. Single application of procedure envisaged

VI *Policy options (joint actions) available*

160. Number (small, large, continuous)
161. Number (fixed or variable)
162. Agenda (fixed or variable)

VII *Fractionation possibilities (decomposability)*

170. Re: issues (including logrolling)
171. Re: participants
172. Re: actions and policies
173. Re: outcomes

VIII *Information available*

180. Available 'objective' knowledge of conflict situation
181. Participants' knowledge of conflict situation
182. Participants' knowledge of others' knowledge and preference structures

will want to evaluate the relevance of the specific procedure for the specific situation. To do so, we construct Table 12.5. In column 3 of this table we list all the important characteristics that any conflict situation (CS) may possess. These characteristics have been detailed in Tables 12.1–12.4. The first is 'need to have (exercise) power'; the second is 'need to have identity', and so forth. In column 5, we list possible 'corresponding' properties of a conflict management procedure (CMP). That is, given a characteristic of a CS, we set down by its side the one or more properties that it is desirable for a CMP to have to match this characteristic. For

Table 12.3 Participants' characteristics

Psychological

200. Conservative
201. Mixed conservative-wild
202. Wild (bold)
203. Pessimist
204. Mixed pessimist-optimist
205. Optimist
206. Regretter
207. Risk-averse
208. Risk-lover
209. Optimiser
210. Expected payoff calculator
211. Satisficer
212. Altruist
213. Retaliatory
214. Obstinacy
215. Cooperative
216. Competitive
217. Passive
218. Aggressive
219. Myopic
220. Non-myopic
221. Strategy-oriented
222. Outcome-oriented
223. Action-oriented

Level of education (sophistication)

230. Low
231. Medium
232. High
233. Extent and type of misperception

Control of resources

240. Economic (money, jobs, patronage)
241. Political (c-power, etc.)
242. Social (c-respect, etc.)
243. External support

Perceptive

250. Sees many joint actions
251. Sees few joint actions
252. Recognises interdependence of actions
253. Recognises interdependence of outcomes
254. Sees many outcome properties
255. Sees few outcome properties

Receptive

260. Ability to learn
261. Willingness to learn (explore new ideas and thinking)
   - about situation
   - about other participants
262. Ability to change focus
263. Extent of commitment to a position

Preference structure and statement capability

270. Ordinal
271. Relative
272. Cardinal
273. Independent
274. Interdependent

example, the first characteristic of conflict situations, 'need to have (exercise) power', can be said to be matched by, that is, can be met by a conflict management procedure, if that procedure possesses the properties 'inescapable sanctions possible', 'veto power possible', 'leader-follower arrangement possible' and perhaps others. These latter properties are listed in column 5 to the right of the 'need to have (exercise) power' characteristic in column 3. Or, to take another illustration, the general characteristic number 15 of conflict situations, namely, a 'calculating social psychology', may be said to be matched by a conflict management procedure if that

Table 12.4 Properties of conflict management procedures

A *Structural properties*

Information requirements re: preferences

1. Ordinal
2. Relative
3. Cardinal
4. Interdependent or independent
5. Other
6. No knowledge of others
7. Full knowledge of others
8. Partial knowledge of others

Other information requirements

9. Full
10. Relatively large
11. Relatively small

Number of joint actions considered

12. Small
13. Large
14. Continuous

Nature of interaction

15. Participatory
16. Non-participatory
17. Face-to-face
18. Non-face-to-face

Cost

19. High in initial phases
20. Low in initial phases
21. High in later phases
22. Low in later phases

Coalition formation

23. Encouraged
24. Discouraged
25. Changes in allowed

Other

26. Requires weights
27. Does not require weights
28. Objective achievement focused

Other (cont'd)

29. Utility outcome focused
30. Last resort capability
31. Usable in a workshop setting
32. Statable in balance-sheet form
33. Statable mathematically
34. Statable without mathematics
35. Presence of mediator required
36. Pairwise comparisons possible

B *Time related properties*

37. Single application-oriented
38. Multiple application-oriented
39. Able to be coupled with others in a fractionated conflict

Number of rounds

40. One
41. Many
42. Either one or many

Size of steps involved

43. Small (Limited commitment)
44. Large
45. Either small or large

Interactive (feedback effects)

46. With search processes
    - for efficient solutions
    - for better satisficing solutions
47. With knowledge accumulation
48. With learning by participants
49. Encourages invention of new tools and creative thinking by participants
50. Removes misperceptions
51. Develops common language and communication channels

C *Motivating properties*

Guaranteed improvement

52. Overall strong
53. Overall weak
54. On each round strong
55. On each round weak
56. Guaranteed nondeterioration

Concessions

57. From ideal point
58. From overly optimistic point
59. From status quo point (in contraction)

Limited commitment

60. Overall
61. On each round

Fairness

62. Fairness (weighted equity) over microtime
63. Fairness over macrotime
64. Equitable over microtime
65. Equitable over macrotime

Other

66. Can accommodate improvements or concessions or concessions
67. Shifts focus from fixed positions
68. Permits reconstruction (reformulation of a game-type conflict
   - from zero to non-zero sum
   - from action to policy-oriented
69. Eliminates bluffing
70. Reveals preferences

D *Psychological properties*

71. Security preserving
72. Confidence building overall
73. Confidence building on early rounds
74. Trust building in first step(s)
75. Allows leader-follower arrangement
76. Conflict reducing (escalating) potential
77. Induces positive reciprocal action
78. Induces negative reciprocal action
79. Provides an incentive to think of others
80. Recognises strategic potential
81. Involves strategy-oriented behaviour
82. Involves outcome-oriented behaviour
83. Involves action-oriented behaviour

E *Solution properties*

84. Steps to solution defined
85. Inescapable sanctions possible
86. Veto power possible
87. Unstable
88. Stable in both micro and macro time
89. Stable in micro time only
90. Unique outcome
91. Non-unique outcomes
92. Preindeterminate outcome
93. Efficient outcome guaranteed
94. Non-efficient outcome possible
95. Certain outcome
96. Uncertain, risky outcomes
97. Probabilistic outcomes

F *Implementability of solutions*

98. Is the solution consistent with international law
99. Is the solution simple enough for the minds of existing administrators (legislators)
100. Is the solution implementable with existing tools, policy instruments
101. Does the solution require development of new tools, policy instruments for implementation

Table 12.5 Worksheet to calculate index of inadequacy

Presence of characteristic in conflict X	Weight for conflict X w_j	Characteristics of conflict situation (CS)	Relative importance ϱ_{ji}	Corresponding properties of conflict management procedure (CMP)	Presence of property in CMP Y	Calculation of inadequacy score of CMP Y
(1)	(2)	(3)	(4)	(5)	(6)	(7)
		Cultural	0.45	85 inescapable sanctions possible	0	(1–0)9(.45) = 4.05
1	9	1. Need to have (exercise) power	0.45	86 veto power possible	1	(1–1)9(.45) = 0
			0.10	75 leader-follower arrangement possible	1	(1–1)9(.10) = 0
0	0	2. Need to have identity	1	17 face-to-face interaction	1	(0–1)0(1) = 0 (whenever < 0, set equal to zero)
		Social Psychology	.50	93 efficiency	1	(1–1)6(0.5) = 0
1	6	15. Calculating	.03	33 statable mathematically	1	(1–1)6(0.03) = 0
			.47	52 guaranteed improvement	1	(1–1)6(0.47) = 0
		Political System	1	81 involves strategic behaviour	1	(1–1)3(1) = 0
1	3	58. Big Power Involvement				
		Economic System	.30	82 involves outcome oriented behaviour	0	(1–0)(.2)(.3) = .06
1	0.2	31. Developing				
			.40	44 large size of steps	0	(1–0)(.2)(.4) = .08
			.30	96 uncertain risky outcomes	0	(1–0)(.2)(.3) = .06
		Social Group	.30	60 fairness (weighted)	1	(1–1)6(.3) = 0
1	6	52. Division over History	.02	79 provides incentive to think of others	0	(1–0)6(.02) = .12
			.58	74 trust building in first steps	.2	(1–.2)6(.58) = 2.78

procedure possesses the properties 'efficiency' (that is, can lead to an efficient outcome), 'statable mathematically', and 'guaranteed improvement.' These properties are listed in column 5 of Table 12.5 to the right of the 'calculating social psychology' characteristic in Column 3.

Just listing characteristics of a CS and corresponding properties of a CMP is not enough. We need to indicate the relative importance of each of the CMP properties for a given CS characteristic. We do this in column 4 of Table 12.5. For example, with regard to the CS characteristic 'need to have (exercise) power', we have judged the relative importance of the CMP properties 'inescapable sanctions possible', 'veto power possible', and 'leader-follower arrangement possible' to be 0.45, 0.45, and 0.10, respectively.

Having done this part of the table, namely columns 3, 4 and 5, which, as we will later indicate, can be computerised, we then take the next step. This is to fill in column 1 of Table 12.5. To do this, however, we must have a given conflict situation in mind. It may be, for example, the current Iraq-Iran conflict, or the US-Vietnam conflict in the 1960s. We need to specify in column 1 whether or not each characteristic listed in column 3 is relevant for the specific conflict being examined. In the relevant row we place a '1' if the answer is yes and a '0' if the answer is no. For example, the 'need to have (exercise) power' was a relevant characteristic of the Vietnam conflict; so we have placed a '1' in the first column of Table 12.5.

Having filled in column 1 we proceed to column 2. In this column we specify the relative importance of each CS characteristic present (i.e., those characteristics which have a '1' in column 1) in the particular conflict situation we are studying. Here we will need a ranking system or the like, such as the Saaty system which we have discussed elsewhere.[1] If we do so, then we might, for the Vietnam conflict, set down the number 9 for the 'need to have (exercise) power', indicating that we judge this characteristic to be of paramount (overwhelming) importance in this conflict situation. In contrast, we may judge the characteristic 'need to have identity', the second listed in column 3, to be of negligible importance; so we simply transfer the 'O' in column 1 to column 2. We may judge characteristic number 15 in column 3, 'calculating social psychology', to be rather significant; so we give it a weight of '6'.[2]

We next proceed to column 6. To fill in column 6, we must first have identified the specific conflict management procedure to be examined for possible use for the given conflict situation. The conflict management procedure might be a many-round procedure employing the principle of split-the-difference over utility outcomes; or, it may be a veto-incremax procedure; or, it might involve the metagame approach. In each row of the column we indicate whether the corresponding property of a CMP listed in column 5 is present or absent in the specific CMP being examined. If present, we record a '1'; if not, a '0.' For example, the property 'inescapable sanctions possible' at the top of column 5 is not present in a split-the-difference or veto-incremax procedure. So, for each of these specific procedures, we would record a '0' at the top of column 6. (Note the possibility of a modification of our method which would allow the use of a fraction between zero and one to indicate the degree of presence of a CMP property in column 6, or of a CS characteristic in column 1, or both. In this article we do not pursue this modification.)

Once the first six columns have been filled in we can proceed to evaluate the specific CMP for use for the given CS. We do this in a very simple, admittedly inadequate way in column 7. In particular, we evaluate a CMP in terms of its ability to match each of the CS characteristics properly weighted by its importance to the specific situation being examined. For example, with regard to the characteristic 'need to have (exercise) power', we make three calculations, as indicated in the first three rows of column 7. Specifically, in row 1 of column 1 there is a '1' which indicates that the CS characteristic, the 'need to have (exercise) power', is present in the specific CS being examined. In row 1 column 6 there is a '0', which indicates that the property 'inescapable sanctions possible' is not present in the specific CMP being examined. Hence, the term (1–0) in row 1 column 7 represents a first rough index of inadequacy. The significance of this inadequacy is in part determined by the importance of the CS characteristic the 'need to have (exercise) power' for the specific CS. This importance, as given in row 1, column 2, is 9; so we must multiply the term (1–0) by 9, obtaining (1–0)9. Finally, as indicated in rows 1–3 of column 5, 'inescapable sanctions possible' is only one of three CMP properties relevant for the 'need to have (exercise) power' characteristic and among the 3 properties its relative normalised weight, as indicated in column 4, is 0.45. So the inadequacy score on the 'inescapable sanctions possible' dimension is (1–0)9(0.45) = 4.05, as noted in row 1 of column 7. In like manner, we calculate inadequacy scores for the CMP properties 'veto power possible' and 'leader-follower arrangement possible' with regard to CS characteristic 'need to have (exercise) power'. See rows 2 and 3 of column 7. Note that these two properties are fully present in the CMP, say, veto incremax, being considered; so the first term for each is (1–1) and in their respective rows we have (1–1)9(.45) = 0 and (1–1)9(.10) = 0. That is, the veto incremax procedure has zero inadequacy on both these counts. Thus, in terms of matching the characteristics 'need to have (exercise) power', the inadequacy of the specified procedure is in total 4.05 for the given CS.

In the above manner, we proceed with respect to all characteristics of the given CS. We then sum the inadequacy scores in column 7 to obtain a measure of the total inadequacy of the given CMP, say, veto incremax. We can do likewise for every other CMP. Observe that the information in columns 3, 4, 5 is common for all procedures and so such information may be stored in a computer. Further, for every possible CMP we can develop the information contained in column 6, and store that, too, in a computer. Hence, when an analyst (mediator) does confront a specific conflict situation, all he needs to do is to fill in columns 1 and 2. The computer can then print out in order of magnitude the inadequacy score for each processed CMP.

One further point needs to be mentioned in order to understand the calculations in Table 12.5. Frequently, the first term in column 7 is negative. This will always be the case, if (1) the particular CS does not require a characteristic so that the entry in the row corresponding to that characteristic in column 1 is '0', and (2) the CMP being studied has a property that could match that characteristic so that the entry in the row corresponding to that characteristic in column 6 is '1'. Whenever this occurs we set the first term equal to zero so that an actual CMP with more of a given property than is required is not penalised. We shall discuss this convention later.

V EVALUATING ACTUAL CONFLICT MANAGEMENT PROCEDURES

In section IV we illustrated how with Table 12.5 we can carry through a set of calculations to derive an index of inadequacy for each CMP. To describe the method efficiently and in more general terms we need to employ simple algebra. First, we let j, $j = 1, \ldots, n$, represent an important characteristic of any possible CS, there being n such characteristics (as listed in Tables 12.1–12.3) when we consider all possible conflict situations. We also let i, $i = 1, \ldots, m$, represent a possible relevant property of any CMP which may be adopted (as listed in Table 12.4), there being m such properties over all possible conflict management procedures. We next let ϱ_{ji} indicate the relative importance of property i of a CMP in matching the characteristic j of a CS. Typical ϱ_{ji} are listed in column 4 of Table 12.5. Thus, if all characteristics j of a CS had the same weight, namely, one, then the weight of property i of a CMP would be given by $\widetilde{w}_i$ where $\widetilde{w}_i = \sum_{j=i}^{n} \varrho_{ji}$ for all i, indicating that the weight of property i is simply the sum of its 'importances' over all characteristics j. However, not all characteristics j of a CS are of equal importance. Rather, each characteristic j must be given a relative weight $\bar{w}_j$, as recorded in column 2 of Table 12.5. Hence the weight of any property i becomes the sum of its 'importances' over all characteristics j after each of these characteristics has been multiplied by its relative weight $\bar{w}_j$. Accordingly, we obtain

$$w_i = \sum_j \bar{w}_j \varrho_{ij}, \quad i = 1, \ldots, n. \tag{12.1}$$

With these derived weights for i, we then can calculate for each actual CMP its index of inadequacy π. To do so, conceive of an 'ideal' CMP to match a give CS. It has all the properties required by that CS. We designate an ideal CMP for the given CS by the symbol (or vector) p with properties $p_1, p_2, \ldots, p_i, \ldots p_m$. If p is an ideal procedure, then by definition the property p_i of this procedure is always equal to '1' whenever property i is required by any characteristic of a given CS, and for reasons to be discussed later, equal to '0' otherwise.

In practice, we know of no actual conflict management procedure that is 'ideal', except perhaps for a very simple CS. Hence we want to contrast an actual CMP with an ideal CMP to see how far short the actual falls from the ideal. For the purpose of this section we shall designate any actual CMP by $\bar{p}$ with properties $\bar{p}_1, \bar{p}_2, \ldots, \bar{p}_i, \ldots, \bar{p}_m$ with $\bar{p}_i = 1$ or 0, according to whether property i is present or absent in the actual CMP being examined. Thus, $(p_i - \bar{p}_i)$ represents a first crude index of inadequacy with regard to property i. In evaluating any actual CMP on a total basis, we must multiply $(p_i - \bar{p}_i)$ by the weight of property i to obtain $w_i (p_i - \bar{p}_i)$. Hence, over all properties i, the index of inadequacy of the actual CMP might seem to be

$$\pi = \sum_i w_i (p_i - \bar{p}_i) \tag{12.2}$$

However, recall that we set $(p_i - \bar{p}_i) = 0$ whenever $p_i < \bar{p}_i$ indicating that in calculating an index of inadequacy for an acutal CMP with more of any property i than is required, this surplus is not allowed to be used to reduce the CMP's inadequacy on other scores. Thus 12.2 must be rewritten as:

$$\pi = \sum w_i (p_i - \bar{p}_i) \quad \text{for} \quad (p_i - \bar{p}_i) \geqslant 0 \tag{12.3}$$

In this way, given any CS with relative weights $\bar{w}_j$ one can immediately calculate the index of inadequacy π for each CMP that exists or is to be considered. This implies that the best CMP is the one with the lowest index π.

VI THE MATHEMATICAL FORMULATION OF THE METHOD

The use of the index of inadequacy derived above involves a set of most serious shortcomings. However, before we discuss these shortcomings, we need to outline in more rigorous mathematical language the steps in the matching and evaluation processes discussed in sections IV and V. The non-technical reader need only glance over this section.

Recall our first task, namely to list the n important characteristics, $j = 1, \ldots, n$, of all possible conflict situations (theoretical and actual). See Tables 12.1 to 12.3. Taking all possible combinations of these characteristics yields theoretically the set S of all possible conflict situations. See the upper left of Figure 12.1. Hence any particular conflict situation s that we examine is contained in this set, i.e., $s \in S$ where $S \subset R^n$.

The next task is to list all possible properties, $i = 1, \ldots, m$, of conflict management procedures. See Table 12.4. Taking all possible combinations of these properties yields the set P of all possible CMP's. See the upper right of Figure 12.1. We designate any procedure, theoretical or actual, as p, with $p \in P$ and $P \subset R^m$.

Now any actual CMP we have discussed is contained in this set as well as many still to be developed, and presumably still more that are purely theoretical. We designate the subset of actual CMP's as $\bar{P}$ where $\bar{P} \subset P$, and any actual CMP as $\bar{p}$ where $\bar{p} \in \bar{P}$. See Figure 12.1.

Next, we map each characteristic s_j $(j = 1, \ldots, n)$ of a conflict situation s into one or more properties p_i $(i = 1, \ldots, m)$ of a CMP required (or, more weakly, desirable) to match that characteristic s_j. The set of p_i $(i = 1, \ldots, m)$ defines a point $p \in P$. The mapping θ of s into p is shown in the upper part of Figure 12.1. The point p may be said to describe for s the properties of an ideal CMP. As will be discussed below, this CMP is taken

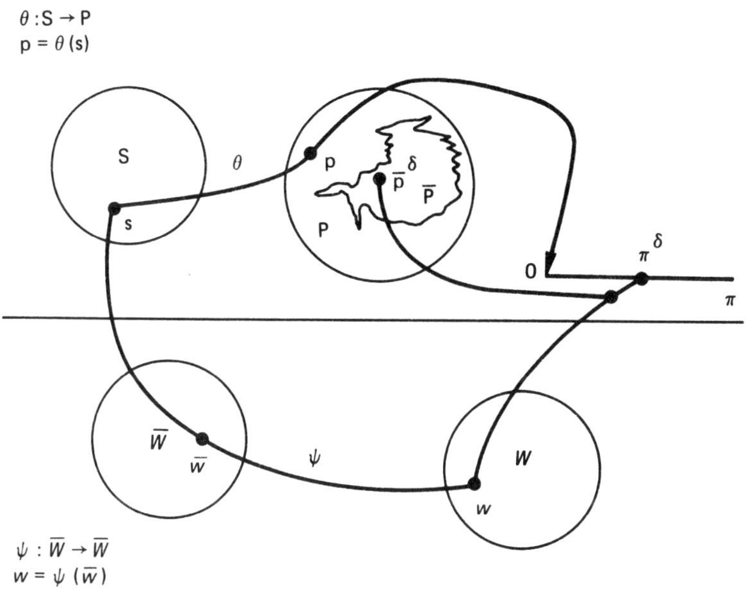

Figure 12.1 Mappings for determining a 'best' practical conflict management procedure

to be the 'minimal property-containing ideal CMP'. In reality, however, we know of no actual procedure $\bar{p}$ that meets or historically has met all the requirements of the characteristics s_j of an actual conflict situation s, except perhaps for some highly simplified situations. Invariably, each existing procedure $\bar{p} \in \bar{P}$ is 'inadequate' (imperfect) for a given s with regard to one or more of the required properties p_i. Therefore, we need to distinguish among existing procedures on the basis of their adequacy to match the characteristics s_j of the given s. As discussed in section V, we need to develop an index of inadequacy for each $\bar{p} \in \bar{P}$. To do so, we define (establish) relative weights $\bar{w}_j$ $(j = 1, \ldots, n)$ for each s_j of the given s. This yields $\bar{w}$, which is a point in the general space $\bar{W}$ of all possible sets of relative weights, i.e., $\bar{w} \in \bar{W}$ where $\bar{W} \subset R^n$. See the lower left of Figure 12.1. The mapping of s into $\bar{w}$ is a perceptual mapping – that is, is the importance of characteristics as perceived by one or more participants, the mediator, or both.

As discussed in section V, the relative importance ϱ_{ji} of every property i $(i = 1, \ldots, m)$ for each given characteristic j $(j = 1, \ldots, n)$, and hence the matrix

$$\begin{bmatrix} \varrho_{11} & \varrho_{12} & \bullet & \bullet & \bullet & \varrho_{1m} \\ \varrho_{21} & \varrho_{22} & \bullet & \bullet & \bullet & \varrho_{2m} \\ \bullet & \bullet & \bullet & \bullet & \bullet & \bullet \\ \bullet & \bullet & \bullet & \varrho_{ji} & \bullet & \bullet \\ \bullet & \bullet & \bullet & \bullet & \bullet & \bullet \\ \varrho_{n1} & \varrho_{n2} & \bullet & \bullet & \bullet & \varrho_{nm} \end{bmatrix} \equiv [\varrho_{ji}] \tag{12.4}$$

must be specified. Typically the ϱ_{ji} are not technologically determined. Rather, they reflect the perceptions of relevant participants, the mediator, or both. Theoretically, they should be independent of the particular CS being studied and the particular CMP being investigated for relevance.

Given the matrix $[\varrho_{ji}]$ and the relative weights $\bar{w}$, for a given s we derive the weights w_i to apply to the properties of CMP's. That is[3]

$$w_i = \sum_j \bar{w}_j \varrho_{ji} \quad i = 1, \ldots, m \tag{12.5}$$

which defines the point $w \in W$. The mapping Ψ of $\bar{w}$ into w is shown in the lower part of Figure 12.1. The w_i then allow us to map each actual CMP namely $\bar{p}^\delta$ ($\delta = \alpha, \beta, \ldots, \zeta$) into a single dimension along which π, the index of inadequacy, is measured. See the upper part of Figure 12.1. That is[4]

$$\pi^\delta = \sum_i w_i (p_i - p_i^\delta) \quad \text{for} \quad (p_i - \bar{p}_i^\delta) \geqslant 0 \quad \delta = \alpha, \beta, \ldots, \zeta \tag{12.6}$$

Finally, at this point we may set up the trivial programme

$$\begin{aligned} &\min \pi^\delta \\ &\{\bar{p}^\delta \in \bar{P}\} \end{aligned} \tag{12.7}$$

which will not be trivial when relevant constraints are introduced.

VII SHORTCOMINGS OF THE MATCHING AND EVALUATION PROCESSES

We now turn to an evaluation of the method developed in sections II–VI. We shall consider the validity and reasonableness of each step. First, however, note a serious general shortcoming. This shortcoming relates to the fact that our thinking follows that of a rational economist or applied mathematician. It is heavily oriented to traditional natural science methods. For example, it does not meet the challenge of neo-Marxists when they claim, and rightly so, that ideology plays a central role in conflict of human societies as history records these conflicts. However, along one direction we do try to indicate how the natural science and neo-Marxist approaches must and can be complementary in actual conflict management. In effect, what we attempt to do is to cull from the natural sciences a systematic and comprehensive framework for studying conflicts in general and then when we study any particular conflict we start with value judgements (ideology) at the core, and continue to consider explicitly the different values and perceptions of the diverse participants at all stages, in ways that will be detailed later.

In the method presented in the previous sections, the first step is to specify the set of all important characteristics j that any conflict situation may have. We recognise that what one writes down as important today may become rapidly obsolete tomorrow, when we may try to employ our method to some degree. None the less, given our social science and related knowledge, we may be able to do a credible job here – credible in the sense of loosely identifying many, if not all, relevant characteristics. On the other hand, given our different language systems and their different capacities at communications, we may not be able to state precisely and without confusion these characteristics in all languages, let alone in any one language. Further, characteristics relevant for one cultural system may be of zero consequence for another. On the other hand, our list of properties can be extended at little or no cost until we do include all those that are relevant to any culture.

While we list properties in keeping with our social science and related knowledge, it must be recognised that any analyst comes to this task with certain disciplinary biases. He sees most clearly those elements (variables) central to his own discipline, and less clearly and in less disaggregated fashion elements central to other disciplines, and especially those disciplines far from his own. We, therefore, look to significant enrichment of Tables 12.1–12.3 from scholars, particularly those from disciplines other than regional science and economics. For instance, a cultural anthropologist would clearly extend category I (cultural) in Table 12.1; and a cognitive psychologist, category IV of Table 12.3.[5]

The second step in our method is to specify the set of all properties i that any conflict management procedure (theoretical or actual) may possess. We have listed in Table 12.4 what we presently consider relevant properties. This listing not only again reflects our limited knowledge and discipline, language and other cultural biases, but also a major bias in that they focus on properties which have tended to be present in CMP's written about and proposed for actual use, such as GRIT, and the metagame procedure. New concepts, new theories, new operational frameworks – all of which are solely needed – would undoubtedly result in major change in the list of Table 12.4.

The third step in our method is to carry through the matching of all properties i with each of the characteristics j that might be present in a specific conflict situation s. This task is required for the identification of p, an ideal CMP.[6] It is extremely difficult and needs the cumulative wisdom of experienced negotiators in each of the several major conflict areas of society. In a Utopian world, of course, one may assume that all participants and the mediator may be able to agree fully on the matchings for each j.

Note that an s may map into more than one point in p. This is so because more than one CMP may have all the desirable properties required by an s, and still other desirable properties. For example, the 'veto power' property may not be required for a given s; yet one CMP may have all the properties i required by an s and the veto power, while a second CMP has all the properties but not the veto power. Both are ideal in the sense of having all required properties i. Yet, strictly speaking, we consider only the second CMP (without the veto power) as 'ideal' since in subsequent calculations involving actual CMP's, those CMP's which do not have the veto power would be wrongly penalised for not having the veto power, were the first CMP used as the benchmark for the ideal. Accordingly, we map every s into a unique p, the minimally ideal CMP.[7]

The fourth step is taken when it is appreciated that every $\bar{p}$ currently in our tool kit is inadequate for handling any conflict situation s. Such recognition then requires that the relative importance, ϱ_{ji}, of all properties i that are matched with a given characteristic j, be specified (with $\sum_i \varrho_{ji} = 1$, and $0 \leq \varrho_{ji} \leq 1$). Difficulties arise not only because of differences in perceptions among participants to be discussed below, but also because a given property i of a CMP may be required to simultaneously serve (match) several characteristics j of a given s. Our method assumes that the matching of a property i with a characteristic j of any s does not affect its ability to match any other characteristic j' of that s.

Once the ϱ_{ji} are determined, we confront in the fifth step the problem that the simple addition of ϱ_{ji} over all j is not meaningful in interpreting the role of a property i. To give significance to the ϱ_{ji} we must have in mind a particular target, that is a particular conflict situation s. This then poses the next problem, namely, what characteristics j of the given s are important. This again is a highly subjective matter about which there may be differences among analysts. In addition, at least some characteristics may be present (absent) to different degrees – another issue involving subjective perceptions. For the simplest case possible we can assume that the characteristic j is either wholly present ('1') or completely absent ('0') in a conflict situation.

Moreover, once the degree of presence or absence of a characteristic j is identified or estimated, we need in the sixth step to state the relative importance $\bar{w}_j$ of each characteristic j for the given conflict situation s. This, as well as the degree of presence, is required if we are to add meaningfully the several inadequacies (contributions) of any i (via the different j) for the resolution of the conflict s. How to obtain the $\bar{w}_j$? A first problem is to settle on an appropriate scale. From a technical standpoint, there is no commonly accepted scale by which to measure the $\bar{w}_j$. At one extreme, each property j may be placed into one of two categories, say, of *major* importance and given a weight of '9', or of *minor* importance and given a weight of '1'. If a more discriminating weighting is possible, its weight may be given by one of the odd integer numbers 1, 3, 5, 7 and 9.[8] But why place characteristics in only 5 categories of relative importance. Why not more, or less? This question needs to be decided by participants and the mediator. Note that where a characteristic j is overwhelmingly dominant (or close to indispensable), it might be given a weight of 9 and all others given a weight of 1 or close to 1. If such relative weights do not adequately portray j's dominance, the upper limit to the scale, namely, 9, can be increased to be sufficiently large, or the lower limit sufficiently decreased, or both, so that an adequate portrayal of dominance is achieved. Note that all characteristics of conflict situations which are not important for the actual conflict situation s are not involved in the assignment of relative weights. In effect, their relative weights are zero.

Whatever the difficulties in establishing the $\bar{w}_j$ for the calculations to follow, it is essential that precise relative weights be derived, that is, relative weights which are measurable on a ratio scale.

With the j, i, ϱ_{ji} and $\bar{w}_j$ defined, we can now proceed in the seventh step to examine actual CMP's for potential use in managing the given conflict s. But before an analyst or others can effectively do so, he must establish what properties are present in any actual conflict management procedure $\bar{p}^\delta$ ($\delta = \alpha, \beta, \ldots, \zeta$). This has difficulties similar to those in identifying the j that are important to any given conflict situation s. Once again, this is a subjective matter, but here the subjective perceptions of an experienced mediator may be of greater importance, relatively speaking, than participants' perceptions. At one extreme, all that can be observed or envisaged is the presence ('1') or absence ('0) of a property i in any given p^δ. Consequently, only a nominal scale is possible; that is, $p_i^\delta = 1$ or 0. A somewhat better situation may exist if it becomes possible through the use of participants, experts, or both to establish the precise degree of presence p_i^δ of a property i in the given p^δ (e.g., using Saaty-type procedures). Hence, we would have $0 \leqslant p_i^\delta \leqslant 1$. Or it may be possible to use scaling techniques, such as Thurstone's to establish a seemingly more precise measure along a ratio and interval scale of the degree of presence (or absence).

Having established the degree of presence $\bar{p}_i^\delta$ of property i in a given $\bar{p}^\delta$ ($\delta = \alpha, \beta, \ldots, \zeta$), we then may proceed in the next step to the measure of inadequacy. The measure employed in the previous section is[9]

$$\pi^\delta = \sum_i w_i (p_i - \bar{p}_i^\delta) \quad \delta = \alpha, \beta, \ldots, \zeta \quad \text{for} \quad (p_i - \bar{p}_i^\delta) \leqslant 0 \qquad (12.8)$$

where

$$w_i = \sum_j \bar{w}_j \varrho_{ji} \tag{12.9}$$

We have all the necessary elements to do so, namely, the $\bar{w}_j$, the ϱ_{ji}, the $\bar{p}_i^\delta$ and the p_i.[10] However, there are many problems involved in the use of 12.8 as a measure of inadequacy of $\bar{p}^\delta$.

First, recall that whenever $\bar{p}_i^\delta > p_i$, by (12.8) we in effect set $(p_i - \bar{p}_i^\delta)$ equal to zero. That is, we do not allow an excess of an actual CMP on one property i over what is required by a given s to reduce the inadequacy of that CMP with regard to other properties i'. We can imagine that for certain s, such reduction might be permissible.

Second, π^δ, as defined by (12.8), employs a specific way, a specific 'distance', for measuring inadequacy (namely, the city-block distance). Many other possible ones could be used. One frequently used alternative is[11]

$$\pi^\delta = \Big(\sum_i w_i (p_i - \bar{p}_i^\delta)^2\Big)^{1/2} \quad \text{for} \quad (p_i - \bar{p}_i^\delta) \geqslant 0 \tag{12.10}$$

which has the effect of penalizing larger inadequacies more than smaller ones. But then why should the inadequacies be evaluated in the particular way suggested by (12.10) or (12.8)? Again this is an issue that must be confronted by participants and the mediator.

The problem of measuring the inadequacy of $\bar{p}^\delta$ is still more complex. The formulae of (12.8) and (12.10) simply add inadequacies over each property and, importantly, assume that an inadequacy of a CMP on one count does not influence its inadequacy on any other count. But frequently such is not the case in reality. One possibility is to add to (12.8) interaction terms

$$\prod_{i} \prod_{i'} a_{ii'} (p_i - \bar{p}_i^\delta)(p_{i'} - \bar{p}_{i'}^\delta) \quad \text{for} \quad (p_k - \bar{p}_k^\delta) \geqslant 0, \quad k = i, i' \tag{12.11}$$
$$i \neq i'$$

where $a_{ii'}$ can be either positive or negative to indicate inadequacy as a result of interaction between properties i and i' and where the above term can be restricted to selected pairs of properties and not cover all pairs.

There are obviously many other possible formulae for measuring inadequacy. One which is difficult to set down mathematically would state that a CMP should be rejected if it fails to achieve for any given s a critical mass, complex or bundle of required (matching) properties. What we are unable to define is the nebulous concept of a critical mass, complex or bundle. However, an operational way to attack the problem of achieving a critical mass, complex or bundle, as well as to ensure that an 'indispensable' property i is present, is to introduce constraints into the trivial programme 12.7, which now becomes the non-trivial programme.

$$\min_{\{\bar{p}^\delta \in \bar{P}\}} \pi^\delta \tag{12.12}$$

$$\text{such that } \bar{p}_h^\delta \geq c_h \tag{12.13}$$

where $h \in H$, where H is a set of selected properties, and where c_h equals 1 for any 'indispensable' property. Still other kinds of constraints are possible.

Additionally, if there are interdependencies among the properties i, are there not likely to be interdependencies among the characteristics j? If so, as is frequently the case in reality, such independencies would need to be taken into account in at least the specification of the $\bar{w}_j$. For example, if one characteristic j implies the existence of another characteristic j', then, perhaps the $\bar{w}_j$ and the $\bar{w}_{j'}$ need to be adjusted to avoid overcounting or undercounting. This, in turn, would lead to a new value for some of the w_i.

Hitherto, we have assumed that all participants and the mediator have the same language, concepts, perceptions, cultural background, and they are in all other respects identical so as to permit the problem to be symmetric. However, in actuality we know that this is not so and that major asymmetries exist in most, if not all, situations. We now address the problems which arise because of asymmetries. These problems involve a sequence of secondary conflicts that must be managed before attacking the main problem of selecting an actual CMP to resolve the primary conflict. We shall stress that the diverse participants with their different values and perceptions, and not the analyst or mediator, should in the main determine relevant variables and magnitudes.

(i) The first problem is the language problem – that is the problem of developing a common language among the participants so that they come to agree, at least to some extent, on the meaning of words, terms and sentences. However, associated with this problem is also the problem of differences among participants in relevant concepts, theories and ideas – a problem which is not as easily managed. This latter problem leads participants to have differences in perceptions concerning the relevant characteristics of the given conflict situation.

The secondary conflict at this point, like a number of others to be noted below, needs to be resolved before the analyst can move on to the next step. In this secondary conflict, as well as the others to follow, there is the problem of attaching different weights to the views, opinions, suggestions and the like of the several participants. This problem itself generates more conflict which needs to be resolved, etc. Moreover, debate and discussion that may be undertaken to resolve these secondary conflicts may often be required to eliminate differences in perceptions; but at the same time such debates and discussions start to introduce interdependence of perceptions which will then affect the ease with which subsequent conflicts are resolved.

(ii) Once the first secondary conflict is resolved, a next one which may arise concerns the relevant scale to use in measuring the degree of presence or absence of a relevant characteristic j in the given conflict situation. This issue and many others mentioned below may become so involved that the participants and mediator may reach the conclusion to forget it (put it aside), i.e., assign either a '1' (relevant and present) or '0' (irrelevant or absent) to any characteristic j. If this is not the case, one participant might be sold on an extension of one approach, say, the Saaty approach, a second on another approach, say, the Thurstone single dimensional scaling and a third on still another approach, say, a simple hunch or a Delphi method. A skillful mediator, here and in other places noted below, may be able to get the participants to use simple procedures to resolve their differences quickly – for example, a simple *action compromise* procedure involving not the most sophisticated, nor the most simple approach, but one falling in between these extremes.

(iii) A third secondary conflict arises when the participants and mediator, again because of differences in language or perception, try to settle on relevant properties i for each given characteristic j.[12]

(iv) A fourth secondary conflict concerns the problem of determining for any given j of s the relative importance of the i, that is, the ϱ_{ji}. As with the previous problem, each participant's notions may be very fuzzy and he may need much assistance from the analyst or mediator. The participants may also follow a suggestion by the mediator that they engage in a give and take discussion – that is, that they interact recursively, perhaps relying heavily on the mediator's or analyst's knowledge, experience, perceptions and the like.

(v) A fifth secondary conflict arises when it is necessary for the participants to agree on what properties any given actual CMP has. Again they may recursively interact and rely heavily on the knowledge and experience of the mediator and experts to resolve this problem.

(vi) A sixth secondary conflict concerns the degree of presence or absence of any property i in an actual CMP being examined. Again the participants may choose to avoid this problem by employing a nominal '1' (presence) – '0' (absence) scale. Here, however, the participants may be more receptive to statements by the mediator and experts concerning the degree of presence or absence of a property i than they would be to statements on the degree of presence or absence of a characteristic j.

(vii) A seventh secondary conflict arises over the need to set up criteria for selecting a 'best' CMP. For some participants this may be a problem of finding a simple formula for determining an index of inadequacy π, which must also involve a choice of values for the several relevant coefficients and exponents. For more technically-minded participants, this may be a problem of choosing an objective function. In the case of the latter, one participant may wish to employ the ideal p as a target and select that actual

CMP which minimises the sum of weighted deviations (distances) from this ideal. A second may want to identify what he considers to be the two chief properties of the ideal CMP – say, efficiency and security – and then to min the max deviation over these two criteria. A third, perhaps a Nash convert, may be enamoured with the multiplication process and want to maximise the product of efficiency and security, after each has been standardised. Finally, another may simply want to minimise search time, subject to satisficing on each relevant property i of a CMP. After recursive interaction and suggestions, etc., from the analyst or mediator, or both, or by the use of some last resort compromise procedure, the participants may reach agreement on a relevant objective function.

(viii) If, however, the objective function agreed to is of the type (12.8), then it becomes necessary to establish relative weights $\bar{w}_j$ for those characteristics j which the participants and mediator have agreed on as important for conflict situation s. Inevitably, a conflict will arise here since participants will have different perceptions on the relative importance of the characteristics. Again, recursive interaction may be employed in resolving these differences. However, in practice there may need to be several occasions for discussion among participants to avoid inconsistency. For example, in the first round of discussions they may agree to set a high relative weight ($\bar{w}_j$) on the characteristic 'calculating social psychology', but later discover that the score on the 'information available' characteristic is extremely low. When this is the case, they will need to enter another round of discussions in order to adjust (lower) the $\bar{w}_j$ attached to the characteristic 'calculating social psychology', since this characteristic demands that a high level of information be available.

Additionally, in the use of any objective function, there is the problem of establishing values for the parameters and relevant constraints to which all participants are agreeable. For example, consider the simplest objective function – namely that of the satisficer which aims to minimise the search time in finding a $\bar{p} \in \bar{P}$ whose properties i exceed a set of binding magnitudes (constraints). The conflict here concerns the relevant i on which to set constraints. Moreover, there is the problem earlier noted on how the i should be measured, which is presumably solved so that the participants can confront the question: what should the level be for each of these i?

When all the above secondary conflicts are managed, we come to the problem of selecting an actual CMP to resolve the primary conflict.[13] We employ, say, the formula or objective function agreed upon in the above processes, and lo and behold the computer churns out a ranking of actual CMP's from best to worst. But does it?

As is obvious in the workshop approaches of Burton [2] and Kelman [5], and the 'yesable proposition-fractionation' approach of Fisher [3], the actual conflict resolution process, successful or unsuccessful, is a dynamic process. It almost invariably involves learning, especially during the recursive interaction and related activity of the participants. The static framework presented in this chapter ignores completely this basic aspect. Even more fundamental is the recognition that the conflict situation itself is constantly changing, and hence, too, the set of relevant characteristics j against which a conflict management procedure needs to be matched.[14]

VIII CONCLUSIONS

We have just noted a major deficiency of our framework – its static character. In subsequent manuscript, we will be able to remove this deficiency to a small extent – to the extent which is possible with all economic, regional science, and other social science models when they attack other than the most simple problems. Like those who put forth these models, we can claim that until better models are constructed which incorporate the dynamic aspects of reality more effectively, we must make do with the anemically dynamic models that we have, or just rely on intuition in attacking problems. But at least for some practitioners, as for us, intuition plus anemic dynamic models is better than intuition alone.

A more telling ground on which the reader may be inclined to reject our model is that it involves so many factors (variables) that it cannot be made operational. But we would claim otherwise. To support our claim we could go through a parallel situation involving just as many factors (variables), and which in a real sense is operational and leads to clear-cut recommendations. Unfortunately, limited space precludes this. However, we present the supporting materials to this claim in full as an appendix to a chapter in a forthcoming book. There, in considering the decision on the location of an integrated iron and steel works, we have a table which is almost identical to Table 12.1 on Systemic Structural Characteristics. For example, all the items listed under 'Economic System' in Table 12.1 pertain, except that we state the relevant entrepreneurship characteristic differently. Instead of Table 12.3 on Participants' Characteristics we would have a table on Structural Properties of Labour Force (including Entrepreneurship) and Associated Institutions. Many of the items would be similar, such as those under the heading 'Psychological' and 'Level of Education'. Others would be dissimilar, such as those relating to the 'Productivity of Existing Labour Force', 'Wage Rates', and 'Institutions'. The number of items in this table would be at least as many as in Table 12.3. In place of Tables 12.2 and 12.4 there would be tables on

1. Structural properties of the transportation system
2. Structural properties of the market
3. Structural properties of raw material supply

4. Structural properties of the energy system
5. Structural properties of production
6. Structural properties of the capital market
7. Structural properties of impacts on external system

The items listed in these tables are even more numerous than those in Tables 12.2 and 12.4.

All the factors just mentioned are considered by a wise industrial location practitioner when he is hired as a consultant to identify the 'best' location for an integrated iron and steel works. Despite the fact that there are so many factors, as numerous as those that a skilled international mediator must consider when he is assigned the task of identifying the 'best' conflict management procedure(s) to handle a given situation, the skillful and wise industrial location practitioner is able to proceed. He looks for what he would designate the major locational cost differentials among alternative sites for the integrated iron and steel works (including existing locations of works and those outside the region or nation that are likely to go into operation in the future because of their efficiency characteristics). For the United States' situation, these major cost differentials would comprise transport costs on coal, transport costs on ore, transport costs on scrap, and transport costs on finished product to the market. Although labor costs constitute by far the largest cost in United States' iron and steel production, labor costs do not differ from site to site (existing or potential) because of national unionisation of steel labour. On the other hand, labor costs do differ markedly from nation to nation. For the United States, a fifth major cost differential would arise from scale economies, wherein competition from small, non-integrated electric steel operations would be taken into account.

With a half dozen or so major cost differentials in mind, the skilled industrial location practitioner constructs a core framework by which to evaluate alternative locations. Onto this framework he grafts many adjustments and qualifications which come to be additional costs and benefits at each location, often, at best, only crudely estimated. This he does when his mind, sometimes unawaringly, races through the innumerable factors already mentioned and many more. Ultimately, he comes up with a recommendation, as scientific as any that can be achieved for complex problems.

In the same manner we believe that the future international conflict mediator should be able to identify the half dozen or so major factors in a conflict situation, as Fisher [3], for example, does in his balance sheet analysis. These factors then help him identify a set of alternative conflict management procedures to be considered for use in a given conflict situation. Onto each of these procedures he grafts additional evaluative elements, here and there, to take into account a number of less important factors as his mind races through the possible innumerable factors at play. Ultimately, he reaches a conclusion on the procedure he recommends or uses, at least at the start, along with a number of back-up ones.

Notes

1. See Glugiewiez and Isard [4] and Isard and C. Smith (forthcoming book manuscript), *Conflict Analysis and Practical Conflict Management Procedures: An Introduction to Peace Science* (Ballinger, Cambridge, MA).
2. In order to keep our presentation as simple as possible we do not consider in Table 12.5 the possibility that the presence of certain CMP properties may have a negative influence in the use of a given CMP; these properties may escalate a conflict. Such influence may be incorporated in a straightforward manner in the construction of an index and, hence, is left to the reader.
3. In matrix form we have: $\mathbf{w} = \bar{\mathbf{w}}\,[\varrho_{ji}]$ where $\mathbf{w}$ and $\bar{\mathbf{w}}$ are $1 \times m$ and $1 \times n$ row vectors, respectively.
4. In matrix form we have: $\pi = \mathbf{w}\,[\mathbf{p} - \bar{\mathbf{p}}]$ where π is a $1 \times \zeta$ row vector and where the elements in the δ column of the $m \times \zeta$ matrix $[\mathbf{p} - \bar{\mathbf{p}}]$ are $(p_1 - \bar{p}_1^{\delta}), (p_2 - \bar{p}_2^{\delta}), \ldots, (p_m - \bar{p}_m^{\delta})$.
5. Note that the problem of elimination of characteristics (errors of commission as against errors of omission) is not a critical one. For example, if an analyst were to set down characteristics which he considers important, but which participants and mediators in conflict situations never do consider important, they will almost invariably, if not always, be given zero weights. However, there could conceivably be a nuisance value in having such characteristics included in a table.
6. At this point, all that we need to know is that a property i is desirable for characteristic j; we do not need to know its relative importance ϱ_{ji}, which is required in the next step.
7. In effect, we choose among potential candidates for the designation 'ideal CMP' as if we were attaching a cost to the use of any p which has unnecessary properties. Nonetheless, certain problems in the use of this convention may arise, which we do not discuss, since they are relatively minor compared to others we cannot adequately handle.
8. See Saaty [6] and Chapter 5 in Isard and C. Smith (forthcoming book manuscript) for one interesting rationale for the use of such numbers. Or, to achieve still more discrimination, the participants and the mediator may use one or more single dimensional scaling techniques such as Thurstone scaling. See, for example, Baird and Noma [1], ch. 7.
9. Note that when a property i of a CMP is not needed to match any important characteristic j for a given conflict situation s, this means that the characteristics j' that property i does match (i.e., for which it is required) have zero weights for s (i.e., each $\bar{w}_{j'} = 0$) and, hence, by 12.5, $w_i = 0$.
10. A useful step might be to test the sensitivity of the ranking of CMP's along the index of inadequacy to changes in the $\bar{w}_j$ and the ϱ_{ji}. Clearly, it would be desirable to happen upon a conflict situation wherein the ranking of CMP's was invariant to the $\bar{w}_j$ and ϱ_{ji} that different analysts and participants might propose.
11. Equations (12.8) and (12.10) are special cases of the general Minkowski metric

$$\pi^{\delta} = \left(\sum_i w_i (p_i - p_i^{\delta})^q\right) \frac{1}{q} \quad \text{for } (p_i - \bar{p}_i^{\delta}) \geq 0$$

where q is a parameter taking any positive value.

12. Formally speaking, the problem here is to settle on an appropriate mapping θ from characteristics j to properties i.
13. We have indicated that before participants choose a CMP to resolve the primary conflict, they must manage a sequence of preceding conflicts. In effect one might claim that there are many stages involved in managing the entire conflict situation. At each stage there is an outcome o which is, in part, a function of the demands, proposals, or the like of the participants. In our general framework this can be viewed as the joint action a. At each stage, too, there is an environment e which consists mostly of information, setting, and the like, largely set by the mediator and analyst. If the conflict at any stage is managed, then the outcome o (which, for example, might relate to relevant properties, weights, or the like), and the information available become inputs into the outcome function at the next stage to be considered, together with the joint action a at that stage and the e. In this sense, we have a series of outcome functions: $o = f(a, e)$.
14. Again, we note that the above presentation is largely in conformity with traditional scientific method and rational thinking dominant in our Anglo-Saxon culture. But, with ideologically-oriented participants, for example, religious fanatics, risk lovers, and revolutionary Marxists, the sequence of steps in our thinking process may be completely wrong-headed. For situations involving these kinds of participants, the framework and sequence of thought-stages may be totally different.

References

[1] Baird, J. C. and E. Noma, 1978. *Fundamentals of Scaling and Psychophysics* (John Wiley, New York).

[2] Burton, J. W., 1972. 'Resolution of Conflict', *International Studies Quarterly*, vol. 16, no. 1, pp. 5–30.

[3] Fisher, R., 1979. 'Coping with International Conflict', Unpublished Manuscript, February.

[4] Glugiewicz, E. and W. Isard, 1981. 'Conflict Analysis in Regional Science', *Regional Science and Economic Development*, vol. 1, no. 1.

[5] Kelman, H. C., 1976. 'The Problem Solving Workshop: A Social Psychological Contribution to the Resolution of International Conflicts', *Journal of Peace Research*, vol. 13, pp. 79–90.

[6] Saaty, T. L. and M. W. Khouja, 1976. 'A Measure of World Influence', *Journal of Peace Science*, vol. 2, no. 1.

13 Integration of Multiregional Models

I INTRODUCTION

The research proposed here addresses the myriad issues surrounding the design, implementation and evaluation of a system of multiregional models. While the research also aims to make significant contributions to the development of individual models, the major focus will be directed towards the *integration* of these models. In this context, the research will provide many different perspectives, but with a clear desire to move towards the strengthening and articulation of model-to-model linkages.

The field of regional science has long been characterised by the attention it has focused on the development of methods of regional analysis and the concomitant building of operational models. Regional and interregional input-output models, regional and multiregional econometric models, commodity flow and linear programming models have all become familiar features of the accumulated literature of the past twenty-five years. Until rather recently, another prominent feature of this literature was the independence of these models. This independence was viewed not so much in terms of the models' theoretical development and foundations (for, clearly, many of them had common ancestry) but rather in terms of their being regarded as separable for use in forecasting and in impact and policy analyses. Often in attacking any particular problem, one of these models would be judged to be best and the others ignored.

Recently, however, it has become deeply appreciated that decisions of firms and governments with regard to one problem affect the outcomes of decisions taken with regard to other problems, and in turn the outcomes of the former are affected by the latter. This is so whether we consider energy and environmental regulations, foreign competition, unemployment or other issues. In effect, then, the several models, each perhaps most useful for one or a highly restricted set of problems, must be interconnected so that there can be a more effective attack on any given problem facing private and/or public decision makers.

In short, there is an urgent need to consider the effects of decisions upon each other, and to be able to link together the several models which would permit this. Hence, a major objective of this research is to investigate,

---

This chapter was written with David Boyce, Gerald Adams, Steven Caldwell, Ronald Miller and T. R. Lakshmanan. Reprinted from an unpublished Research Proposal submitted to the United States' National Science Foundation in 1983.

formulate and implement linkages among the models, the ways the outputs from one model could be used as inputs into another, and the ways each model could provide consistency checks on estimates derived from another model. A specific objective of this last type is the consistency between regional model outputs and national model inputs, in effect a 'top-down, bottom-up' consistency check.

The philosophy of the proposed research may be traced to Isard's call, some twenty years ago, for consideration of 'Channels of Synthesis'; see Isard (1960). These early notions, while undoubtedly precursors of the present proposal, were general and involved consideration of models at a much more primitive stage of development. In recent years, linkages between two models have been explored by a small, but growing number of authors. For example, see Bargur [3], Mathur [39], Ledent [34], Ledent and Gordon [35], Madden and Batey [38], Schinnar [48], L'Esperance [37], Conway [15], Amano and Fujita [1], Wilson [53], Leontief and Strout [36], Russell, et al. [47] and Golladay and Haveman [26].

These important developments have heralded an important new initiative in regional science: the richness of the linked models has been clearly demonstrated in almost all cases. What is proposed here is a process which would link together all or some of the following *six* component models to form a flexible family of both small and large multiregional models for the United States. The particular models employed would be dependent on the problem or set of problems being addressed:

1. a national econometric model (NATLEC);
2. a multiregion econometric model (REGLEC);
3. an integrated comparative cost, industrial complex, input-output, linear programming model (CICIOP);
4. a commodity flow-transportation network model (TRANS)
5. a micro-simulation demographic migration model (DEMO);
6. a factor demand-investment supply econometric model (FACTIN).

Each of the separate models has reached a rather high stage of development; for operation, however, each depends very crucially on exogenous inputs or very limiting *ceteris paribus* assumptions about the external-to-the model world. One of our major foci in this research endeavor is the improvement in the quality of the functioning environment in which these models operate. The achievement of this improvement will be derived, in part, through new initiatives in model *development* undertaken in the context of research on model *integration*, the two processes being viewed as mutually dependent and supportive. In addition, research on feasible and desirable conflict management (compromise) procedures will be undertaken to provide a sound basis for formulating basic model input assumptions concerning state and federal expenditures and policies.

II RESEARCH OBJECTIVES

The major research objectives of this proposal may be stated as follows:

1. The development of an integrated multiregional model of the United States' economy.
2. The articulation and specification of the linkages among the various models comprising the integrated model, including bottom-up and micro/macro consistency relationships.
3. The continued development of the structuring and functioning of the separate models in response to the linkages available to and from other models.
4. The development and standardisation of a data base necessary to implement a multiregional modelling effort on different spatial levels of aggregation (e.g., Census regions, state, BEA regions), and identification of critical missing data.
5. Evaluation of the integrated model in terms of its strengths and weaknesses, and identification of research problems for high priority consideration.
6. The dissemination to university, industrial and government research and analysis groups of the data, algorithms, computer programs and documentation for application of the models to specific problems.

Clearly, the first four objectives are interdependent. Since the structure will be modular in character, it will not preclude the integration of other models into the system, or the substitution of one module for another competitor of a similar type (e.g., in the choice of the national econometric model which is used). In the development of the linkages among the modules, care will be taken to specify them in the most flexible way possible so as not to render impossible the type of substitution just mentioned.

The other feature of the linkage system, apart from the joining of the models, will be the provision of consistency checks. Given the mixed bottom-up/top-down nature of the econometric modules, consistency checks are of paramount importance. For example, flows of energy products derived from the combined input-output/linear programming model will be compared with those derived from the commodity flow module. Outputs of regional micro-simulation models will be confronted with outputs (regional and national) of the macro econometric models. Procedures for reconciling differences among model outputs will be investigated as a means of imposing greater consistency on the system of models. In the next section, an overview of the operation of the integrated model is provided together with some discussion of the individual models.

III RESEARCH METHODOLOGY

The general outlines of the research methodology will be presented in this section, with major interconnections by a series of figures.

The Overall Structure

The six basic component modules have already been indicated in Section I. There are many ways in which these components may be linked in either a full or partially integrated model. What follows then is a description of just one of the ways. A listing of specific steps to be taken in following this way is presented in Appendix A.

Consider a national econometric module as the starting point. It is proposed to use the Wharton Economic Forecasting Associates model since it already has a strong indirect linkage with a prototype REGLEC model. (The use of other national models may be investigated in later phases of the project).

For a given projection year, NATLEC generates the national magnitudes for the major components of final demand (household expenditures, investment, government expenditures, foreign exports and imports, and inventory change). See Figure 13.1. (At this point the reader may also wish to refer to Figure 14.2 in this volume and the listing of specific steps in Appendix A.) These magnitudes will feed into CICIOP where they will be disaggregated by region to help drive the multiregional input-output model. NATLEC also provides information on material prices, interest rates and other price variables which feed into the REGLEC and FACTIN modules. Finally, this module provides national aggregates against which checks can be made of the sum of regional estimates for similar variables such as output, employment and investment by sector. In this way, top-down/bottom-up consistency will be examined.

In the first round CICIOP provides three basic functions. First, it disaggregates national magnitudes from NATLEC so that they can be used as inputs in its several submodels. Second, via comparative cost and industrial complex studies, it conducts locational studies for key industrial sectors, takes into account their changing locational patterns and thus establishes non-traditional final demand by region. Through the combined operation of an input-output/linear programming submodel, it determines output, labor demand, value added by sector and region, and in the case of energy the optimal pattern of energy production, conversion, interregional transmission and shadow prices. The information generated flows to TRANS, REGLEC, DEMO, and FACTIN.

TRANS takes from CICIOP sector outputs, final demand by region and technical production coefficients by sector and region (in commodity

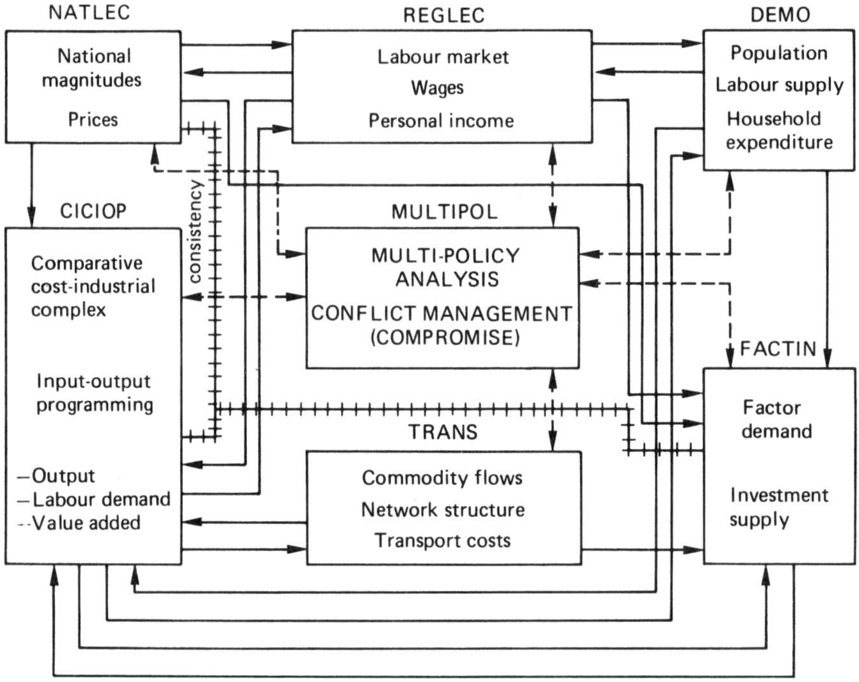

Figure 13.1 Outline of an integrated multiregion model: its components and linkages

terms), which it converts into a commodity classification. TRANS then generates interregional commodity flows using a reasonable constraint regarding the crosshauling of commodities observed to occur in aggregated commodity flow data. The flows are specified by carrier and by mode. Transportation costs for the various commodities are also generated in this module. The information generated flows to CICIOP and FACTIN as indicated below.

To determine regional wage rates, REGLEC takes from CICIOP labour requirements by sector and region, and non-wage income by region, and, for the first round only, uses standard demographic projections of labour force by occupation and skill (by demographic group). However, in the case of manufacturing industries which are highly unionised and where the labour market is predominantly a national market, REGLEC receives from NATLEC wage rates by manufacturing sector and passes them onto DEMO. Employment and unemployment are still, however, regionally determined with the use of labour requirements from CICIOP. Also, REGLEC takes prices from NATLEC and generates, to the extent possible, regional price vectors. REGLEC reflects a bottom-up approach to econometric modelling. Potential extensions to this approach would encompass a regional government sector, and investment and sector output projections by region.

DEMO takes from CICIOP regional industrial structure and labour requirements by sector by region and from REGLEC wage rates by region and occupation, and available regional price vectors. It approaches the estimation of population characteristics (including migration) through the use of micro-modelling and simulation techniques. As a result, labour supply by region is obtained according to a highly disaggregated set of specifications, and the associated elements of household income, expenditures and savings are generated.

Given sector outputs and energy shadow prices from CICIOP, and relevant factor prices from NATLEC and REGLEC, FACTIN determines by sector and region the demand for capital, labour, energy and materials, employing a translog cost function. Taking investment magnitudes in national industries from NATLEC, the investment in each production sector by region is determined based on a combination of factors from location and investment theories and, in part, on the particular characteristics of each sector. Its regional investment estimates are in part conditioned by household savings estimates from DEMO.

Finally, there is added to the six modules a multidecision analysis and conflict management (compromise) submodule which mediates conflicts among the different interest groups and regions. This is a new exploratory area and only a small amount of resources will be invested here. However, this research is essential to (1) CICIOP, in estimating (in its first and subsequent runs) the government expenditure vectors as components of final demand by regions; (2) TRANS, for specifying assumptions as to the nature and extent of the federal programme for transportation in the future; (3) DEMO, for estimating tax, wage and income policies, and future housing and social infrastructure investments as they affect migration; (4) REGLEC, for specifying assumptions on unemployment compensation, social security, and like policies; (5) FACTIN, for estimating future energy and other prices influenced by governments at all levels; and (6) NATLEC, for specifying reasonable assumptions regarding federal and state policies.

Once the first runs of the above six modules have been carried through, it is necessary to perform consistency checks on the estimates of certain key magnitudes. Regional outputs by sector, as generated by CICIOP, must be consistent with national outputs; hence adjustment in both national and regional outputs may be necessary. Regional employment by sector as generated by CICIOP and FACTIN must be consistent with national employment by sector; again adjustment in all these estimates may be necessary. Regional investment in non-national industries generated by FACTIN and other information on investment by region generated by CICIOP, TRANS and possibly REGLEC must also be consistent with national investments; again adjustments in all these data may be necessary.

The above discussion pertains to the first runs and the consistency checks of the several modules. However, once the modules have been run, additional inputs to each become available. First CICIOP receives from NATLEC the first round data appropriately modified for consistency. It receives from FACTIN information on factor demands by sector and region that permit changes in its technical coefficients. From TRANS it obtains information on transportation costs, and from REGLEC wage differentials to utilise in its comparative cost and industrial complex studies projecting the spatial pattern of basic manufacturing activities. It also receives estimates of interregional commodity flows (interregional trade coefficients) from TRANS and makes adjustments to its own projections to achieve consistency. From REGLEC, it receives personal income estimates; and from DEMO population and labour force estimates, and regional household expenditure to help determine its final demand vector. Other diverse useful inputs are also received.

In the second run, TRANS receives from CICIOP the first run data modified, but also particular interregional flows more accurately determinable by CICIOP and other modules, or as required to achieve consistency. REGLEC in the second round receives labour supply data from DEMO and modified data from NATLEC and CICIOP, the labour demand information by sector and region generated by FACTIN already having altered the input-output labour coefficients of CICIOP. Also, in the second round, DEMO gets modified data on wage rates and prices from REGLEC and on outputs and labour requirements from CICIOP. FACTIN also receives first round data appropriately modified for the second round. At the end of the second round there must also be a consistency check such as those carried out at the end of the first round.

Third and subsequent rounds proceed in the same manner as the second round. A more systematic statement of the sequence of operation depicted above is presented in the set of steps listed in Appendix A.

The Details of the Modules

We now describe in more detail the basic elements of each of the six modules and the multipolicy formation research. Consistent with the sequence of operations described in the previous section we begin with CICIOP.

CICIOP

As already indicated, the comparative cost, industrial complex, input-output, programming model (CICIOP) provides three basic functions in the overall integrated multiregional model: (1) to disaggregate national variables from the national econometric module (NATLEC) by region so that they can be used for multiregional analysis; (2) to provide locational studies of key industrial sectors and corresponding adjustments in final demands (and total outputs) by region; and (3) to generate economic output, labour demand, and value added by sector and region. The overall structure of CICIOP and inputs from other modules is shown in Figure 13.2.

The core of CICIOP is a multiregional input-output model, which will ultimately be specified as a full fifty-region (state), eighty-five sector model of the US. (The current version of the US Multiregional Input-Output Model (MRIO) developed by Polenske [44] is envisioned as the starting point. The nine-region, nineteen sector version can serve as an initial 'pilot model.' The MRIO model is being updated by Jack Faucett Associates and will be specified as the full fifty-state, eighty-five sector model, but can be aggregated according to policy scenarios under study.) To the extent possible, the 'Chenery–Moses type' of interregional trade coefficients (ignoring origination region and destination sector of commodity shipments) assumed in the MRIO model and preserved in the TRANS module will be replaced with full interregional trade coefficients (IRIO).

Dantzig and Parikh [18] suggest embedding an input-output model in the constraint matrix of a linear programming model as a means of (1) permitting technological substitution and (2) accounting for environmental, energy, and other technological constraints, both of which are ignored in the traditional Leontief model. In CICIOP we follow Dantzig's paradigm and, further, add constraints specifying additional sectoral detail in certain key industry sectors, e.g., energy flows in the economy.

In using CICIOP in the context of the full multiregional model, three major considerations are important: (1) specification of regional final demands as an exogenous input to the multiregional input-output model, (2) modification of technology coefficients as a function of relative prices of factor inputs, and (3) modification of interregional trade coefficients. The information for all three of the inputs come directly from other modules, the first from NATLEC, DEMO and REGLEC; the second from FACTIN, and the third from TRANS. In the following we discuss these three steps and the ultimate structure of CICIOP in the context of these other modules.

Specification of Regional Final Demand

Figure 13.2 indicates that the national econometric model (NATLEC) produces initial estimates of magnitudes for both 'traditional' and 'nontraditional' final demand. For the former these consist of capital investment, federal government (military and civilian) and household expenditure as well as foreign imports and exports and changes in industry

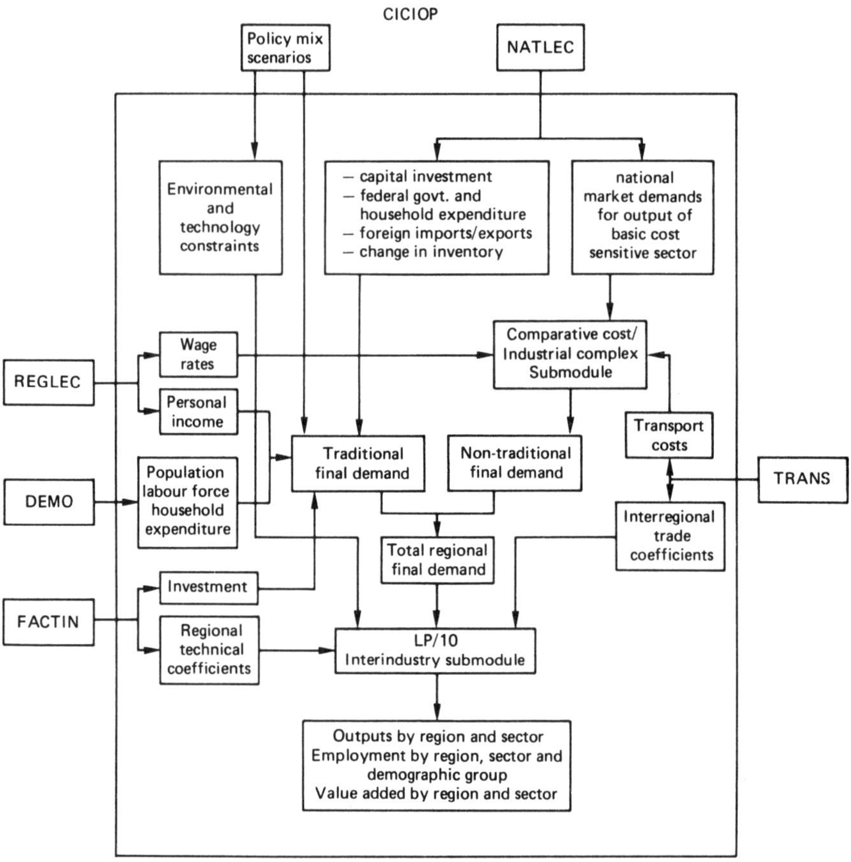

Figure 13.2 Inputs to and internal structure of CICIOP

inventories. These are supplemented after the first run by population and labour force estimates by region from DEMO and personal income estimates by region from REGLEC. The regional final demands must also be stated after adjustments to (1) factor out imports expected to come from other regions and (2) increase the level of deliveries a region is expected to export to other regions. Isard *et al.* [44] and Polenske [29] discuss this process in detail.

Magnitudes from NATLEC for non-traditional final demands consist of national markets (demands) for the output of basic cost-sensitive sectors (iron and steel, aluminum, petrochemicals, etc.). These are supplemented by transport costs by commodity and mode from TRANS and wage rate differentials by REGLEC, thereby permitting the conduct of comparative cost and the industrial complex studies. The input requirements for these basic cost-sensitive sectors (whose levels are specified by region) are recorded in new columns in the final demand section for each region. Blair and Miller [7] show that non-traditional final demands can be posed equivalently as a modification of the technical coefficients (e.g., a new sector) and a new final demand for the new activity.

Input-Output Technology Coefficients

Linkage of CICIOP with the factor substitution submodule of FACTIN permits a specification of input-output technical coefficients as a function of the relative prices, which addresses a principal weakness of the traditional static Leontief model.

Further, in order to focus on specific regional policy issues (or the impact of national policy on regions), it may be desirable to remove from the standard input-output technology matrix of input coefficients, certain sectors of particular interest, i.e., those that might vary in response to specified policy initiatives, and respecify them. By posing the input-output impact analysis problem as a linear program, as in Dantzig and Parikh [18] or Goettle *et al.* [25], we can add additional constraints to account for regional production capacity, environmental load, and employment (manpower) constraints, as well as consistency in specifying, for example, energy flows. For example, an LP problem may be to Min $Z = c'X$ subject to the traditional input-output constraints $[(I-A)X \leqslant Y]$, capacity constraints $[X \leqslant k]$, pollution emission constraints $[PX \leqslant B]$, and others.

Incorporation of Interregional Trade Coefficients

The TRANS module generates estimates of commodity flows between regions. These estimates are then to be used to modify or replace the estimates in the MRIO framework employed in the first-run of CICIOP. Both estimates use the Chenery-Moses assumption, namely,

$$a_{ij}^{LM} = t_i^{LM}\, a_{ij}^{M}$$

where t_i^{LM} is the fraction of the pool of the input good i in region M (regardless of consuming sector) coming from region L. Additionally, for some flows complete sector origin and destination information may be available from other sources which permits construction of full or 'Isard Ideal' IRIO coefficients:

$$\tilde{a}_{ij}^{LM} = t_{ij}^{LM}\, a_{ij}^{M}$$

where t_{ij}^{LM} is the origin/destination specific trade coefficient (sector i located in region L to sector j located in region M). Such flows may be entered as constraints in the TRANS module.

Basic Research Issues

In addition to the issues involved in integrating CICIOP with the other multiregional models, which are discussed elsewhere in this proposal, two basic research issues in the assembly of CICIOP itself will be addressed in the proposed research.

(1) A first basic research issue concerns alternatives for incorporating dynamic characteristics in CICIOP. As noted, CICIOP will be solved as a LP problem with the hybrid MRIO-IRIO input-output coefficients matrix posed as the principal set of constraints. Additional constraints, as suggested earlier, include environmental, employment, and other technology constraints which, along with the inputs from FACTIN and NATLEC and other modules can be specified as a function of time. In order to incorporate consistently these dynamic inputs, the LP can be expanded to a number of planning periods ($t = 1, \ldots, \tau$). The objective function then becomes Min $Z = c' [X(1) + X(2) + \ldots + X(\tau)]$ subject to a set of constraints for each planning period.

(2) A second basic research issue concerns the development of a hybrid MRIO-IRIO input-output coefficients matrix. Neither the validity of the use of IRIO coefficients nor the adequacy of MRIO coefficients as a suitable approximation of full IRIO coefficients has ever been addressed. It is an important research area which we wish to investigate in a systematic fashion. Also as the assembly of CICIOP progresses, several additional research issues may become important, depending upon availability of data and/or the specific policy issues selected for study. These topics include the role of commodity-by-industry accounts, procedures for non-survey estimation of data that are unavailable, the appropriate level of aggregation for specific policy problems, and the development of methods for incorporating additional sector-specific considerations (e.g., the LP approach to dealing with energy flows discussed earlier).

TRANS

The commodity flow-transportation network module provides three basic functions in the overall integrated multiregional model: (1) a consistent set of interregional commodity flows (interregional trade coefficients) for CICIOP; (2) estimates of interregional shipment costs and service characteristics by commodity for CICIOP; and (3) information on investment requirements in the transportation sector and on transport costs to FACTIN. The specific objectives are:

1. represent the commodity transportation system from the viewpoints of both industrial sectors (shippers) and the transportation industry (carriers);

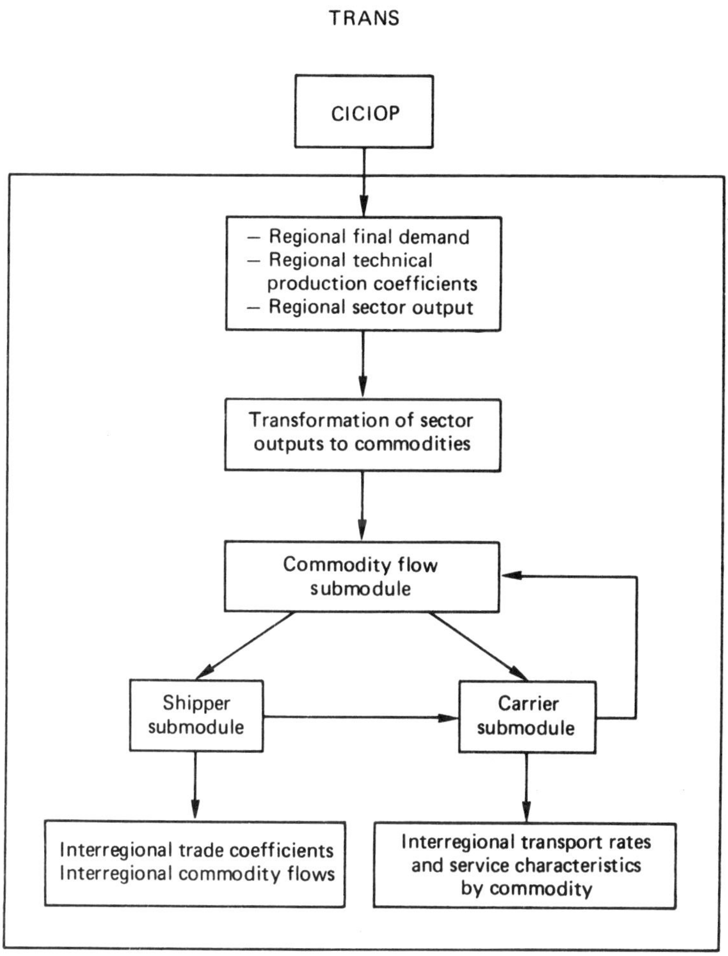

Figure 13.3 Inputs to and internal structure of TRANS

2. represent the pattern of interregional commodity flows corresponding to sales and purchases of industrial sectors;
3. estimate interregional shipment costs (rates) and service characteristics from the viewpoint of industrial sectors;
4. estimate the demand for transportation services from the viewpoint of the transportation industry.

The inputs required for TRANS are from CICIOP, and consist of regional technical input-output coefficients by region, regional output and regional final demand (see Figure 13.3).

In contrast to its predecessor described in Boyce and Hewings [11], the

TRANS module described in this proposal seeks to represent in a behaviorally realistic but aggregated manner the pattern of interregional commodity flows and the underlying transportation system. The key features of the present module shown in Figure 13.3 may be described as follows:

1. Representation of *flows as commodities* using the commodity-industry input-output model, and of *crosshauling* as the dispersion of interregional flows.
2. Specification of *shipper networks* for each commodity class representing the multimodal transportation system as viewed by industrial sectors at the regional level; industrial sectors are assumed to be cost minimisers with respect to the shipper network.
3. Specification of *carrier networks* representing the transportation system from the viewpoint of major individual carriers or groups of smaller carriers; each carrier is assumed to be a profit maximiser with respect to its network.
4. Specification of the relationships between the shipper and carrier networks: shippers specify demands for transportation services on the carrier network; carriers specify rates and service characteristics on the shipper network.

The concept of shipper and carrier networks as a way of representing more appropriately the complexities of the transportation system was devised by Friesz and Viton [21]. The concept is generalised here to more than one commodity and reinterpreted in the input-output context.

Shipper and Carrier Networks

Each shipper network corresponds to a commodity class and consists of (1) one or more nodes for each region and each available mode (truck, rail, inland water, pipeline); (2) nodes representing the connections between modes or major carriers within each mode; (3) links connecting the nodes between which transportation services are available over the more detailed carrier network.

Paths connecting an origin region to a destination region show feasible shipment possibilities in terms of modes or combinations of modes, but not in terms of actual routes. Corresponding to each link in the shipper network are a freight rate and service characteristics (transit time, delay, loss and damage, etc.), as specified or determined by a carrier or group of carriers. A detailed shipper network for the US might correspond to the 510 state economic areas or the 175 Bureau of Economic Analysis regions.

Each carrier network consists of (1) nodes corresponding to the regional centroids on the shipper network; (2) nodes corresponding to intersections of all significant facilities or services; (3) interline points between this carrier and other carriers or modes. Links connect pairs of nodes which are physically connected on the actual transportation system. For each link in a carrier network, a cost function states the relationship between total operating cost and link operations, including the total flow by major commodity type, if appropriate. This total cost function is assumed to be an increasing function of flow, although average cost may decrease with increasing flow until congestion causes costs to increase.

Shipper and Carrier Models

Shippers in the aggregate are assumed to select the path over the shipper network that minimises a generalised function of rate and service characteristics. These shipper path decisions, which correspond to mode and carrier alternatives, not to physical routes, in turn determine the demand over the more detailed carrier network.

Each major carrier or group of carriers represented by a carrier network is assumed to route the demands for its transportation services so as to maximise its profits. In addition to determining the routing through the carrier network, the carrier model determines the rates and levels of service on the shipper network. Rates are determined as a function of variable cost from shipper origin to destination, variable costs on competing modes or routes, flow on the route or routes and commodity type. Regulatory relationships of rates to variable costs may be applied in the absence of competition from other modes.

Commodity Flow Model

The two network and two behavioural models stated above provide the basis for restating the Leontief–Strout–Wilson input-output commodity flow model; see Wilson [53]. Unlike its predecessors, however, this model is stated in commodity terms using the commodity-industrial input-output system; see Gigantes [24].

The commodity flow model consists of three elements plus an assumption about shipper behaviour. The first element concerns regional conservation of flow. The second concerns the representation of crosshauling in the model. The third concerns output constraints by commodity and region to be jointly determined with CICIOP. Conservation of flow at the regional level requires that shipments of a commodity into a region plus regional production of the commodity must equal regional intermediate and final demand for the commodity plus shipments out of the region. If regional production of each commodity is to be determined by the model, then

$$\sum_{I=1}^{U} s_h^{IJ} = \sum_{j=1}^{m} \left(\sum_{i=1}^{n} b_{hi} d_{ij} \right) \sum_{K=1}^{U} s_j^{JK} + y_h \quad \text{for all } h, J$$

shipments of commodity h into J	=	requirements of commodity h for production of all commodities $j = 1, \ldots m$	+	final demand for commodity h

In the above simplified equation, production of commodity h in region J implicitly balances shipments of h out of region J less the region's own contribution to demand. The input-output coefficients $b_{hi}d_{ij}$ represent the input of commodity h required to produce one unit of commodity j by all industrial sectors $i = 1, \ldots$ n.

A phenomenon, often referred to as *crosshauling*, is the apparent flow of a commodity in both directions between a given pair of regions. Such observed flows can arise in survey data for two reasons. First, different commodities in the same commodity class may well flow in both directions. If the level of commodity aggregation is high, then the observed amount of crosshauling will also be high, as compared with more detailed commodity flows. Likewise, as the level of spatial aggregation is increased, the level of apparent crosshauling will also increase. Second, crosshauling may actually occur because of a variety of market and institutional factors including established trading relationships, pricing policies and attempts to insure a stable source of supply.

An important question for the commodity flow model is how to insure that the appropriate level of this phenomenon is represented in the model. If crosshauling is viewed as primarily the result of aggregation, then a statistical approach may be readily justified to impose a similar effect on the model. The level of crosshauling, or more generally the dispersion of flows from a cost minimising solution, may be described by the entropy function:

$$s_h = - \sum_{J=1}^{U} \sum_{K=1}^{U} s_h^{JK} \ln(s_h^{JK}) \qquad \text{for all } h$$

By constraining the commodity flow model to the observed level of crosshauling, the model will more adequately describe the observed flow phenomena.

These two relationships, together with output constraints and the shipper behavior assumption of cost minimisation, provide the basis for deriving the commodity flow model. If the transportation costs on the shipper network are independent of flow, then the solution to this problem will provide a commodity flow model with the desired characteristics.

Basic Research Issues

The TRANS module is perhaps the least developed of all the modules represented in the proposal. A large body of results from network equilibrium theory and models, as well as recent advances in freight network modelling, can be readily applied to bring this module to an operational status. Nevertheless, the first major research issue is the formulation and solution of an operational model; many details related to this subject are provided in Fisk and Boyce [20].

A second major research issue concerns appropriate levels of aggregation of the submodules. Because of the way that the shipper submodule is defined, regions as large as Census regions or as small as state economic areas could be defined, while utilising a detailed definition of a carrier network. Thus, the issue of the best level of aggregation in relation to the other modules will also be a central research issue.

REGLEC

One of the shortcomings of the modules presented so far is that they fail to incorporate the impact of market processes which determine wages and the prices of goods and services. A basic function of REGLEC (the multiregional econometric module) is to overcome this deficiency by explicitly deriving wages and prices which largely, but not entirely, equilibrate supply and demand in labour and products markets. REGLEC also functions to provide personal income estimates by region.

This module describes the labour market and the determination of prices. Drawing heavily on the work of Milne *et al.* [42], Courbis [16], and Fromm *et al.* [22] a submodel is to be constructed which includes labour demand for each sector, labour supply and unemployment, and the wage, labour cost and price determination mechanism. If desired, additional breakdowns by skill class for the relevant magnitudes can be estimated at the level of each sector. Input requirements (see Figure 13.4) are wage rates for manufacturing (national) industries from NATLEC, levels of output from CICIOP and NATLEC, labour requirements (demands) from CICIOP (as modified by FACTIN), the regional labour force structure supply from DEMO, and relevant prices from NATLEC. These inputs are provided on an annual basis in order to yield estimated wage rates and prices by region for each year of the projection.

To investigate the operation of the US labour market in a realistic manner, we recognise that for a number of key manufacturing industries the labour model is a national one oriented to national union activity and national companies and/or industrial operations. Hence, for these industries, the determination of the wage rate and employment and unemployment levels must be examined in the context of NATLEC, our national

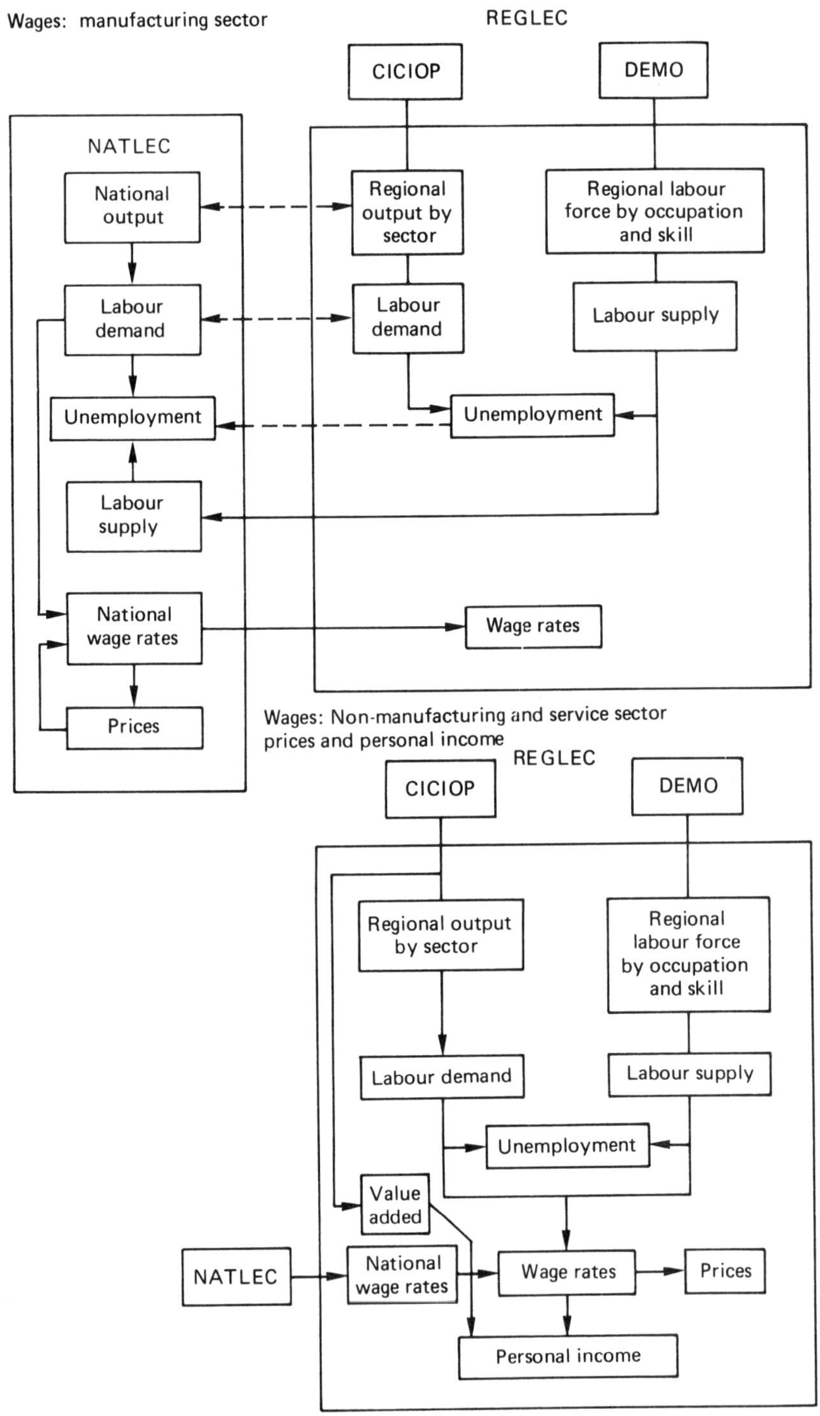

Figure 13.4 Inputs to and internal structure of REGLEC

model. This is also the case for the determination of a number of prices. Accordingly, the REGLEC module must be more closely linked with the NATLEC module than the other modules; it must be so to furnish wage rates, wage rate differentials, employment and unemployment to the DEMO module.

A framework embodying a two-tier labour and product market is used. One tier, nationally oriented, covers the major manufacturing industries. Product demand is determined on the national level. Employment and wage decisions are made largely on a national level. The second tier, locally oriented, encompasses some of nondurable manufacturing, agriculture, and services where labour is likely to be nonunionised and where wage determination tends in the long run to equilibrate the market. Price determination is also incorporated in some of the same two-tier market processes with prices of agricultural and mineral products and most manufactures determined on the national level and prices of services determined on the regional level. This produces a mixed top-down/bottom-up model which reflects market processes with a substantial degree of realism.

Approach

For manufacturing industry i, wages for labour in skill group d for each year are determined in NATLEC as a function of (1) wages in previous years, (2) a lagged structure of the ratio of demand for that labour class by all manufacturing industries in the given year to total national supply of that labour in the several previous years, (3) change in average productivity in industry i over the last several years, and (4) change in average productivity over the last several years in all manufacturing industries. That is,

$$w_{i,d}(t) = \alpha_{0,d} + \alpha_{1,d} \sum_{\theta=t-1}^{t-\tau} w_{i,d}(\theta) + \alpha_{2,d} \sum_{\theta=t-1}^{t-\bar{\tau}} \frac{\sum_i D^L i,\, d^{(\theta)}}{S_d^L(\theta-1)}$$

$$+ \alpha_{3,d} \sum_{\theta=t-1}^{t-\tilde{\tau}} \left(\frac{x_i(\theta)}{E_i(\theta)} - \frac{x_i(\theta-1)}{E_i(\theta-1)} \right) \Big/ \tilde{\tau}$$

$$+ \alpha_{4,d} \sum_{\theta=t-1}^{t-\hat{\tau}} \left(\frac{\sum_i x_i(\theta)}{\sum_i E_i(\theta)} - \frac{\sum_i x_i(\theta-1)}{\sum_i E_i(\theta-1)} \right) \Big/ \hat{\tau}$$

where

$w_{i,d}(t)$ = average wage of skill group d in manufacturing industry i in year t.

$D^L i,\, d(\theta)$ = total requirements (demand) of labour of skill group d by industry i in year θ.

$S^L_d(\theta)$ = total supply of labour in skill group d in year θ.

$X_i(\theta)$ = total output of manufacturing industry i in year θ.

$E_i(\theta)$ = employment in industry i in year θ.

$\alpha_{0,d}, \alpha_{1,d}, \alpha_{2,d}, \alpha_{3,d}$ and $\alpha_{4,d}$ are parameters to be estimated.

For non-manufacturing and service industry j, wage rates for labour in skill group d in region J for each year are determined in REGLEC as a function of (1) wage rates over the past years, (2) the average of wages in that group over national industries weighted by their presence in region J, represented by X^J_i in the following equation, and (3) the ratio of the demand for the labour in industry j to the total supply of that labour in region J in the previous year.

$$w^J_{j,d}(t) = \beta^J_{0,d} + \beta^J_{1,d} \sum_{\theta=t-1}^{t-\tau} w^J_{j,d}(\theta) + \beta^J_{2,d} \frac{\sum_i X^J_i w_{i,d}(t)}{\sum_i X^J_i}$$

$$+ \beta^J_{3,d} \frac{\sum_j {}_D L^J_{j,d}(t) + \sum_i {}_D L^J_{i,d}(t)}{{}_S L^J_d(t-1)}$$

where all symbols represent variables already defined, except that they have the superscript $J(J = A, \ldots, U)$ and relate to that region; and where $\beta^J_{0,d}, \beta^J_{1,d}, \beta^J_{2,d}, \beta^J_{3,d}$, are estimated parameters. In addition to the above, REGLEC may also estimate its own labour demand equations (by sector, or by sector and skill group) using nested CES functions, or by translog functions, where the arguments are capital, value-added and perhaps energy or raw materials. This will provide a consistency check with FACTIN and, upon summing across regions, for NATLEC.

In estimating personal income by region, REGLEC receives value added data (excluding wages) by sector from CICIOP, as reference magnitudes to check its allocation of nonwage income elements. To these elements it adds wage income derived by summing the product of wages estimated by skill group and employment, all adjusted appropriately for taxes and transfer payments. Also, unemployment equations, as appropriate, will be estimated.

Basic Research Issues

(1) *Wage Interaction, National and Regional.* As already mentioned, localised employers cannot make wage decisions independently of the competition of the manufacturing industries. Because they can pay higher wages, the latter have first claim on the available labour force. The

interactions between the regional labour market and the national labour market call for extensive additional study. Research on price variables, their determination and their effects on other variables will be undertaken.

(2) *Aggregation and Model Solution.* The model system proposes a unique blend of linear and nonlinear specifications involving several major but closely interrelated blocks of equations. The approach involves, as appropriate, top-down relationships operating at an aggregate level as well as bottom-up highly detailed specifications, the latter to be aggregated for consistency and solution purposes. Parts of the system call on linear programming solution algorithms. While the solution of such a system is now well within the limits of modern software and hardware technology, solution procedures must be devised which meet the following aims:

1. integration of data base and data base upkeep with model simulation and forecasts;
2. solution of the system using principally convergent iterative procedures integrated with matrix inversion approaches;
3. solution algorithms which will permit maximum transparency and ease of modification. (each module should be approachable by itself for separate solution and adjustment);
4. maximum flexibility in use, for adjustment, introduction of alternative assumptions, and flexible simulation of alternatives;
5. ease of display capability for major aggregates and for detailed data;
6. rapid solution capability on time share terminals and maximum use of distributed micro computer facilities.

The development of the solution procedures, principally using program modules now available from a number of sources, will represent one of the major research areas to be carried on simultaneously with the estimation of the model system itself.

DEMO

The specific objectives of the population module (DEMO) arc:

1. to represent the multiregional population and link it to the multiregional economy;
2. to simulate the household distribution of income (and other distributional outcomes);
3. to estimate regional population aggregates (e.g., consumption, labour supply, fertility, and migration) while taking into account the behavioral heterogeneity of each regional population.

In attaining these objectives DEMO provides regional labour supply by occupation, skill and demographic composition (age, race, sex, household status) to REGLEC, regional population and labor force to CICIOP, population and labour force and household by region to FACTIN, and household income and expenditures for CICIOP. In turn, DEMO receives from CICIOP outputs by sector by region, labour requirements by sector and region (as modified by FACTIN), possibly pollution emissions by type by region, and unearned income elements, and from REGLEC, wage rates and prices (see Figure 13.5).

Approach

The DEMO model will directly simulate the basic demographic behaviours – birth, death, marriage, divorce and interregional migration – for each individual and household in a sample that has been constructed to represent the full multiregional population. In addition, DEMO will incorporate a broader set of individual and household choice processes: education, occupation, labour supply, earnings, consumption, and savings.

By tabulating individual and household outcomes from the model according to region of residence and other socioeconomic characteristics, DEMO will yield detailed breakdowns of the multiregional population; hence DEMO will have the capability to address a wide range of distributional issues. By aggregating outcomes over all members of the sample and weighting up to population totals, DEMO also will be able to estimate aggregate quantities by region (e.g., aggregate labour supply and household consumption). Consequently, DEMO will enable the integrated system of models to connect regional economic policies or unguided trends to both aggregate and distributional outcomes at the household and individual levels.

Many issues that can be addressed only by a micro model such as DEMO involve heterogeneous responses. In so far as behavioural responses of individuals or households to an economic constraint are heterogeneous, the aggregate population response to the economic environment will depend on the composition of the population, as well as its size. In such cases, DEMO is a singularly appropriate tool with which to model aggregate population behaviour, as well as its distribution.

We will not attempt to build a wholly new microanalytic population model. Rather, our strategy will be to adapt an existing microanalytic model to our specific purposes. Since several of the more promising micro models are rapidly evolving, it would be preferable to make the final selection when the project is about to begin. However, it seems clear that one of the three currently operating versions of MICROSIM (Policy Research Group, [45]; [46]) will be most suitable for adoption as the basis for DEMO.

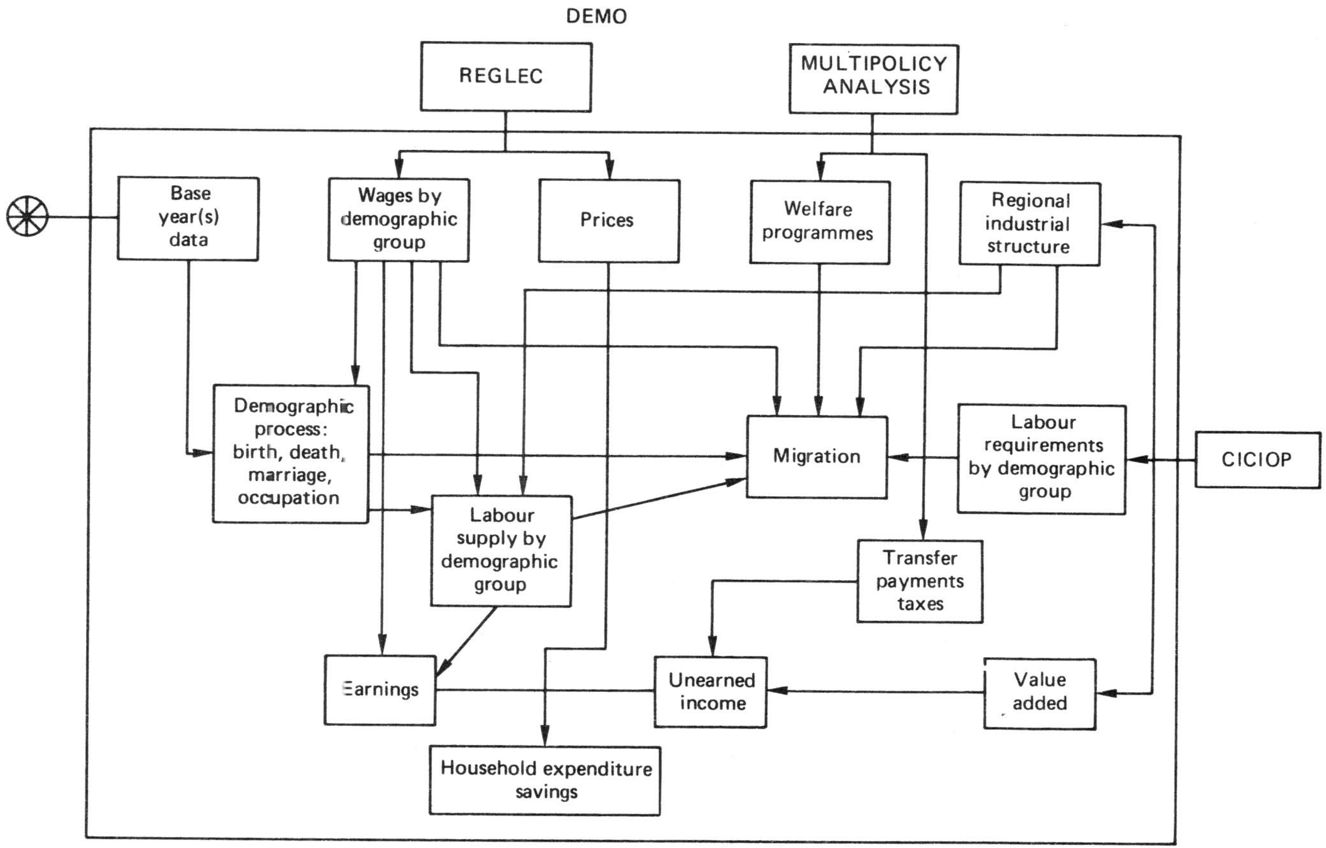

Figure 13.5 Inputs to and internal structure of DEMO

Labour Supply

Labour supply at the micro level is assumed generally to be a function of family income, wage rates and a vector of personal socioeconomic characteristics in a specification based on a utility maximisation approach involving a labour versus leisure tradeoff. Although the various attempts at developing labour supply components for microsimulation models are significant, they do have certain shortcomings, some of which we propose to overcome. First, none of these supply models contain any significant feedback effects from their labour supply module to the macro model environment in which they operate. For example, although a macro model may supply the relevant micro model with, say, labour demand and wage rates, there is no equilibrating process which further adjusts these wage rates as a result of continuing labour demand and labour supply adjustments. This shall be done in our integrated model. Second, in order to enhance the accuracy of predictions from the proposed system of integrated models, there will be additional disaggregation of labour supply response functions by class of workers (e.g., elderly, teenagers, married women, etc.) and by region of the nation.

Household Expenditures

In developing the household expenditure module for our dynamic microsimulation model, we do not propose to 'reinvent the wheel' by starting from scratch to develop a new module. To the extent possible, we propose to draw on work that has already been accomplished. We will do intensive testing of the best available expenditure module, which apparently has not yet been done, and modify only those equations which do not perform well or appear to be inadequate in other ways. The household expenditures model proposed for inclusion in DEMO will provide CICIOP and NATLEC with consistent sets of estimates for personal consumption expenditures. These estimates will be sensitive to socioeconomic distributional impacts that might occur from a variety of policy alternatives (e.g., tax policies, wages and income policies, prices, etc.) which will be generated by the Multipolicy Analysis Submodule. The necessary socioeconomic household data (e.g., age, race, sex, family composition, assets, etc.) will be obtained directly from the micro model.

Migration

The linking of CICIOP, FACTIN, REGLEC and TRANS leads to changes in the spatial distribution of production and the spatial pattern of employment demand, wages, pollution, and other factors. The population then adjusts to these new signals, in large part by means of interregional migration.

Evidence is accumulating on the magnitude of migration responses to the relevant socioeconomic factors. Rather than undertake extensive new research, we intend to draw from the available literature in so far as possible. Our approach would be to specify a discrete choice (logit or probit) model in which the probability for each household of leaving its current region of residence would be a function of a vector of household characteristics (e.g., age, sex, family cooperation, race, skill) and a vector of characteristics describing the socioeconomic environment of the current region (e.g., industrial structure, unemployment, relative wage rates). Given a pool of households from this first stage that migrate from their current region, the second stage would assign each household characteristics and the characteristics of the competing regions. Parameter estimates for the two sets of leaving and destination functions would be drawn in so far as possible from the current literature.

Basic Research Issues

Individuals seldom respond homogeneously to economic constraints and opportunities. For example, the underlying utility maximisation hypothesis with linear budget constraints does not recognise certain existing nonlinearities in the labour market. Important discontinuities or other nonlinearities exist in work related income transfer programmes, in the structure of progressive federal and state income taxes, and in social security payroll taxes, to name just a few. Not recognising these nonlinearities can be expected to introduce a bias into a model's estimated labour supply responses. Similarly, household consumption and savings behaviours exhibit nonlinearities in response to income and price constraints. Finally, individual demographic and social behaviours, such as marriage, fertility, divorce, education and migration exhibit such nonlinearities. By representating the joint distribution of individual and household characteristics, the DEMO model will be capable of capturing these nonlinearities as they determine both aggregate and distributional population behaviour.

FACTIN

The factor demand and investment module (FACTIN) has two submodules (see Figure 13.6) each with separate functions in the overall integrated multiregional model.

1. The dynamic factor demand module estimates (at regional/sectoral levels) demands for labour, capital, materials and energy (and categories thereof) in a dynamic adjustment framework, to be used to modify the regional coefficients of CICIOP and its labour requirements projections.

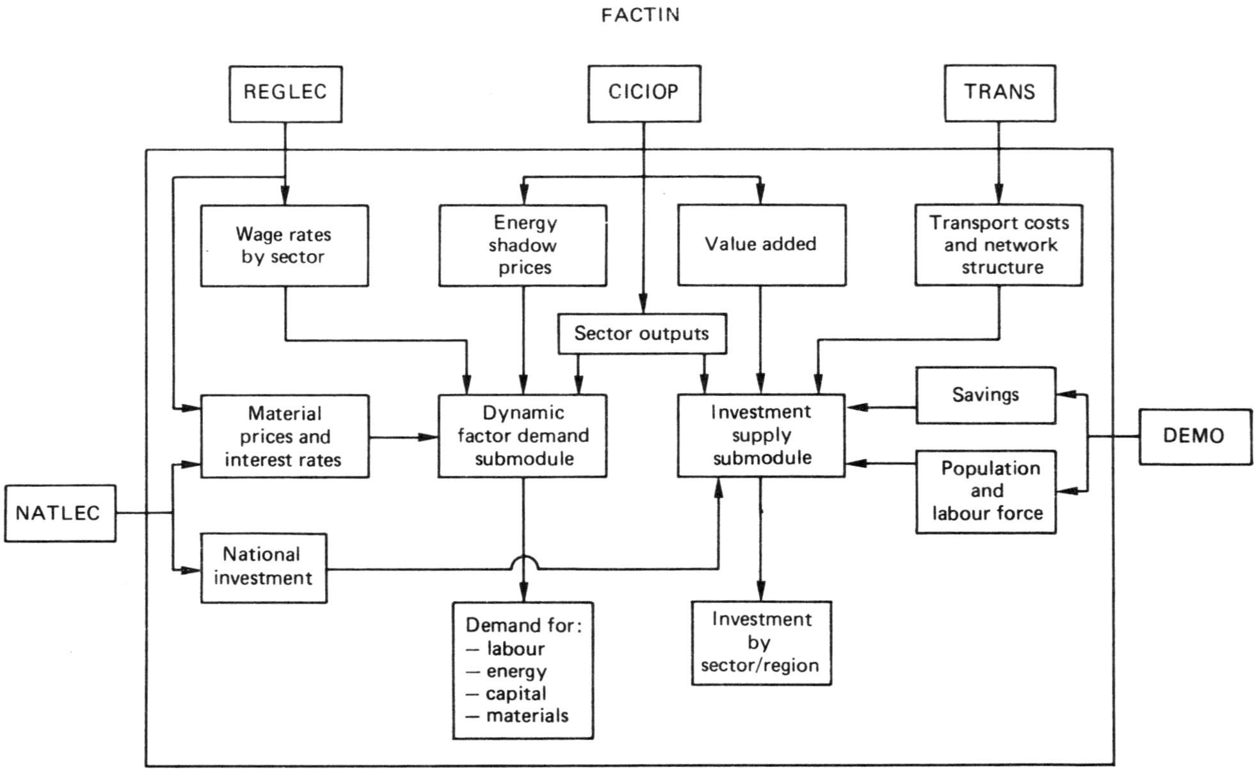

Figure 13.6 Inputs to and internal structure of FACTIN

2. The investment supply module estimates regional investment supply by sector for CICIOP.

Inputs for the factor demand module are energy shadow prices from CICIOP, wage rates from REGLEC, and material prices and interest rates from NATLEC and REGLEC. For the investment supply module, value added is provided by CICIOP and REGLEC, transport costs and network structure by TRANS, investment by national industry by NATLEC, and savings and other region specific characteristics by DEMO. Sectoral output – for both sub-modules – is obtained from CICIOP.

Factor Demand Module

The factor demand model described below avoids the technological rigidity of regional input-output models by deriving their structures from neoclassical production theory, which allows for factor substitution. The fundamental technological relationship of neoclassical theory is the *production function*

$$X = F(K, L, E, M) \tag{13.1}$$

which relates the level of output (X) to levels of capital (K), labour, energy (E) and material (M) inputs.

It can be demonstrated that the production technology represented by (13.1) may also be represented by a *cost function*

$$C = g(P_K, P_L, P_E, P_M, X) \tag{13.2}$$

where C represents the minimum cost at which output level X can be produced given the prices of capital (P_K), labour (P_L), energy (P_E) and materials (P_M) (McFadden, [40]). Input substitution occurs when these prices change relative to one another. The effect of prices on the demand for individual factors of production is made explicit by Shephard's Lemma which states that the level of any factor employed in the production of output level X is equal to the first partial derivative of the cost function (13.2) with respect to the price of that input.

To estimate the parameters of the cost function without imposing prior restrictions on the Allen elasticities of substitution, we propose to use a class of flexible functional forms – the translog; see Christensen, Jorgenson and Lau, [14] and Lakshmanan [32]. The cost function in translog form is written:

$$ln\ C = \alpha_0 + \alpha_1\ ln\ X + \sum_i \alpha_i\ ln\ P_i + \beta_{XX}(\ln X)^2$$

$$+ \frac{1}{2} \sum_i \sum_j \beta_{ij} \ln P_i \ln P_j + \sum_i \beta_{xi} \ln X \ln P_i \tag{13.3}$$

where $i, j = K, L, E, M$ $\quad P_i$ = factor prices

Applying Shephard's Lemma, the share of production costs attributable to factor i is

$$SH_i = a_i + \sum_i \beta_{ij} \ln P_j + \beta_{xi} \ln X \tag{13.4}$$

Using pooled cross-sectional data for states, the parameters of the above instantaneous adjustment model will be estimated using econometric techniques for 2-digit and 3-digit (where possible) SIC code manufacturing industries (Lakshmanan *et al.* [33]).

Dynamic Specification

Then we attempt dynamic versions of the model in two stages – first a partial adjustment model, and then a fully dynamic adjustment model. In dynamic models, in general, there is an assumption of an optimal or 'desired' level of any input $X_i^*(t)$ at any given time t, to which the firm adjusts over some period. The process of adjustment of input levels over time may be represented as:

$$\dot{X}_i(t) = \lambda_i \left[X_i^*(t) - X_i(t)\right] \qquad 0 \leq \lambda \leq 1 \tag{13.5}$$

where $\dot{X}_i = dX_i(t)dt$. The static model is the case when $\lambda = 1$ for all i.

In the Treadway version of the dynamic adjustment model, the cost of adjustments to new levels of input stocks is viewed as the sum of the market price of the input and the cost of allocating resources to the process of adjustment in terms of output foregone (Treadway [50]; [51]). Berndt, Fuss and Waverman ([5]; [6]) adopted Treadway's internal adjustment cost concept but maintained the distinction between variable and quasi-fixed inputs in developing a model in which the firm minimises the present value of an infinite stream of production costs. In addition to determining the optimal or desired level of quasi-fixed inputs, this model determines an optimal partial adjustment parameter [λ_i in (13.5)].

We propose to try initially a quadratic functional form, though we may experiment with other functional forms later.

Investment Supply Module

The investment supply model integrates elements of a top-down approach, where a national investment pool is distributed among regions depending upon local costs, and a bottom-up approach where regional investment is generated purely by regional demands. In the case of industries whose investment depends upon spatially invariant variables, whose outputs and prices are determined in national markets, and whose technology is easily transferred (multiplant manufacturing industries), a top-down approach consistent with recent work by Crow [17] is used. In this framework, the probability of a nationally oriented firm making an investment in a particular region depends on profitability, access to markets, access to inputs, local taxes, land prices and amenities in that region relative to other regions. This qualitative choice is represented by a multinomial logit model. The predominance of cost and profitability determinants in this model indicates a supply side orientation. In the case of industries which are more closely tied to the local economy (housing, retail, services) a 'hybrid' approach which takes both regional demand and supply determinants and selected national variables into account is used. For example, investment in housing is represented as:

$$I_h^J = f(PY^J, GHH^J, r^J, FHA) \tag{13.6}$$

where

I_h^J = investment in housing in region J
PY^J = income in region J
GHH^J = growth rate of households in region J
r^J = mortgage rate in region J
FHA = Federal Housing Administration interest rate (below market).

Research Issues

The use of a large number of production inputs and flexible production functional forms (that impose no a priori restrictions on elasticities of substitution as traditional forms such as Cobb–Douglas and CES do) in the factor demand module allows us to portray a more general representation of the regional growth process. Further, the dynamic version of this model permits the incorporation of the costs of adjustment to regional growth and change. Using this approach, it is possible to estimate distinct long-run and short-run own and cross-price elasticities for labour, energy and materials where capital is assumed quasi-fixed. Since a major policy issue of the coming decade is how best to manage the consequences of industrial restructuring, particularly in the Northeast US, this dynamic model will help in exploring regional adjustment processes and flexibility to change.

The investment supply model is specified as an interactive regional-national submodel where top-down and bottom-up approaches are combined and is expected to contribute to the growing literature on hybrid multiregional modelling (Bolton [9]).

Multipolicy Analysis and Conflict Management Submodule

As already suggested, each of the modules requires as inputs for the year of projection certain basic assumptions, e.g., on state and federal government expenditure vectors (programmes) for CICIOP, on transportation programmes for TRANS, on wages and income transfers for DEMO, on social security for REGLEC and energy development for FACTIN. Since typically these decisions are the outcome of the mediation of different conflicting objectives among regional and private interest groups, it is essential to analyse this process of mediation. The research on multidecision analysis confronts this process, bearing in mind the give and take that occurs in reality. It will utilise the systematic classification of conflict management (compromise) procedures in Isard and Smith [30].

Over 100 conflict management procedures have been identified as practical in one situation or another, or in one context or another. These range from simple split-the-difference procedures through equal increment (decrement), veto incremax, game-theoretic, demand-revealing, and priorities-determining methods.

With regard to each key decision issue or complex of interconnected issues, key characteristics of the conflict situation among interest groups will be identified. Leaving aside systemic structural properties, such characteristics relate to (1) nature of disagreement among interest groups and participants (issues and stakes, qualitative and quantitative), (2) participants involved (few or many, representing themselves or large constituencies, with or without technical support teams), (3) coalition characteristics (ease of formation and stability), (4) options available (large numbers, or small), (5) fractionation possibilities, (6) information available, and so forth. Characteristics of participants (interest groups) are also of critical importance – whether they are conservative or bold, myopic or not, strategy-oriented or not, highly educated or not, etc., etc.

Additionally, the characteristics of a conflict situation among interest groups are to be matched by relevant properties of conflict management (compromise) procedures – their structural, time-related, motivating, psychological, and solution properties. Against this background of relevant data and characteristics, a set of reasonable, workable or likely-to-be-used conflict management procedures will be identified for each situation or complex of conflict issues. We will be particularly interested in sets of procedures that yield compromise decisions relatively insensitive to which procedure is used.

Currently, considerable research is being conducted by many scholars on coalition analysis in a highly mathematical form, using technical concepts such as the core, nucleolis and fibre and making such strong assumptions that the resulting analysis has little relevance to the real world. At the other extreme, the highly particularistic historical case studies lead to few useful principles. We intend to examine systematically various factors present in the formation, disruption and reformation of coalitions to isolate a few basic ones whose interaction are most critical for understanding. In particular, we should like to identify conditions governing the formation of stable coalition patterns with regard to different kinds of conflict situations. In doing so, we shall employ concepts from the traditional field of location theory which have recently been found to yield much insight into coalition formation when viewed as an agglomeration process in policy space. Other basic research issues concern the processes of learning and diverse dynamic phenomena in interactive decision making.

Multiregional Data Base Acquisition

Given the scope and complexity of the various modules that will be linked in the integrated multiregional model, a major issue in its implementation is the feasibility of assembling the requisite multiregional data base. There is no compelling need to initiate the development of special data bases for the model. The several large multiregional data bases generated by the statistical activities of Federal agencies, when appropriately drawn upon, are quite impressive. In terms of content, detail (sectoral, temporal and spatial), reliability and accessibility, available data bases are adequate for the most part for the development of the integrated multiregional model. Where shortfalls exist for some variables, in terms of availability for selected points in time or for geographical scales other than desired, the problems are not intractable. What is striking is that these large regional data bases are so underutilised by the regional modeling community. These observations on data bases can be illustrated in the case of three modules – FACTIN, TRANS and DEMO.

The data base for the two FACTIN modules is drawn from several regional data bases. For the dynamic factor demand module, the 2 and 3 digit industry data on sector output, value added, factor quantities and prices are drawn from *The Census of Manufacturers Series*, *The Annual Survey of Manufacturing* and *The Energy End Use Data Base*. Further special research reports such as those pertaining to book value of capital stock, are utilised. In the case of constructed variables (e.g., sector depreciation rate) necessary data will be obtained from special studies such as Bulletin 2034, US Department of Labor, on *Capital Stock Estimates on I–O Industries: Methods and Data 1947–76*. For the investment supply modules, special data files on local taxes, assessed land values, and amenity variables will be drawn upon. For accessibility, the transportation network descriptors will be obtained from TRANS file.

The interregional commodity flow data required for TRANS can be obtained from several sources: Census of Transportation, Corps of Engineers, US DOT one per cent waybill sample and TSC Compilation of Composite Commodity Flow Table. The commodity-industry system can be built up from recent I–O table data. Finally, the multimodal network system, delineated in terms of nodes and different types of links can be characterised from the data available from the Transportation System Center network model of the US.

A variety of public use sample data on households have been made available and used in microsimulation models such as MATH, TRIM, DYNASIM, and MICROSIM. These have been enlarged by special data file mergers (e.g., 1973 IRS/SSA/CPS match file) or studies such as income maintenance experimental panel data (SIME/DIME). The DEMO module will tap into such data bases as are appropriate to its concerns with labour supply and migration.

It must be clear that as the integrated model is developed over the three year period, the data requirements for maintenance of the model will become evident. Indeed, one of the outputs of the project is the delineation of a future plan for acquisition of data – secondary and primary – for the operation and maintenance of a large integrated multiregional model.

IV CONCLUSIONS

The proposed research can, to reiterate, be of critical importance in several directions. First, it would advance the state of the art in the most basic areas of regional science – industrial location, transportation, resource use, population and income distribution, employment, and regional growth and decline. Such advances would come not only from basic research in each of these areas, but also from the input of relevant parameters and information from other areas which have impact on any given area. Hence, the more restricted models designed to attack the problems of regional science in each of these areas would be significantly enhanced in quality.

Even more significant is the deeper insight that would be gained on how decisions in any particular sector of the national or a regional economy, both private and public, come to affect policies in other sectors of the national and regional economies. Accordingly, the overall impact of a decision, in terms of costs, benefits, and welfare to people in different groups can be much more accurately identified. In this regard, the proposed research is of critical importance for the decision making process of major industries, as well as the federal and state levels of government, where the need to evaluate impacts of one decision area on another is so urgent.

References

[1] Amano, K., Fujita, M., 1970. A Long Run Economic Effect Analysis of Alternative Transportation Facility Plans – Regional – National, *Journal of Regional Science* vol. 10, pp. 297–324.

[2] Ballard, K., Wendling, R., 1980. The National-Regional Impact Evaluation System: A Spatial Model of US Economic and Demographic Activity, *Journal of Regional Science*, vol. 20, pp. 143–58.

[3] Bargur, J., 1969. A Dynamic Interregional Input-Output Programming Model of the California and Western States Water Economy, Water Resources Center, University of California, Berkeley.

[4] Berndt, E. R., Christensen, L. R., 1973. The Internal Structure of Functional Relationship: Separability, Substitution and Aggregation, *Review of Economic Studies*, vol. 60, pp. 403–10.

[5] Berndt, E. R., Fuss, M. A., Waverman, L., 1977. Dynamic Models of the Industrial Demand for Energy, Electric Power Research Institute, Report EPRI EA–580, Palo Alto.

[6] Berndt, E. R., Fuss, M. A., Waverman, L., 1981. Dynamic Adjustment Models of Industrial Energy Demand: Empirical Analysis for US Manufacturing, 1947–1974. Electric Power Research Institute, Report EPRI EA–1613, Palo Alto.

[7] Blair, P., Miller, R. E., 1983. Spatial Aggregation of Multiregional Input-Output Models, *Environment and Planning A*, vol. 15, pp. 187–206.

[8] Bolton, R. E., 1960. *Defense Purchases and Regional Growth*, Brookings Institution.

[9] Bolton, R. E., 1980. Multiregional Models in Policy Analysis, in Adams F. G. and N. Glickman, (eds), *Modeling the Multiregional System* (Lexington Books, Lexington, MA), pp. 255–84.

[10] Bolton, R. E., 1980a. Multiregional Models: Introduction to a Symposium, *Journal of Regional Science*, vol. 20, pp. 131–42.

[11] Boyce, D. E., Hewings, G. J. D., 1980. Interregional Commodity Flow, Input-Output and Transportation Modelling: An Entropy Formulation, presented at the First World Regional Science Conference, Cambridge, MA, revised 1981.

[12] Boyce, D. E., LeBlanc, L. J., Chon, K. S., Lee, Y. J., Lin, K. T., 1983. Implementation and Computational Issues for Combined Models of Location, Destination, Mode and Route Choice, *Environment and Planning A*, forthcoming.

[13] Caldwell, S., Greene, W., Mount, T., Saltzman, S., Boyd, R., 1979. Forecasting Regional Energy Demand with Linked Macro-Micro Models, *Papers of the Regional Science Association*, vol. 43, pp. 99–113.

[14] Christensen, L. R., Jorgenson, D. W., Lau, L. S., 1973. Transcendental Logarithmic Production Frontiers, *Review of Economics and Statistics*, vol. 55, pp. 28–45.

[15] Conway, R. S., 1979. The Simulation Properties of a Regional Interindustry Econometric Model, *Papers of the Regional Science Association*, vol. 43, pp. 45–57.

[16] Courbis, R., 1980. Multiregional Modelling and the Interaction Between

Regional and National Development: A General Theoretical Framework, in F. G. Adams and N. J. Glickman (eds) *Modelling the Multiregional Economic System*, Lexington Books, D. C. Heath and Co., pp. 107–130.

[17] Crow, R. T., 1979. Output Determination and Investment Specifications in Macroeconomic Models of Open Regions, *Regional Science and Urban Economics* 9, 141–158.

[18] Dantzig, G., Parikh, S., 1976. On a PILOT Linear Programming Model for Assessing Physical Impact on the Economy of a Changing Energy Picture, in F. Roberts (ed.) *Energy: Mathematics and Models*, SIAM, Philadelphia.

[19] Diewert, W., 1971. An Application of the Shephard Duality Theorem: Generalized Leontief Production Function, *Journal of Political Economy*, vol. 79, pp. 481–507.

[20] Fisk, C. S., Boyce, D. E., 1983. Optimal Transportation Systems Planning with Integrated Supply and Demand Models, submitted to *Transportation Science*.

[21] Friesz, T. L. and Viton, P. A., 1982. Economic and Computational Aspects of Freight Network Equilibrium Models: A Synthesis, submitted to *Transportation Research*.

[22] Fromm, D., Loxley, C. J., McCarthy, M., 1980. The Wharton EFA Multiregional Econometric Model, a Bottom-Up Approach, paper presented at the First World Regional Science Conference, Cambridge.

[23] Fuss, M., 1977. The Demand for Energy in Canadian Manufacturing, *Journal of Econometrics*, vol. 5, pp. 89–116.

[24] Gigantes, T., 1970. The Representation of Technology in Input-Output Systems, in A. Brody and A. P. Carter (eds), *Contributions to Input-Output Analysis* (North-Holland, Amsterdam).

[25] Goettle, R., Cherniavsky, E., Tessmer, R., 1977. An Integrated Multiregional Energy and Interindustry Model of the US, Brookhaven National Laboratory, BNL 22728.

[26] Golladay, F., Haveman, R., 1977. *The Economic Impacts of a Tax Transfer Policy* (Academic Press, New York).

[27] Harris, C. C., Jr., 1980. New Development and Extensions of the Multiregional Multi-Industry Forecasting Model, *Journal of Regional Science*, vol. 20, pp. 159–71.

[28] Isard, W., Anselin, L., 1982. Integration of Multiregional Models for Policy Analysis, *Environment and Planning A*, vol. 14, pp. 359–76.

[29] Isard, W., Bramhall, D. F., Carrothers, G. A. P., Cumberland, J. H., Moses, L. N., Price, D. O., Schooler, E. W., 1960. *Methods of Regional Analysis: An Introduction to Regional Science* (MIT Press, Cambridge, MA).

[30] Isard, W., Smith, C. A., 1982. *Conflict Analysis and Practical Conflict Management Procedures. An Introduction to Peace Science* (Ballinger Publishing Co., Cambridge, MA)

[31] *Journal of Regional Science*, 1980. *Symposium on Multiregional Forecasting and Policy Simulation Models* 20, 129–206.

[32] Lakshmanan, T. R., 1983. A Multiregional Model of the Economy, Environment and Energy Demand in the United States, *Economic Geography*.

[33] Lakshmanan, T. R., Anderson, W., Jourabchi, M., 1982. Regional Dimensions of Factor and Fuel Substitutions in US Manufacturing, in A. Anderson, T. Puu, W. Isard and U. Schewizer (eds) *Structural Economic Analysis and Planning in Time and Space*, forthcoming (North Holland, Amsterdam).

[34] Ledent, J., 1978. Regional Multiplier Analysis: A Demometric Approach, *Environment and Planning A*, vol. 10, pp. 537–60.

[35] Ledent, J. and Gordon, P., 1979. Modeling Interregional Population Distribution and Economic Growth with a Focus on Rural-Urban Systems, IIASA, Laxemburg.

[36] Leontief, W. W., Strout, A., 1963. Multiregional Input-Output Analysis, in T. Barna (ed.) *Structural Interdependence and Economic Change*, (MacMillan, New York), pp. 119–50.

[37] L'Esperance, W. L., 1977. Optional Stabilization Policy at the Regional Level, *Regional Science and Urban Economics*, vol. 7, pp. 25–48.

[38] Madden, M., Batey, P. W. J., 1980. Achieving Consistency in Demographic-Economic Forecasting, *Papers of the Regional Science Association*, vol. 44, pp. 91–106.

[39] Mathur, P. N., 1972. Multiregional Analysis in a Dynamic Input-Output Framework, in A. P. Carter and A. Brody, (eds), *Input-Output Techniques* (North-Holland, Amsterdam), pp. 208–30.

[40] McFadden, D., 1978. Cost, Revenue and Profit Functions, in M. Fuss and D. McFadden (eds), *Production Economics: A Dual Approach*, 1, The Theory of Production, North Holland, Amsterdam.

[41] Miller, R., Blair, P., 1985. *Input-Output Analysis: Foundations and Extensions*, Prentice-Hall, Englewood Cliffs, NJ.

[42] Milne, W., Glickman, N., Adams, F. G., 1980. A Framework for Analyzing Regional Growth and Decline: A Multiregion Economic Model of the United States, *Journal of Regional Science*, vol. 20, pp. 173–89.

[43] Orcutt, G., Caldwell, S., Wertheimer, R. II, 1976. *Policy Exploration Through Microanalytic Simulation* (The Urban Institute, Washington, DC).

[44] Polenske, K., ed., 1980. *The US Multiregional Input-Output Accounts and Model*, Lexington Books, Lexington.

[45] Policy Research Group, 1982a. *An Evaluation of Forecasting Accuracy in the MICROSIM Model.* Task 1 Interim Report, Washington, DC.

[46] Policy Research Group, 1982b. *An Assessment of the Macroeconomic Forecasting Model in MICROSIM and Use of Alternative Forecasts.* Task 3 Interim Report, Washington, DC.

[47] Russell, C. *et al.*, 1976. Environmental Quality Management: An Application to the Lower Delaware Valley, Resources for the Future Research Paper R-1, Washington, DC.

[48] Schinnar, A. P., 1976. A Multidimensional Accounting Model for Demographic and Economic Interactions, *Environment and Planning A*, vol. 8, pp. 455–75.

[49] Smith, C., 1983. Integration of Multiregional Models for Policy Analysis, PhD dissertation, Cornell University, Ithaca.

[50] Treadway, A. B., 1971. The Rational Multivariate Flexibility Accelerator, *Econometrica*, vol. 39(5), pp. 845–55.

[51] Treadway, A. B., 1974. The Globally Optimal Flexible Acceleration, *Econometrica*, vol. 39, pp. 845–55.

[52] Treyz, G., 1980. Design of a Multiregional Policy Analysis Model, *Journal of Regional Science*, vol. 20, pp. 191–206.

[53] Wilson, A. G., 1970. *Entropy in Urban and Regional Modelling* (Pion, London).

Appendix A
A feasible set of steps for integrating the modules*

This appendix lists one of many possible feasible sets of steps for operating the integrated multiregional model. This set of steps is consistent with the interconnections of the modules depicted in Figures 13.1–13.6. The steps are:

1. From NATLEC, the national magnitudes for the year of projection (say 2000) of the major components of final demand – household expenditure, investment, government expenditures, foreign exports, foreign imports, and inventory change – and the national markets (demand) for the output of the basic cost-sensitive sectors are obtained for CICIOP.
2. With the use of standard procedures, the national magnitude for traditional final demands are disaggregated by region.
3. With the use of comparative cost and industrial complex methods, the national outputs of basic cost-sensitive sectors are distributed among the regions.
4. From steps 2 and 3 and using the MRIO (Multiregional input-output) framework or an LP/MRIO framework with energy or another sector given special significance, the programming format determines outputs by region by sector.
5. With the use of sector and regional data on outputs per labour by sector and region, labour requirements are determined by sector and region (adjusted for expected increase in average productivity to year of projection).
6. From steps 2 and 3, and from the MRIO study, regional output, regional final demands and regional technical production coefficients are entered into the TRANS module. With the use of the commodity-industry input-output transformation, the data are converted to commodity terms.
7. For TRANS, the shipper networks and carrier networks by commodity classes are specified.
8. With the use of the results of steps 6 and 7, the system of submodels (interregional flow, network and carrier) in TRANS are solved simultaneously to yield interregional commodity flows, interregional shipment costs, and service characteristics by commodity.
9. The changes in the spatial pattern of outputs from 1980 to 2000 (yielded by CICIOP) are converted to equal absolute annual changes. These in turn are converted to increments (decrements) in labour requirements by sector by region.
10. Labour requirements by sector and region are transformed into labor requirements by relevant age-sex-class demographic group, using standard demographic procedures.
11. With the use of the labour requirements (from step 10) and appropriate average productivity data (from CICIOP and NATLEC), the REGLEC module is run to estimate regional wage rates for non-manufacturing and service sectors by year up to year of projection, to supplement NATLEC's estimates of national wage rates by demographic class for manufacturing industries.
12. With the use of wage rates, employment levels and elements of value added data (from CICIOP), personal income by region are estimated by the REGLEC module.
13. With the use of national price data from NATLEC, and estimated regional wages, REGLEC generates price vectors by region.
14. The results of step 11 and data on regional industrial structure and labour requirements from CICIOP are fed into DEMO to yield labour supply by region by demographic group by year up to year of projection.

*This Appendix draws heavily upon the work of Christine Smith [49].

15. The projected wage rates and prices from REGLEC, other value added elements by sector from CICIOP, and transfer payments and taxes from multipolicy analysis are fed into DEMO to yield household personal income and in turn household expenditures and savings.

16. Projected wage rates and prices from REGLEC, energy shadow prices and sector outputs from CICIOP, material prices and interest rates from NATLEC are fed into FACTIN to yield factor demands by sector by region.

17. Household savings by region and other region specific characteristics from DEMO, value added and sector outputs from CICIOP, network extension and transport cost from TRANS, and investment in national industries are fed into FACTIN to yield investment estimates by sector by region.

18. A consistency check is conducted with respect to the sum of regional investment estimates by sector with national investment estimates. Necessary adjustments in both regional and national estimates are made.

19. A consistency check is conducted with respect to the sum of regional outputs by sector from CICIOP and NATLEC's national output by sector. Necessary adjustments in both regional and national estimates are made.

20. Consistency checks are conducted with respect to the sum of regional employment by sector from CICIOP and NATLEC's national employment by sector. Necessary adjustments in both regional and national estimates are made.

21. From both NATLEC and FACTIN, the modified national and regional estimates of investment are obtained for CICIOP, to be adjusted in the light of multi-policy analysis.

22. From personal income estimates from REGLEC, and household expenditures from both NATLEC and DEMO, an appropriately modified household consumption vector by region is obtained for CICIOP to be adjusted in the light of multi-policy analysis.

23. From NATLEC and multipolicy analysis, modified level of government expenditures, foreign exports and foreign imports are obtained by region for CICIOP.

24. From NATLEC, modified national market (demand) for the output of the basic cost-sensitive sectors are obtained from use of a national final demand vector modified in the light of the top-down, bottom-up checks. The modified national market (demands) are fed to CICIOP.

25. With the use of transport costs from TRANS and wage rate differentials from REGLEC, the modified national outputs of basic cost-sensitive sectors are distributed among the regions using comparative cost and industrial complex methods.

26. On the basis of factor demands by sector and region estimated by FACTIN, CICIOP's technical production coefficients by region are adjusted.

27. On the basis of interregional commodity flows generated by TRANS, CICIOP's interregional trade coefficients are adjusted.

28. With the use of the new estimates of traditional final demand (steps 21–23) and the new estimates of non-traditional final demand (step 25), with both adjusted in the light of multipolicy analysis, the MRIO framework (with adjusted coefficients from steps 26 and 27) is run to determine new (2nd round) levels of outputs by region by sector. Alternatively, an LP/MRIO framework, with energy or another sector given special significance, is run with the use of a standard programming format.

29. With the use of sector and regional data on outputs per labour by sector and region, 2nd round labour requirements by sector and region are determined (adjusted for expected increase in average productivity to year of projection).

30. From steps 21–23, 25 and 26, 2nd round regional final demands and regional

technical production coefficients are entered into the TRANS module. With the use of the commodity-industry input-output transformation, the data are converted to commodity terms.

31. With the use of the results of step 30, and the shipper and carrier networks identified in step 7, the system of submodels (inter-regional flow, network and carrier) are solved simultaneously to yield 2nd round interregional commodity flows, interregional shipment costs, and service characteristics by commodity.

32. The 2nd round estimates of changes in the spatial pattern of outputs from 1980 to 2000 (yielded by CICIOP), are converted to equal absolute annual changes. These in turn are converted to increments (decrements) in labour requirements by sector by region.

33. The above labour requirements by sector and region are transformed into labour requirements by relevant age-sex-class demographic group.

34. With the use of (1) labour requirements (from step 33), (2) appropriate average productivity data (from CICIOP and NATLEC), and (3) labour supply by demographic group from DEMO, the REGLEC module is run to estimate regional wage rates for non-manufacturing and service sectors by year up to year of projection, to supplement NATLEC's 2nd round estimates of national wage rates by demographic class for manufacturing industries.

35. With the use of the above wage rates, employment levels and elements of the 2nd round value added data (from CICIOP), personal income by region is estimated by the REGLEC module.

36. With the use of modified national price data from NATLEC and modified regional wages, REGLEC generates 2nd round price vectors by region.

37. The wage rate estimates of step 34 and 2nd round data on regional industrial structure and labour requirements from CICIOP are fed into DEMO to yield labour supply by region by demographic group by year up to year of projection.

38. The 2nd round wage rates and prices from REGLEC, other value added elements by sector, interest and dividends from CICIOP, and transfer payments and taxes from multipolicy analysis are fed into DEMO to yield 2nd round household personal income and in turn household expenditures and savings.

39. The 2nd round wage rates and prices from REGLEC, energy shadow prices and sector outputs from CICIOP, material prices and interest rates from NATLEC are fed into FACTIN to yield 2nd round factor demands by sector by region.

40. The 2nd round household savings by region and other region specific characteristics from DEMO, value added and sector outputs from CICIOP, network extensions and transport costs from TRANS, and investment in national industries are fed into FACTIN to yield 2nd round investment estimates by sector by region.

41. A consistency check is conducted with respect to the sum of 2nd round regional investment estimates by sector with 2nd round national investment estimates. Necessary adjustments in both regional and national estimates are made.

42. A consistency check is conducted with respect to the sum of 2nd round regional outputs and employment by sector from CICIOP and NATLEC's 2nd round national output and employment by sector. Necessary adjustments in both regional and national estimates are made.

Iterations repeating steps 21 to 42 are conducted until all major discrepancies are eliminated and until the changes from round-to-round are sufficiently small to invoke a prespecified stopping rule.

14 Linked Integrated Multiregional Models at The International Level

I INTRODUCTION

This paper reports some of our present thinking on future research directions in regional science, particularly as it is developing at Cornell University. We are also currently involved in the very practical problem of projecting the growth of certain regions of a nation, for example the East Texas–Louisiana region of the United States. We shall show how these two objectives are intimately linked. Since there will be insufficient space to present all the relevant material, this paper represents one part of a much longer report (see Isard and Anselin [8], Isard and Smith [9]).

II TOWARDS A REFORMULATION OF THE CONCEPT OF A SOCIETY

To begin, let us go far back in regional science and redefine the concept of a society S. This society comprises many elements, and has been partially defined elsewhere (see Isard *et al.*, 1969, pp. 597–767). Briefly put, a society can be represented by the following equation:

$$S = \left\{ R^{\phi}_{+}, \varrho, \left[\zeta^{J}, (X^{J}_{i}, \succcurlyeq^{J}_{i}, \alpha^{J}_{i}, g^{J}_{i})^{m}_{l}, (Y^{J})^{n}_{l}, (s^{J \to L}_{+f})^{\bar{f}}_{l} \right]^{U}_{A} \right\} \tag{14.1}$$

First, it involves ϕ ($h = 1, \ldots, \phi$) commodities, the first ℓ of which are economic, the remaining being noneconomic. For example,

$\ell + 1$ − c solidarity
$\ell + 2$ = c-power
$\ell + 3$ = c-respect

The ϕ commodities are indicated by the symbol R^{ϕ}_{+}, representing the nonnegative quadrant of a ϕ dimensional Euclidean space. Any bundle of goods consumed by an individual is a vector in this space.

This chapter was written with Christine Smith. The body of the chapter is reprinted from *Papers of the Regional Science Association*, vol. 51, (1983), pp. 3–19. The appendix was written especially for this volume.

Second, our society S consists of U regions. For simplification we assume that each region J ($J = A, \ldots, U$, as indicated at the right-hand bracket) has one and only one government whose policies are set forth (after proper determination) by a single leader. For the moment we assume that each leader (region) like the traditional consumer, trader and firm is interested in maximising a relevant payoff. In this case, we take this payoff to be GRP (Gross Regional Product). The action space of government J is ζ^J, and consists of policies and programmes along the lines discussed elsewhere (see Isard *et al.*, [5], pp. 617–620, 706–9).

Third, S contains a set of rules ϱ, where each rule ϱ ($\varrho = 1, \ldots, \bar{\varrho}$) is a way of resolving conflicts such as frequently occur when a joint policy needs to be adopted after each government proposes that set of policies which is optimal for the constituency in its own region.

Fourth, S contains a set of m consumers in each region J. Each consumer i ($i = 1, \ldots, m$) in region J has a combined action (consumption) set X_i^J (see Isard *et al.*, [5], pp. 620–4 and 709–29) over all his roles; a preference structure $\succcurlyeq_i^J$; equities in firms j, J ($j = 1, \ldots, n$) and trading establishments f, $J \rightarrow L$ ($f = 1, \ldots, \bar{f}$), summarised by α_i^J, which entitle him to a fixed share of profits of any firm or trading establishment; and an initial stock of goods g_i^J including his labour $g_{\ell i}^J$. Each consumer faces the customary budget constraint.

Fifth, S contains a set of n producers in each region J. Each producer j, J ($j = 1, \ldots, n$) has a production set Y_j^J and confronts the customary transformation function as a constraint.

Finally, S contains a set of $\bar{f}$ trading establishments in each region J. Each trader f ($f = 1, \ldots, \bar{f}$) is able to ship an amount $s_{hf}^{J\rightarrow L}$ of commodity h to every region L ($L = A, \ldots, U$), $J \neq L$. His shipment space $S_{+f}^{J\rightarrow L}$ consists of non-negative shipments only.

Among equilibrium conditions for S are the following (the full set appears in Isard *et al.*, [5], chapters 13 and 14).

1. $\tau_h^{J\rightarrow L} \leq 0; \quad \tau_h^{J\rightarrow L} s_{hf}^{J\rightarrow L} = 0 \quad h = 1, \ldots, \phi; \quad f = 1, \ldots, \bar{f};$
   $J, L = A, \ldots, U; \quad J \neq L$
   where $\tau_h^{J\rightarrow L} = p_h^L - p_h^J - p_{\ell-2}^J w_h d^{J\rightarrow L}$

   p_h^L, p_h^J = prices of commodity h in regions L and J, respectively,
   $p_{\ell-2}^J$ = the transport cost per unit weight per unit distance,
   w_h = weight of a unit of h, and
   $d^{J\rightarrow L}$ = the distance betwen regions J and L.

2. $z_h^J \leq 0; p^J z_h^J = 0 \quad h = 1, \ldots, \ell; \quad J = A, \ldots, U$
   where z_h^J is the excess demand for commodity h on the market in region J.

In addition to equilibrium conditions, our society S needs to select (its interest groups need to agree upon) a rule $\varrho \in \varrho$ by which any given policy conflict among the several regions (regional authorities or leaders) is to be resolved. Under appropriate conditions, one rule may be to reach a solution with the use of the concept of a Shapley value (see Shapley, [14]); another, with the use of a Fair Division scheme (see Lucas, [11]; and Englebrecht–Wiggans, [1]); still others, with the use of an incremax procedure (see Isard *et al.*, [5]); or a Nash equilibrium allocation (see Nash, [13]). For such cases, we shall talk of a Shapley, Lucas, Isard-Smith and Nash society, respectively.

III TOWARDS A MEANINGFUL SET OF SUBSYSTEMS FOR ANALYSIS OF SOCIETY

With this background of an abstract society S, we wish to focus on a meaningful set of subsystems for analysis (for a detailed discussion of the contents of these subsystems, see Isard, [6]; [7]). We depict this set in Figure 14.1. At the extreme left and right of the bottom layer are the P^J and P^L boxes, representing the production subsystems of region J and L, respectively. The production subsystems of other regions can be added. These production subsystems are interconnected with each other via imports, exports, pollution diffusion, etc. This interdependence is summarised by the R box which, in effect, provides the subsystem coupling functions on the 'real' level. The P^J ($J = A, \ldots, U$) and R boxes together form society's production system P.

The boxes C^J and C^L in the middle layer of Figure 14.1 are concerned with cognitive activity and decision-making in regions J and L, respectively. From cognitive activity and interaction of interest groups in regions J and L come commands, orders and decisions, as indicated by arrows from the C^J and C^L boxes directly to the production subsystems P^J and P^L of regions J and L, respectively. The cognitive subsystems of other regions can be added, with respective connections to their production subsystems. All these cognitive subsystems are inter-connected via flows (to and from the I box) of ideas, regulations and other statements (information linkages). Exchange of ideas, debate and discussion, and other cognitive interaction take place within the I box which furnishes the subsystem coupling functions on the cognitive level. The C^J ($J = A, \ldots, U$) and I boxes together form society's decision-making system C.

Finally, the top layer in Figure 14.1 refers to society's information subsystem M. We have insufficient space to review the numerous vertical and horizontal flows among the full set of boxes that represent a society. The purpose of Figure 14.1 is simply to enable us to represent our conception of an abstract society S, as defined by equation 14.1, in the diagrammatic form used in systems analysis as an intermediate step to developing a meaningful and balanced operational model.

IV OUTLINE OF AN OPERATIONAL INTEGRATED MULTIREGION MODEL

What is a meaningful and balanced operational model? One such model for a multiregional system within a nation is presented in Figure 14.2 We have already discussed most of the figure's modules in Isard and Anselin [8] and Isard and Smith [9]. The modules involved are:

1. NATLEC (a national econometric module);
2. CICIOP (a multiregional comparative cost, industrial complex, input-output, programming module);
3. TRANS (a transportation module);
4. FACTIN (a factor demand-investment supply module, one for each region);
5. REGLEC (a regional econometric module, one for each region) and
6. DEMO (a multiregion demographic module)

To proceed to our objective of designing a multinational modeling system, we must discuss the mapping of Figure 14.1 onto Figure 14.2.

First, it is clear that when we use Figures 14.1 and 14.2 to represent the system of a single nation which is highly industrialised like the United States, we in effect map the *I* box and part of the C^J ($J = A, \ldots U$) boxes in Figure 14.1 onto the INPOL box in Figure 14.2 Within the *I* box, the cognitive interaction required for joint decision making takes place; so that we need to arrange for the choice of a rule $\varrho \varepsilon \varrho$ to take place within the INPOL box, which is in effect the operational counterpart of the *I* box. This choice is the result of debate and discussion, deliberations and negotiations that occur among the different regional leaders. Moreover, we specifically indicate in the INPOL box sequences of steps (where each step resolves a subconflict), each sequence corresponding to the requirements of a rule that might be chosen. For example, a first step may be to choose a rule to resolve differences among leaders on an appropriate agenda; a second step to choose a rule to reach agreement on weights that might be attached to different objectives; a third step to choose a rule to determine minimal food requirements and other binding magnitudes regarding constraints, and so forth (see Isard and Smith, [9]). This selection activity is an extremely important one. As we can see from the integrated multiregional model in Figure 14.2, this rule must generate a specific objective function for the programming submodel which then drives the multiregional system through the first iteration. With the objective function specified, and with the command or order to the production system $P = (P^A, \ldots, P^U; R)$ to optimise on this function, we move to the particular module in Figure 14.2 which represents that system. The order to optimise an objective function (whose form, weights and other parameters may have been decided via a

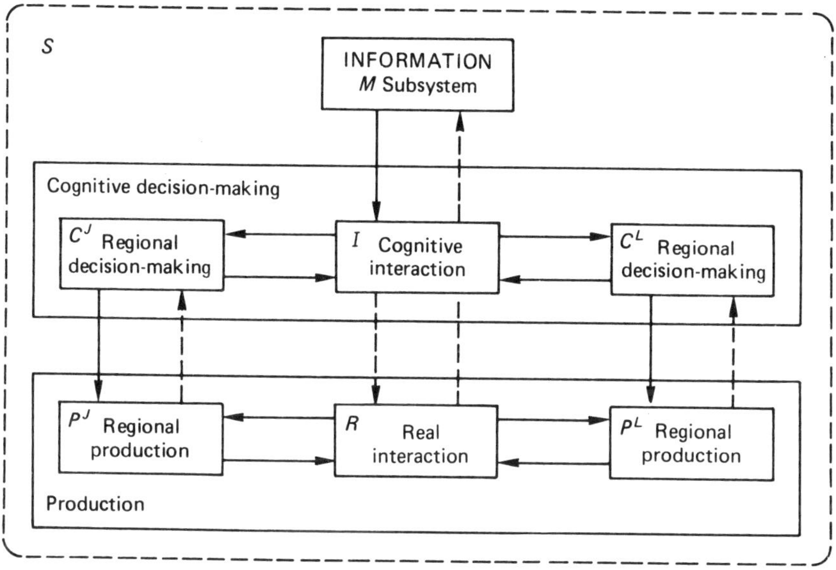

Figure 14.1 A system's structure

series of compromises in the INPOL box) is in itself only a fraction of the input from the I box into the production system P. Whereas the regional leaders may make decisions which guide the functioning of the system at the macro level, there are innumerable other decisions made by numerous actors (consumers, producers, traders) on the micro-level – decisions to generate a birth, to expand output or shipments, to make an investment, etc. These micro-decisions are largely independent of actions taken and agreed upon by leaders, yet such decisions have a major impact on the production system. There are a whole series of decisions (orders, commands, policies) made; not only at the system level but also at the individual and group levels (taking into account all kinds and sizes of groups). They all stem from the cognitive $C = (C^A, \ldots, C^U, I)$ subsystem and impinge on the production system; their understanding is necessary for better projection of the production system. Because the production system incessantly receives such orders and commands from the C subsystem, we have placed the INPOL box at the centre of Figure 14.2.

V CONSTRUCTION OF AN INTEGRATED MULTINATION MODEL

Because of shortage of space we cannot further review Figure 14.2. We must proceed to the main subject of our paper, namely the preliminary

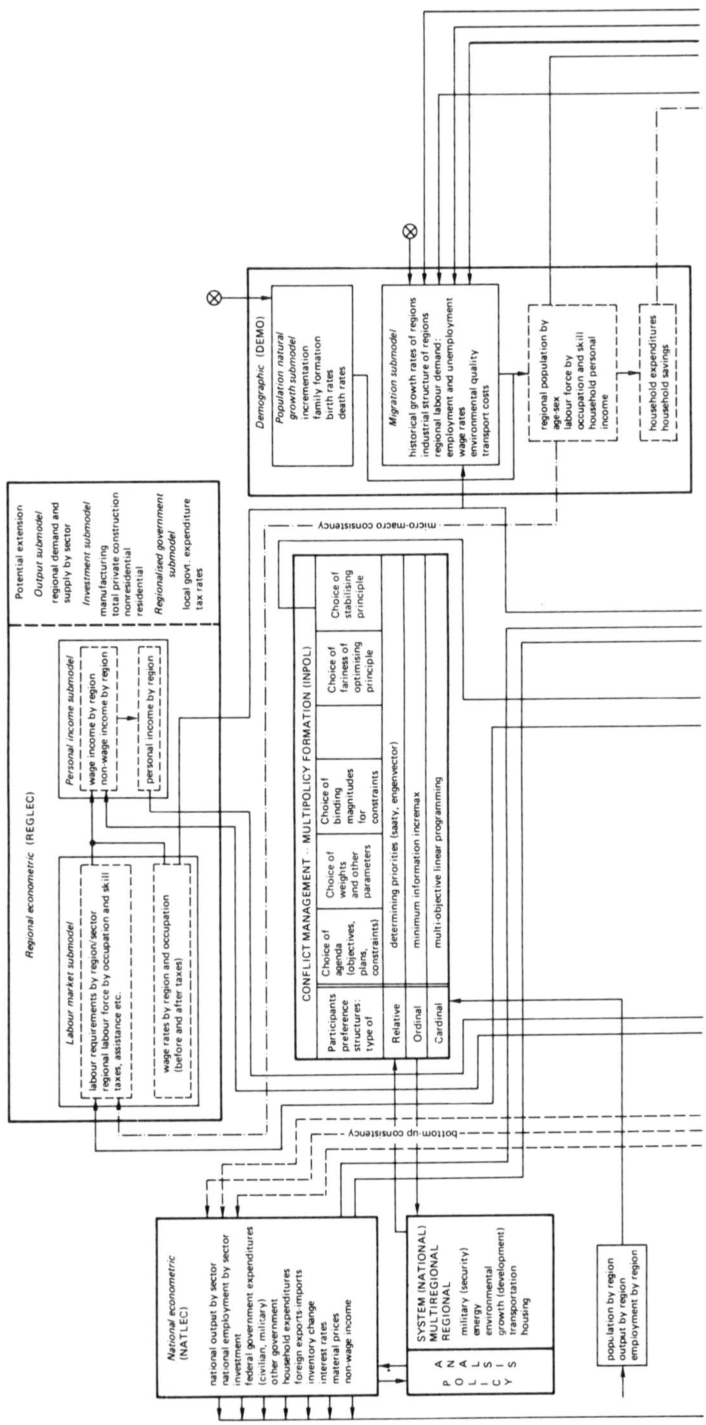
Demographic (DEMO)
Population natural growth submodel
incrementation
family formation
birth rates
death rates
Migration submodel
historical growth rates of regions
industrial structure of regions
regional labour demand:
employment and unemployment
wage rates
environmental quality
transport costs
regional population by age-sex
labour force by occupation and skill
household personal income
household expenditures
household savings
Potential extension
Output submodel
regional demand and supply by sector
Investment submodel
manufacturing
total private construction
nonresidential
residential
Regionalised government submodel
local govt. expenditure
tax rates
Regional econometric (REGLEC)
Personal income submodel
wage income by region
non-wage income by region
personal income by region
Labour market submodel
labour requirements by region/sector
regional labour force by occupation and skill
taxes, assistance etc.
wage rates by region and occupation (before and after taxes)
micro-macro consistency
CONFLICT MANAGEMENT – MULTIPOLICY FORMATION (INPOL)
Participants preference structures: type of
Choice of agenda (objectives, plans, constraints)
Choice of weights and other parameters
Choice of binding magnitudes for constraints
Choice of fariness of optimising principle
Choice of stabilising principle
Relative
determining priorities (saaty, engenvector)
Ordinal
minimum information incremax
Cardinal
multi-objective linear programming
bottom-up consistency
National econometric (NATLEC)
national output by sector
national employment by sector
investment
federal government expenditures (civilian, military)
other government
household expenditures
foreign exports-imports
inventory change
interest rates
material prices
non-wage income
SYSTEM (NATIONAL) MULTIREGIONAL REGIONAL
POLICY ANALYSIS
military (security)
energy
environmental
growth (development)
transportation
housing
population by region
output by region
employment by region

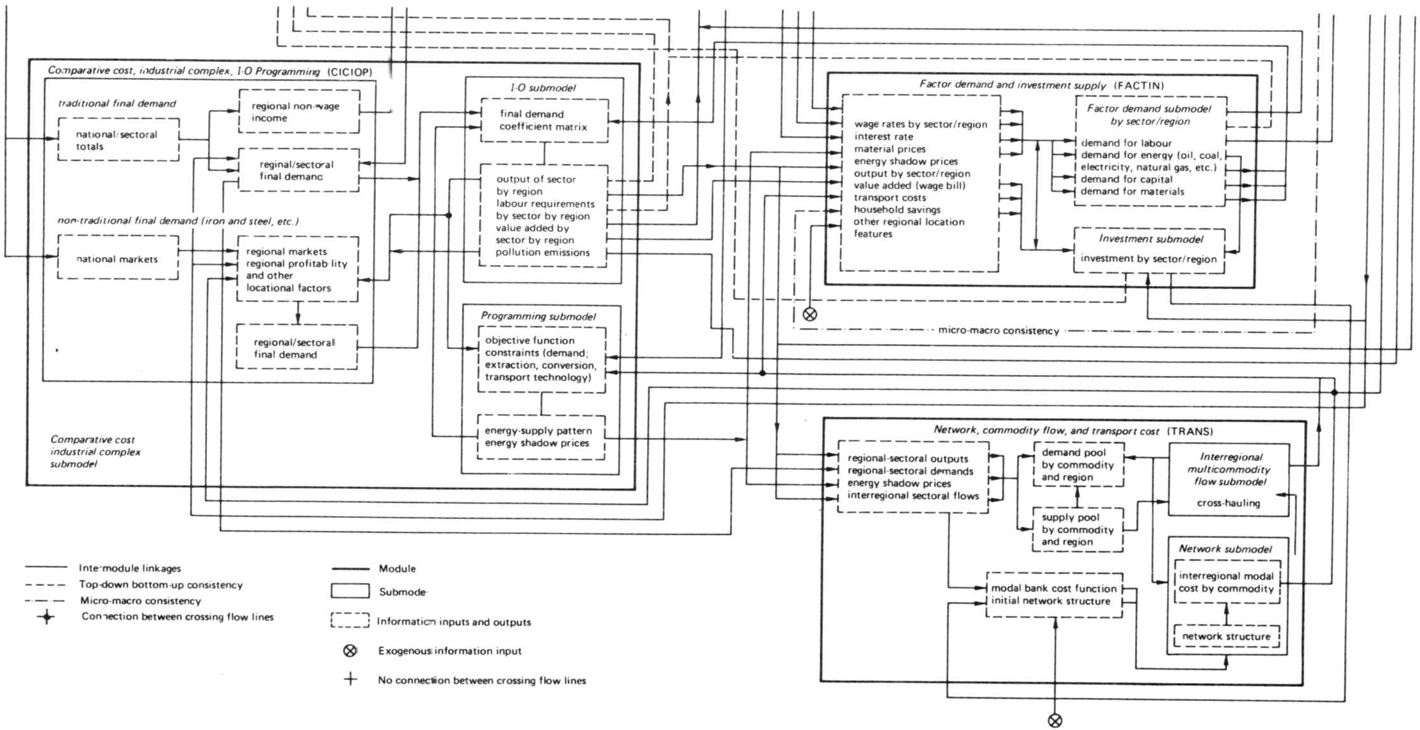

Figure 14.2 Conflict and policy analysis within an integrated multiregional model

construction of a multinational integrated model. Unfortunately, the state of the art in global modelling is such that we cannot go as far at the multinational level as we can within a nation on a multiregional basis. To facilitate our presentation, we regard each nation as a point, each point being a society which (as outlined in section II) consists of a multiregional system, wherein each region is regarded as a subpoint. We thus have a hierarchical system of points in the global system G, as depicted in Figure 14.3. In several important respects, a system of nations is unlike a multiregional system within a nation. Our multinational system therefore requires a different set of modules. First, the transportation system may be viewed as relatively less important to model; for the present exercise, we eliminate it. Second, the migration submodel of a DEMO module would be exceedingly difficult, and possibly misleading, to develop. Political, social and other non-economic factors governing international migration are so important that we cannot hope to present an operational model at this time; it must capture diverse feedbacks from other modules that affect birth and death rates and propensities to migrate. Next, we do not have an econometric model for the globe as a whole that corresponds to our national econometric model. On the other hand, do we need one? Nor do we have on the international scene a parallel to our FACTIN module, though this might not be too difficult to develop. So we are left with:

1. an econometric module for the world system of the LINK type (see Klein, [10]), which parallels REGLEC;
2. a global Input-Output module of the Leontief [12] type, which parallels the interregional input-output submodel of CICIOP;
3. a World Programming module of the Novosibirsk type (see Granberg and Rubinshtein, [2]), which parallels the interregional programming submodel of CICIOP; and
4. an Industrial Location-Trade CC-IC module (see Isard and Anselin, [8]), since the industrial location, comparative cost and industrial complex analysis of CICIOP can be conducted for the world system as well as for each national system.

In the LINK model, each nation J constructs an econometric model of its own economy. Each of these models is based on assumptions about the volume of exports and import prices together with assumptions about other input values associated with domestic policies and other exogenous variables. They thus obtain the endogenous variables, total value of imports M^J and prices of exports PE^J, as functions of exports E^J and import prices PM^J, that is

$$M^J = f^J(E^J, PM^J, \ldots) \qquad J = A, \ldots, U$$
$$PE^J = g^J(E^J, PM^J, \ldots,) \qquad J = A, \ldots, U$$

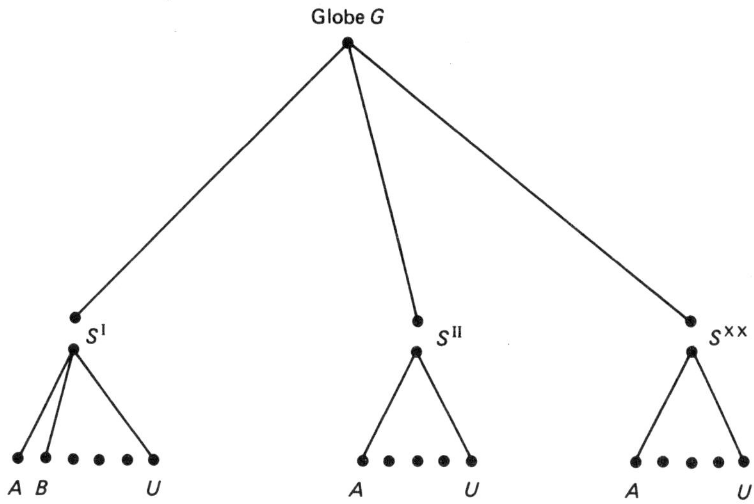

Figure 14.3 The hierarchical system

The purpose of the LINK model is to derive consistent estimates of exports E^J and prices of imports PM^J for each nation. It does this by substituting the import values M^J and export prices PE^J over all nations into the two equations:

$$E = AM$$
$$PM^J = \sum_J PE^J a^{JL} \qquad J = A, \ldots, U$$

where E is a $U \times 1$ column vector of exports (in dollars) by nation;
M is a $U \times 1$ column vector of imports (in dollars) by nation; and
A is a $U \times U$ matrix of coefficients where any $a^{JL} = x^{JL}/\sum_J x^{JL}$, and
x^{JL} = export of nation J to nation L.

The Leontief World Input-Output model is strictly a multiregion model, and not a pure interregional model, since all exports go to a common pool to which each nation contributes in specified proportions and from which each nation draws some of its inputs as imports. The model divides the world economy into 15 regions. Each region is described in terms of 45 sectors of economic activity, including 5 types of pollution abatement activities to treat the emissions of 8 types of major pollutants. The 15 regions are interconnected through exports and imports of some 40 classes of goods and services, capital flows, aid transfers, and foreign interest payments. The model is designed to be a general purpose economic model applicable to the analysis of the evolution of the world economy from various points of view – designed to attack such problems as food and agriculture, mineral resources, pollution and pollution abatement, and structural changes in the economies. The mathematical formulation of the model is the familiar pool model first developed by Chenery and Moses. The basic trade coefficient matrices are equivalent to:

$$T_k = \underset{UxU}{(t_k^{JL})} \qquad\qquad k = 1, \ldots, n$$

where regions are denoted by $A, \ldots, J, L, \ldots U$ and commodities by $k = 1, \ldots, n$ and where t_k^{JL} is the amount of good k which L indirectly receives from J via the pool of k to which J has contributed. Standard input-output equations characterise the rest of the model which is concerned with commodities which are region-specific.

The Novosibirsk programming model follows the customary format. It has lower and upper bounds on the output levels of relevant sectors. For sectors producing transportable outputs, it has a typical market clearing condition:

$$x_i^s - \sum_{j \varepsilon J_1^S} a_{ij}^s x_j^s - \sum_{k \varepsilon J_2^S} a_{ik}^s x_k^s - h_i^s u^s - v_i^s + w_i^s - \lambda^s c_i^s \geqslant q_i^s$$

where

x_i^s = output of industrial sector i in region (nation) s;
a_{ij}^s = output of *ith* sector consumed to produce a unit output of the *jth* sector in region s ($i, j \varepsilon J_1^s$);
J_1 = the set of industrial sectors;
x_k^s = output of pollution abatement sector k in region s;
a_{ik}^s = output of *ith* industrial sector necessary to produce a unit output of the *kth* pollution abatement sector in region s ($k \varepsilon J_2^s$);
J_2 = the set of pollution abatement sectors;
h_i^s = input of the *ith* sector per unit capital investment in region s;
u^s = amount of capital investment in region s;
v_i^s = output of *ith* sector to be exported from region s;
w_i^s = output of *ith* sector to be imported into region s;
λ^s = degree to which a given consumption target is reached in region s;
c_i^s = government and variable household consumption target for sector i in region s;
q_i^s = fixed consumption target for sector i in region s.

For sectors producing nontransportable outputs, the terms v_i^s (exports) and w_i^s (imports) drop out. Since the model is concerned with the pollution problem (environmental quality objective), we have the customary constraints for abatable pollutants:

$$-\sum_{j \varepsilon J_3^s} a_{\ell j}^s x_j^s + \sum_{k \varepsilon J_2^s} a_{\ell k}^s x_k^s - \lambda^s c_\ell^s > - q_\ell^s (\ell \varepsilon J_1^s)$$

where

a_{lj}^s = output of abatable pollutant ℓ generated per unit of the *jth* industrial sector output in region s ($\ell \varepsilon J_3^s$),
J_3 = the set of abatable pollutants;
$a_{\ell k}^s$ = output of abatable pollutant ℓ destroyed per unit of *k-th* pollution abatement sector output in region s ($k \varepsilon J_2^s$);
c_ℓ^s = variable consumption target pollution emissions of type ℓ in region s; and
q_ℓ^s = fixed consumption target for pollution emissions of type ℓ in region s.

There are other constraints pertaining to the transportation sector, non-abatable pollutants, labour, capital stock and capital investment, and the necessary equality of imports and exports worldwide. One objective function examined maximises the percentage of consumption targets realised, where these percentage levels are the same over all sectors and regions. Other objective functions maximise world consumption levels, where regional and sectoral shares are prespecified. Still other variations of the model take into account balance of payments constraints.

Comparative Cost-Industrial Complex analysis essentially involves a determination of the profitability or non-profitability of different production sectors or selected complexes of sectors. The comparative cost approach is well known and does not require description here. The industrial complex approach can be summarily stated as follows. For that part of a complex characterised by linear relations, we calculate the net advantage Z as:

$$Z = \varrho A X$$

where

ϱ = a $1xl$ row vector of cost and revenue differentials by commodity between the location under study and the location which is most competitive with it;
A = an activity matrix with ℓ rows (each corresponding to a commodity) and n columns (each corresponding to a production activity);
X = an $nx1$ column vector of the different production activities.

For that part of a complex characterised by non-linear relations (in effect scale economies in the use of any input), we make use of standard engineering equations of the form:

$$C_k = \left(\frac{O_k}{O_i}\right)^{\alpha} C_i \qquad L_k = \left(\frac{O_k}{O_i}\right)^{\beta} L_i$$

where

O_i = the level of output for some standard (efficient) industrial plant currently in operation;
O_k = the proposed level of output of the new plant;
C_i and L_i = capital investment and labour requirements, respectively, of the existing plant (known data); and
C_k and L_k = the capital investment and labour requirements to be determined for the new plant.

The parameters α and β are based on past experience, and are notably stable (see Isard, Schooler and Vietorisz, [3]).

From these estimates we can determine, for the location under study, its net advantage or disadvantage for an industrial complex with specific reference to nonlinear inputs.

In Figure 14.4, we depict these four analytical approaches surrounding a central INPOL box, which is going to play a much more important role than in Figure 14.2. We intend (later) to connect the central INPOL box to additional appropriate modules or submodules.

Before indicating what we can and cannot do, let us return to our conception of a general equilibrium for a world society G, and refresh ourselves on some of the conditions that should be fulfilled by our operational model. First we need:

1. supply-demand equilibrium conditions for each commodity in the world system, taking into account changes in inventory, and a similar balance for certain commodities in certain regions or smaller geographic areas where, for various reasons, the commodity is not shipped beyond particular regional boundaries.
2. a system of prices for the several nations such that when a good is exported from one location to another, the delivered price is equal to the price at the originating location plus transport cost – subject however, to a relaxation (tolerance) factor to recognise diversity, imperfect knowledge, inefficiency, heterogeneity of commodities placed in any class, exploratory actions and other factors which give rise to cross-hauling (real and fictitious) which can conveniently be captured by an entropy-type term.

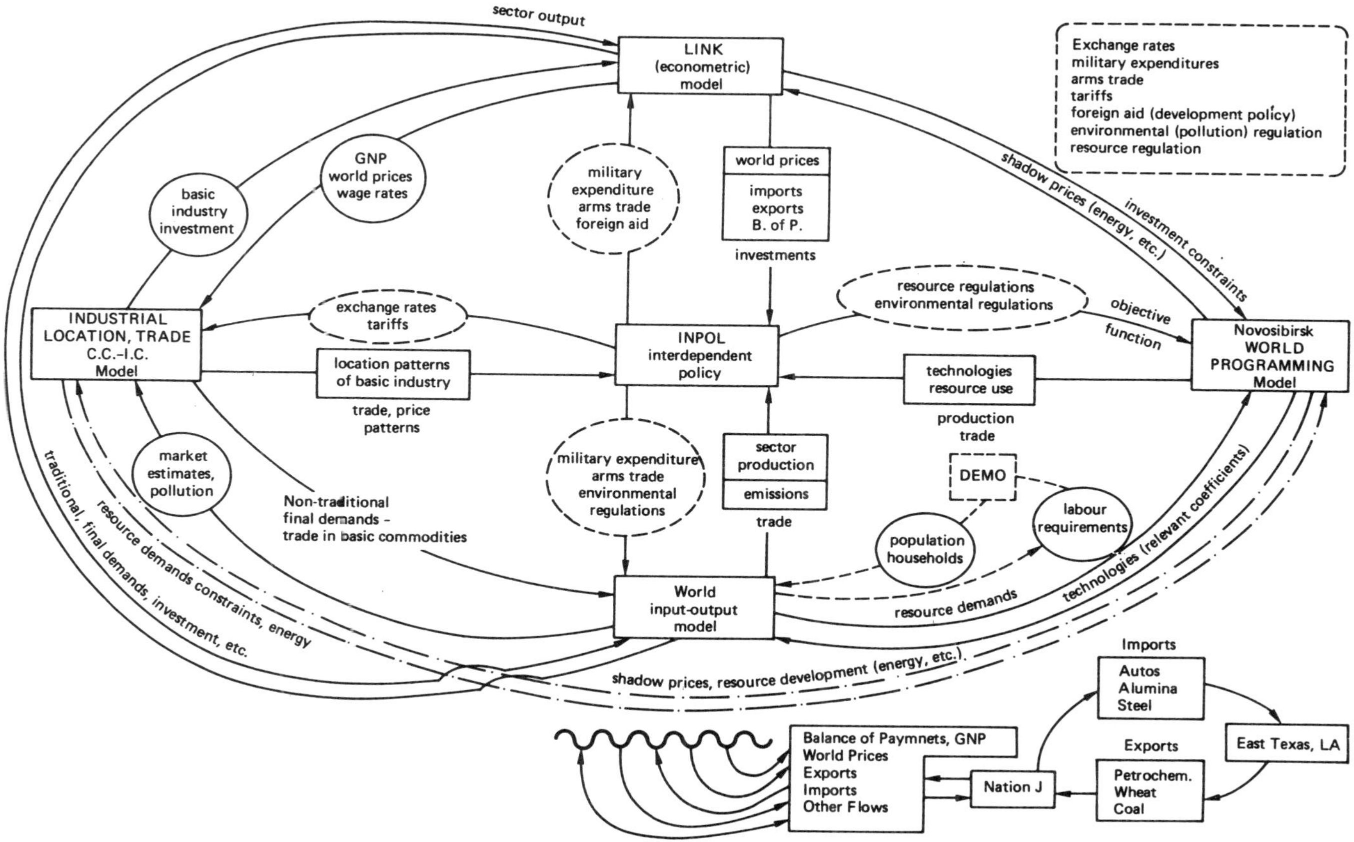

Figure 14.4 An integrated multination model

3. a balance of payments restriction on each nation, again subject to tolerance limits to allow for non-excessive debt and the equivalent, after taking into account interest, rent and dividend payments and capital and other transfers. Over all nations, the flows of money (or the equivalent) must sum to zero.

Against the background of such conditions, let us briefly recall the rationale for each of the modules noted in Figure 14.4. The *LINK (econometric) module* embodies competition on world markets for a number of basic commodities (wheat, coal, oil in part, etc.) and captures to an important extent the effect of world prices on a nation's domestic economy. The *World Input-Output module* reflects the technological requirements for production, and yields projections of output and pollution by world region, for alternative sets of assumptions. The *Novosibirsk World Programming module* reflects the reality of scarce resources and the forces which move the system toward efficiency in their use. So also does the comparative cost and industrial complex analysis of the *Industrial Location-Trade CC-IC module*, when performed on a world basis and when related to the profit motive of entrepreneurs. On the other hand, each approach offers only partial explanation, of the behaviour of the multinational system; each can attack only part of the problem of projecting the world system. It is here that the INPOL box plays a critically important role, much more so than in the integrated multiregional model.

For example, take the question of the determination of exchange rates among national currencies. As is well recognised, the traditional location theory-comparative cost approach rests upon identification of relevant locational cost (revenue) differentials, and the overall net differential, which guide industrial location decisions over the middle-run and long-run. In an intranational setting, these differentials can be calculated at a given moment of time, and changes in them can be forecast with a high degree of reliability. Hence, we are able to project changes in the regional production pattern of iron and steel, aluminum, petrochemical and other basic industries in a fairly reliable way. However, when we move to the international scene we must recognise that locational cost (revenue) differentials among nations are subject to the short-run fluctuations of exchange rates. As a consequence, the reliability of any projection of middle- and long-run differentials is greatly diminished, thereby reducing the validity of any projection of the global distribution of the above mentioned basic industries. None the less, such projections must be made and we must make them as reliable as possible. Thus we must conduct research to determine the pattern of change of exchange rates that is most likely to occur in the middle- and long-run, recognising the reality of short-run fluctuations.

This exchange pattern will of course be in large part an outcome of policies (inclusive of monetary and fiscal policies) determined within the INPOL box of the multinational system. Projections of exchange rates will also affect the results of all other modules depicted in Figure 14.4. For example, because they affect projections of global production patterns for basic industries, they affect the estimation of relevant interregional trade coefficients contained in any World Input-Output model. They also affect the interregional activity analysis matrix of coefficients and the set of relevant finished product prices embodied in an appropriately extended World Programming model, and so affect the optimal pattern of world resource use and production generated by this model. They also affect export and import projections for each nation and thus its GNP; and thus the projections of world prices and associated outcome variables of the LINK module.

So all these models rely essentially on exchange rate patterns and thus policy, and this linkage is indicated in Figure 14.4 by arrows that lead from the INPOL box to each other module. Although it is exceedingly difficult to develop relevant exchange rate forecasts or, more appropriately, a range of likely future exchange rate patterns for the middle and long-run, we must certainly do so; cooperatively with experts in the field who tend to be more concerned with short-run fluctuations and middle-run patterns.

From the INPOL box must come information relating to a host of other policies which greatly affect the output of our integrated model – policies on military expenditures, arms trade, tariffs, foreign aid (development), environmental (pollution) regulations, regulations concerning the development of ocean and other common resources. In Figure 14.4, we indicate the set of policies that flow from the INPOL box to each other box by a dashed oval, which is assumed to include all policies just mentioned as well as many others. These policy areas are indicated in the oval at the upper right corner of Figure 14.4. However, in each oval connecting the INPOL box to another module, we specifically designate one or more policies which are critical in the particular connection. We would consider the most important items to be:

1. exchange rates and the set of specific commodity tariffs in the connection to the Industrial Location Trade CC-IC module;
2. military expenditures, arms trade and environmental regulations in the connection to the World Input-Output module;
3. resource and environmental regulations in the connection to the Novosibirsk World Programming module; and
4. military expenditures, arms trade and foreign aid in the connection to the LINK (econometric) module.

Some of the most important feedbacks to the INPOL box are, as indicated:

1. the world pattern of outputs of selected basic industries (steel, aluminum, petrochemical) from the Industrial Location-Trade-CC-IC module;
2. the world pattern of output and pollution emissions by sector and nation from the World Input-Output module;
3. optimal technologies and resource use patterns for given constraints from the Novosibirsk World Programming module;
4. world prices, imports, exports, balance of payments and investments from the LINK (econometric) module.

In Figure 14.4, we have also indicated some of the main inputs into each module from other modules. For example, levels of non-traditional final demand and trade in selected basic commodities go from the Industrial Location Trade-CC-IC module to the World Input-Output module. Resource demands go from the World Input-Output module to the Novosibirsk Programming module and so forth.

Again, we caution the reader that the above is our current thinking on the most important items for an integrated, operational multinational model given the current state of regional science research. It should be viewed as strictly preliminary.

VI SOME IMPLICATIONS FOR THE PROJECTION OF REGIONAL CHANGE WITHIN A NATION

Let us proceed from the above structure, in which we have emphasised the INPOL box, to the basic problem of projecting the growth of a region within a nation, say the East Texas–Louisiana region of the United States. At the bottom of Figure 14.4, we indicate the link to a nation and then to each region of a nation. It should be clear from the above discussion what these links are – balance of payments, GNP, world prices, imports and exports, etc. We have also indicated links to the specific region of interest, namely the East Texas–Louisiana region. Once again, it is clear what some of the important exports and imports are, and the need to make sound projections for these exports and imports if we are to have reliable projections of the region's growth – as well as implications for the growth of other regions.

VII SOME CONCLUSIONS AND EXTENSIONS

Limited space forces us to conclude at this point. We must reiterate that we are looking ahead. It is just a matter of time before an integrated multiregion model for a nation (like the one described in Isard and Anselin, [8]) will be in operation. It is also just a matter of time before an INPOL module will be added to that operational model; then the model depicted in Figure 14.2 will be in operation. But such a model is still inadequate for projecting regional changes in many nations like the USA and Australia, which have one or more regions with major export-import sectors. Here we need a model that combines Figures 14.2 and 14.4. Although we cannot hope to make that combined model operational at this time, we have outlined it primarily to provoke and stimulate others, especially those in the new generation of regional scientists, into such creative directions.

In pursuing these multiregional and multinational research tasks, we must also make explicit the many policies with which each government is concerned. We must drop the simplifying assumption of one government (interest group) in each region, and recognise the reality of many. It is this area of policy making, containing many interrelated policy issues and governmental authorities, wherein all kinds of coalitions, alliances and blocs form and dissolve, which should be a research area of primary concern to regional scientists. It is one in which we must become increasingly involved, since it requires as input so much of the basic knowledge which only well-seasoned regional scientists can provide.

References

[1] Englebrecht-Wiggans, R., 1977. On the fair and efficient allocation of commodities. Ithaca, New York, USA: Cornell University, School of Operations Research and Industrial Engineering, Technical Paper 356.

[2] Granberg, A. G. and Rubinshtein, A. G., 1979. Some lines in development of the United States global input-output model. Novosibirsk, USSR: Institute of Economics and Organization of Industrial Production.

[3] Isard, W., E. Schooler and T. Vietorisz, 1959. *Industrial Complex Analysis and Regional Development* (MIT Press, Cambridge, MA).

[4] Isard, W., 1960. *Methods of Regional Analysis*. Cambridge, Massachusetts: MIT Press.

[5] Isard, W., 1969. *General Theory, Social, Political, Economic and Regional*, (MIT Press, Cambridge, MA).

[6] Isard, W., 1979. The origins, development and future of regional science, *Papers, Regional Science Association*, vol. 43.

[7] Isard, W., 1979a. A definition of peace science, the queen of the social sciences, part I. *Journal of Peace Science*, vol. 4, pp. 1–47.

[8] Isard, W. and Anselin, L., 1982. Integration of multiregion models for policy analysis. *Environment and Planning A*, vol. 14, pp. 359–76.

[9] Isard, W. and Smith, C., 1984. Incorporation of conflict and policy analysis in an integrated multiregion model. *Regional and Industrial Development Theories, Models and Empirical Evidence*. (North-Holland, Amsterdam).

[10] Klein, L. R., 1977. *Project LINK*. Athens, Greece: Center of Planning and Economic Research.

[11] Lucas, W. F., 1978. The apportionment problem. School of Operations Research and Industrial Engineering, Technical Report 388, (Cornell University, Ithaca, New York).

[12] Leontief, W., 1977. *The Future of the World Economy: A United Nations Study* (Oxford University Press, New York).
[13] Nash, J. F., 1950. The bargaining problem. *Econometrica*, vol. 18, pp. 155–62.
[14] Shapley, L. S., 1950. A value for N-person games. In *Contributions to the Theory of Games, II*, Annals of Mathematical Studies, H. W. Kuhn and A. W. Tucker (eds) (Princeton University Press, Princeton, New Jersey).

Appendix 14A

Some notes on current conceptual thinking regarding world models and progress in their construction

1 INTRODUCTION

In this brief appendix I wish to bring the reader up-to-date on certain developments with regard to world system models. Section 2 draws entirely upon the conceptual thinking of Christine Smith as contained in the paper she presented at the American Economics Association, Chicago, December 1987. Section 3 records some of the general features of the applied model which we are running jointly using Cornell's supercomputer and associated computer facilities.

2 THE SMITH COMPUTABLE GENERAL EQUILIBRIUM (CGE), INPUT-OUTPUT (I-O), SOCIAL ACCOUNTS (SAM) CONCEPTUAL FRAMEWORK

Ideally we would seek a dynamical general equilibrium spatial system. Smith pointed out that in practice we are not at the stage where current computations/findings for the world would permit such a system in continuous space to be implemented. Rather we must be satisfied with treating the world as a set of points wherein each region or relevant spatial unit is viewed as a concentration at a point. The procedure that one must follow is to build up from this set of points to a world system model by adding successively elements of strength to an initial structure that represents a crude approximation to the world system.

Smith observed that historically one of the earliest structures proposed, and one that is probably the most generally accepted as valid, is the input-output structure. Leontief's World Input-Output model is an interregional depiction of this structure. The many shortcomings of an input-output framework are well known – constant production coefficients, constant consumption coefficients alike for each household, constant shares assumption for projecting trade, etc. It is however operational and is subject to steady improvement as the data for different nations become more extensive and of higher quality. This structure is easily extended and has been to probe into environmental problems via the use of pollution coefficients and through the incorporation of pollution abatement activities and technology.

Most world system modelling strategies employ an input-output frame of some type to capture the effects of intersectoral linkages on each region's (nation's) economy. However, as just noted it has many inadequacies. A first improvement upon the Leontief framework might be the recognition that the household sector comprises different socio-economic groups and that at least the constant consumption pattern that might be assumed to be alike for each member of a socio-economic group should or can be different from that of another socio-economic group. Moreover, data are available in many countries for approximating the consumption pattern for each of these socio-economic groups. However, to effect this improvement at the same time one needs to disaggregate the value added component of each input-output sector into different income generating categories – wages and salary incomes by different occupation group, profits by companies and unincorporated enterprises, etc. One would then need to allocate the income generated by each of these value-added items to the several socio-economic groups. Once achieved this SAM type of addition to input-output would then allow feedback between factor markets and final demand in operating a linear structural framework within input-output. An increasing number of nations are developing national accounts data (income and product accounts) which will allow one to perform this disaggregation.

A further improvement, Smith noted, would be to recognise that the assumption of constant coefficients of consumption can be relaxed by use of non-linear consumption patterns for each of the socio-economic groups. Such nonlinear consumption functions may be of the LES, Cobb Douglas, or flexible form type. Similarly, non-linear production functions (CES, Cobb Douglas, translog or other flexible form) can be introduced in order to permit basic substitution among inputs. Already, there have been developments in CGE models which permit such non-linearities to be taken into account, where desirable. Frequently, however, to retain operationality or avoid excessive data requirements for parameter estimation, the input-output constant coefficients form is in part retained so that we have two-level, two-tier production functions. Such permit substitution among factor inputs (labour, capital, etc.) but retain constant relationships for intermediate inputs. One should note however that while the non-linearity extension can be viewed on the surface as a close representation of reality – for we all know that reality is largely non-linear – yet the non-linearity that is generated from the use of CES, Cobb–Douglas and other functions can be argued to pertain to societies wherein behaving units act in accord with first-order conditions for utility and profit maximisation. So the question that needs to be asked is whether non-linearity for a first-order condition society which we know is not a true representation of actual society is a better representation than a linear economy representation that is a mixture of all kinds of behaviour, including first order. Moreover, when we use a CES function there is the additional assumption not only that the first order condition is met but also that substitution is constrained by a constant elasticity parameter. Furthermore, when one employs the translog or other flexible form function, changes are restricted to a small area around that behaviour consistent with a first order condition for society. Clearly when we are treating developing economies looking forward to major growth such cannot be justifiably used.

Smith noted the need for still another improvement. Such stems from the requirement for projecting a bilateral trade matrix that is more satisfactory than the trade matrix that is projected by the constant shares approach carried over from the Leontief model. CGE models have experimented with CES functions but in a most inadequate way. Here improvement is possible, certainly in some respects, by using the multi-method approach suggested by Isard and Dean [1].

Smith contends that yet another important improvement of the above framework is needed. Whether we think of centrally planned or market economies we are aware of all kinds of imperfections in their operation. Neoclassical and pure CGE models have no effective way to incorporate unemployment (an imperfection of the labour market), discrepancies between savings and investment (an imperfection of the capital market), imbalances in balance of payments (an imperfection in foreign exchange markets), and so on. We need to explicitly recognise that these imperfections exist in the short-run and that despite our policies to reduce the severity of these imperfections they will be constantly with us. Hence when there are stable patterns in these imperfections over time we should resort to econometric estimation of parameters that would capture these imperfections. Thus valid econometric estimations should be added to the extended CGE framework sketched above.

There are of course still other problems that should be addressed: for example, changes in tastes, socio-economic groups, regions, industries, credit creation, and capital flows. For the present, however, these must be put on the back-burner, according to Smith.

3 FEATURES OF A PRACTICAL OPERATIONAL WORLD SYSTEM MODEL IN PROGRESS

While the Smith CGE-I-O SAM framework has a number of desirable features, at this time (1988) problems of data availability on a world scale make its operation impractical. To get started on the construction of an operational model, a different approach has been necessary. Given that the Klein LINK model was running on the Cornell supercomputer as of January 1988, Dr Smith and I deemed it highly desirable to employ that model as a base upon which to add successive improvements.

The next logical step was then to add on an input-output framework. At first our thinking was to add the Leontief World Input-Output model. Although problems would arise from the need to reconcile Leontief's set of world regions with the nations of the LINK model, we did not judge that problem to be severe. We did, however, feel that many of the parameters comprising the World Input-Output model were too out-of-date to combine meaningfully with the LINK projections. Moreover, the increasing stock of new data becoming available would necessitate the development of a mechanism for periodic updating of the Leontief World I-O parameters to be incorporated into the modelling framework. Hence, we decided to follow the procedure of adding on, as data and time permit, an input-output model for each nation. Thus far we have done so for the most recent input-output model available for both the United States and Australia. Such a decision has turned out to be a wise one, for it is clear that the particulars of the procedure to splice an input-output table on to projections of LINK magnitudes depend significantly on the type of accounting conventions employed in the input-output table available for the given nation. We know that these tables differ highly among nations, however the splicing mechanism in each case revolves around taking LINK projections of final demand magnitudes for a given year (say 1990) and using these as inputs to the input-output model. The latter is then run to generate sectoral output (GDP) projections for 1990 which are consistent with these LINK magnitudes. (The procedures for achieving this linkage of input-output and econometric models will be discussed in Smith [3].) The advantage of this approach of proceeding on a one-by-one nation basis is further enhanced when the extension to incorporate a social accounting framework is contemplated. As discussed in Section 2, not all countries have the data available currently to produce a meaningful, operational SAM. However, for the United States such data exist due largely to the work of Adam Rose and his associates [2]. We have taken Rose's data relating to consumption and income patterns for each of seven categories of households, and extended the latest United States input-output table to become a SAM. Linkage of the SAM and the LINK frameworks to provide more disaggregated projections for 1990 has been achieved, following an adaption of the method sketched above for I-O and LINK. That is, we take the LINK projections for final demand magnitudes excluding endogenous consumption as inputs to the SAM model and use this latter model to project sectoral outputs (GDP), consumption by household type, and income distribution by household type. (Details of this procedure will once again be provided in Smith [3]).

Finally, experiments are currently being conducted aimed at improving upon the constant share trade projection mechanism employed in the LINK model. These experiments revolve around using a combination of gravity model and constant share approaches, and the initial results look promising.

References

[1] Isard, W. and W. Dean, 1987. 'A Multimethod Approach toward World Trade Matrix Projection', *Environment and Planning A*, vol. 19, pp. 1059–66.

[2] Rose, A., B. Stevens and G. Davis, 1988. *Natural Resource Policy and Income Distribution* (John Hopkins University Press, Baltimore, MD).

[3] Smith, C., 1989. 'On the Integration of Econometric, Input-Output and Social Accounting Models and Frameworks', Paper presented at Workshop on Large-Scale (Global) Social Science Models, National Center for Supercomputing Applications, University of Illinois (Urbana), September.

15 Policy Analysis Using an Integrated Multiregional Model

I INTRODUCTION

In the last few years, it has become increasingly evident to regional scientists that there is a high degree of interdependence involved in understanding and projecting the impacts of different policies, singly and in combination. The interdependence arises from the fact that each policy, say housing, energy, environment, or industrial development comes to affect the operation of any economic or other sector of a region and the region's welfare (state of affairs) which then influences how other policies may affect these sectors and the region's welfare. For example, it is obvious that any policy designed to maintain or improve the environmental quality of a developing region is highly dependent upon what energy policy is chosen, whether all-out nuclear, all-out coal, a combination of coal, oil, and solar, or something else. At the same time, the impact and effectiveness of an energy policy such as an all-out coal policy is highly dependent upon environmental regulations, regarding say allowable sulphur dioxide and particulate emissions since these regulations can significantly affect the cost of operating coal power plants. Similarly, it is easy to show how transportation, housing, urban land use, employment, industrial development, and other policies have impacts which can be projected only when we know the specific policies (and instruments) adopted in other policy areas.

In past decades, regional scientists have developed different kinds of models to understand and project the effects of different types of policy. For example, we have developed interregional linear programming to handle the implications of alternative energy policies; interregional input-output to get at multiplier effects of alternative industrial development policies; comparative cost and industrial complex analysis to gauge the impact of an across-the-board, nation-wide environmental regulation upon the cost of production in different regions and thus the relocation of industry; gravity and entropy-type models to project impacts of transportation development policies; factor substitution models to estimate changes

---

This chapter was written with Christine Smith. Reprinted from M. Chatterji, *et al*, (eds), *Spatial, Environmental and Resource Policy in the Developing Countries* (Gower, Aldershot, UK, 1984), pp. 64–78.

in production practices from wage and price policies; and regional econometric and demographic models to estimate the impact of social welfare policy, unemployment compensation, etc., on migration. Now, however, when we have come to recognise the interdependence of the impacts of different policies, it is essential that we develop models which integrate several or all of the models mentioned above. It is essential that we do so if we are to project accurately the impacts of any policy, whether we are considering a developed or developing region. That is, the impact or the effectiveness of any policy by itself cannot be examined without reference to what other policies are likely to be in effect.

Accordingly, the authors have developed in other manuscripts an integrated multiregional model to help gauge better the impacts of several different types of policies upon a multiregion system for both developed and developing countries (Isard and Anselin, [1]; Isard and Smith [3] and [5]). As indicated in Figure 14.2, this model consists of the following modules:

1. a national econometric module (NATLEC)
2. an integrated comparative cost, industrial complex, input-output and programming module (CICIOP)
3. a transportation module (TRANS)
4. a demographic module (DEMO)
5. a multiregion econometric module (REGLEC)
6. a factor demand-investment supply econometric module (FACTIN)
7. a multipolicy formation module (INPOL).

In a number of articles (Isard and Anselin [1]; Isard and Smith [3], [4], [5], and C. Smith [8]) and in a book (C. Smith [7]) we have established and developed most of the linkages, and, in part, the data base for the operation of such a model. We have also suggested the operation of a partially integrated model for situations often characteristic of developing nations where the data base is insufficient for the operation of a fully integrated model. However, we have yet to indicate in an effective manner the diverse approaches that might be adopted within the multipolicy formation module. Such approaches take into account the consequences of any mix of policies that might be proposed and lead to the identification of a best or most likely set of compromise policies. That is, in terms of Figure 15.1, we seek that set of compromise policies that might result after one or more sets have been fed into the aggregate of non-policy modules, indicated by the long, narrow, rectangular box designated 'Use of the Non-Policy Modules of the Integrated Model' in the center of the figure, and the results examined. (The specific ways and exact points at which a given set of policies[1] link to each module of the model, or of a partially integrated model, are discussed elsewhere: Isard and C. Smith [4] and C. Smith [7].)

In developing this paper, we also wish to point up how a specific

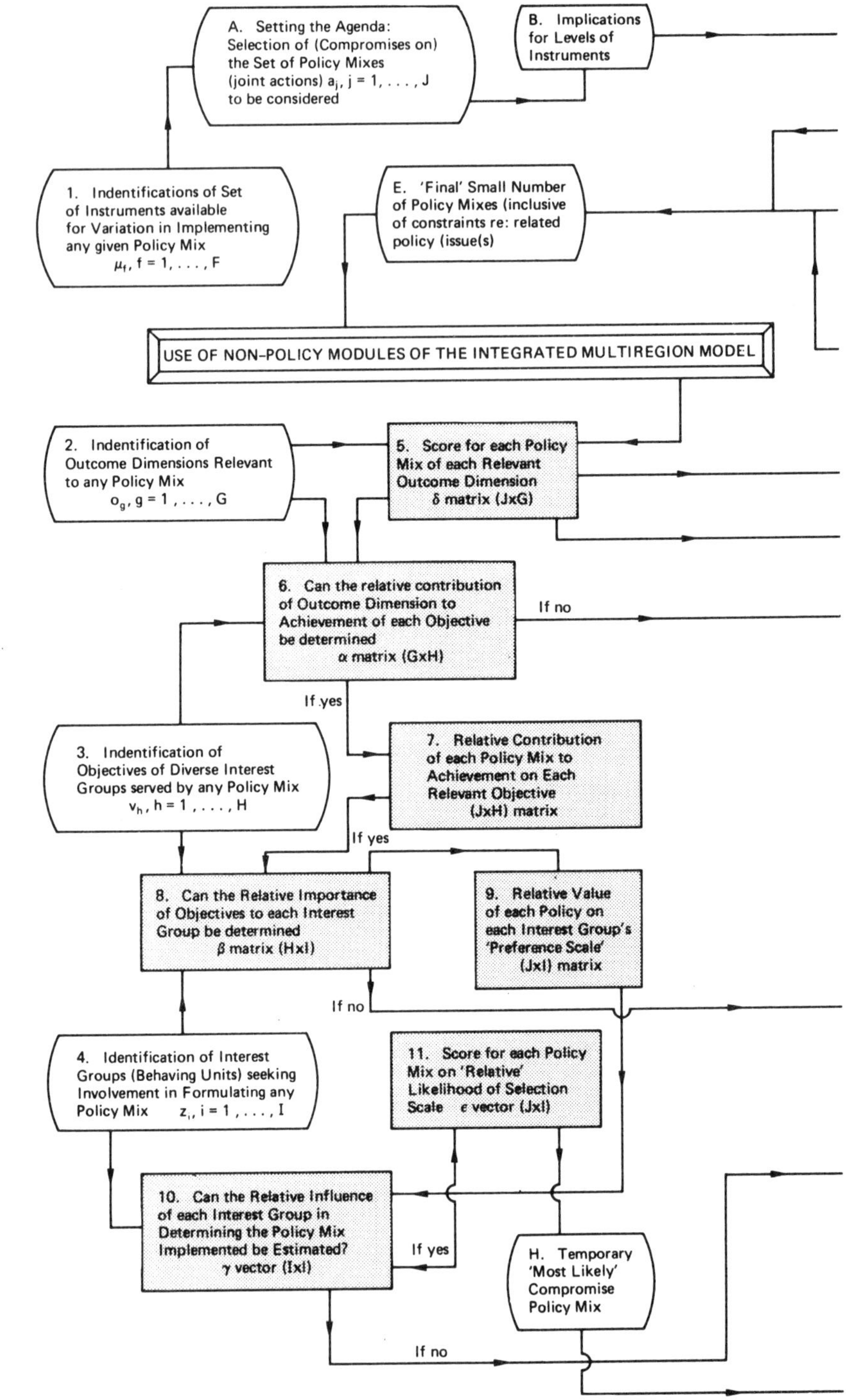

Figure 15.1 A policy projection framework

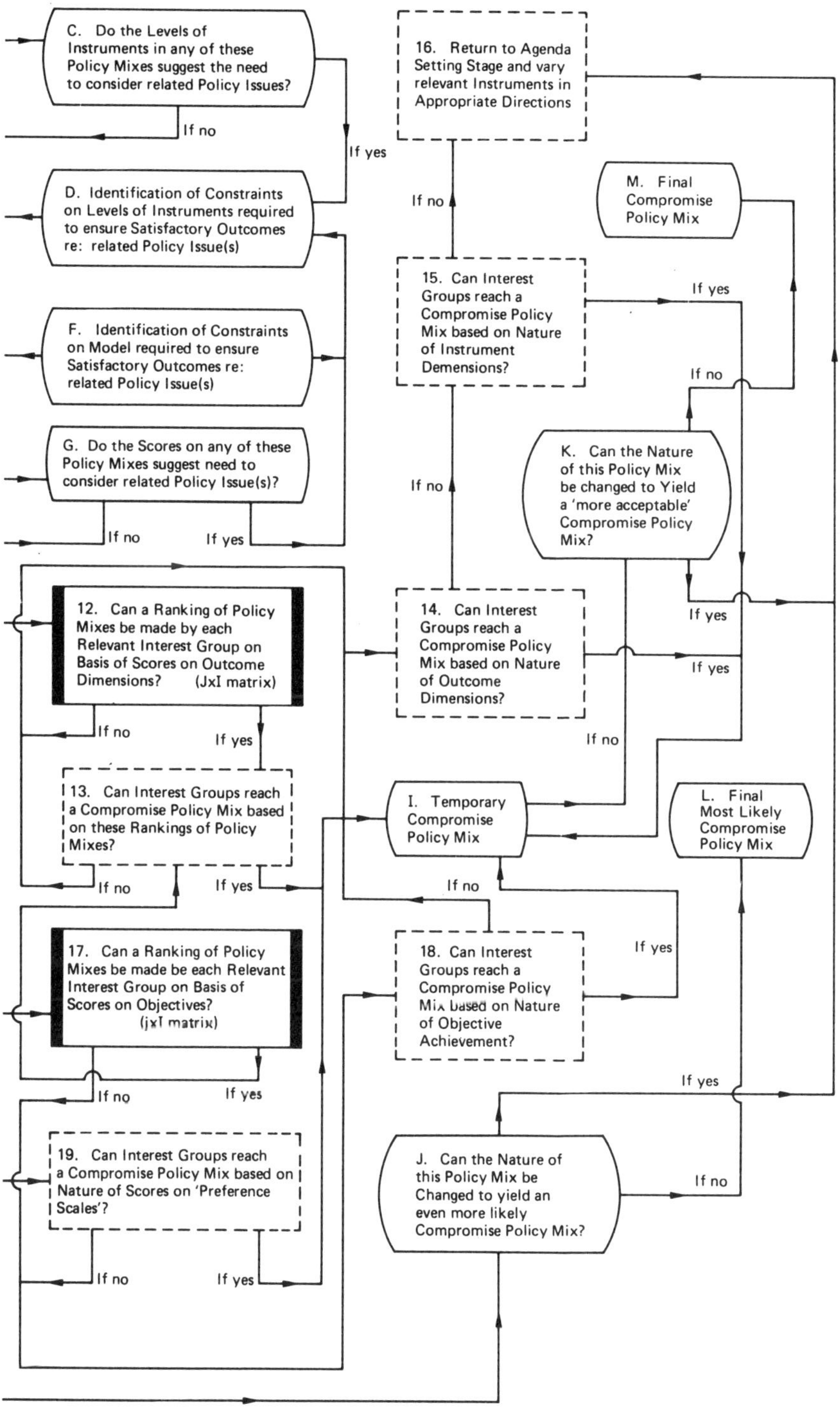
C. Do the Levels of Instruments in any of these Policy Mixes suggest the need to consider related Policy Issues?
If no
If yes
16. Return to Agenda Setting Stage and vary relevant Instruments in Appropriate Directions
D. Identification of Constraints on Levels of Instruments required to ensure Satisfactory Outcomes re: related Policy Issue(s)
If no
M. Final Compromise Policy Mix
15. Can Interest Groups reach a Compromise Policy Mix based on Nature of Instrument Demensions?
If yes
F. Identification of Constraints on Model required to ensure Satisfactory Outcomes re: related Policy Issue(s)
If no
G. Do the Scores on any of these Policy Mixes suggest need to consider related Policy Issue(s)?
If no
If yes
If no
K. Can the Nature of this Policy Mix be changed to Yield a 'more acceptable' Compromise Policy Mix?
12. Can a Ranking of Policy Mixes be made by each Relevant Interest Group on Basis of Scores on Outcome Dimensions? (JxI matrix)
14. Can Interest Groups reach a Compromise Policy Mix based on Nature of Outcome Dimensions?
If yes
If yes
If no
If yes
If no
13. Can Interest Groups reach a Compromise Policy Mix based on these Rankings of Policy Mixes?
I. Temporary Compromise Policy Mix
L. Final Most Likely Compromise Policy Mix
If no
If yes
If no
17. Can a Ranking of Policy Mixes be made be each Relevant Interest Group on Basis of Scores on Objectives? (jxI matrix)
18. Can Interest Groups reach a Compromise Policy Mix based on Nature of Objective Achievement?
If yes
If no
If yes
If yes
19. Can Interest Groups reach a Compromise Policy Mix based on Nature of Scores on 'Preference Scales'?
J. Can the Nature of this Policy Mix be Changed to yield an even more likely Compromise Policy Mix?
If no
If no
If yes

compromise procedure (the method of determining group priorities a la Saaty) (Isard and C. Smith [2], pp. 147–55, 165–68) can be used to identify (project) the most likely mix of policies to be adopted (Isard and C. Smith [4]) in a given year of projection (say the year 2000). Because of limited space, we discuss only in general how other conflict management procedures (as systematically covered in Isard and C. Smith [2]) can be effectively employed to supplement the Saaty approach.

II THE BASIC ELEMENTS OF A POLICY CHOICE

We now list the basic elements in a policy choice by a given government unit (say, the Federal government):

1. Interest groups influencing the decisions of the given government unit z_i ($i = 1, \ldots, I$) (oval 4 of Figure 15.1),
2. Objectives of importance to one or more interest groups v_h ($h = 1, \ldots, H$) (oval 3 of Figure 15.1),
3. Outcome dimensions (as derived from the operation of the integrated model) o_g ($g = 1, \ldots, G$) (oval 2 of Figure 15.1),
4. Policy proposals (joint actions) a_j ($j = 1, \ldots, J$) (ovals A and E of Figure 15.1),
5. Policy instruments (specified in a form amenable for use in the integrated model) μ_f ($f = 1, \ldots, F$) (oval 1 of Figure 15.1).

Given these elements, there exist many ways by which to evaluate the different policy mixes that may be proposed for the year 2000 in terms of their outcome implications as determined via the operation of the non-policy modules of an integrated multiregion model. With reference to Figure 15.1, one possible set of steps may be as follows:

Step 1: Selection of an Initial Set of Policy Mixes to be Considered

While many policy proposals (mixes or joint actions) may be considered, in practice only a few can be examined for outcome implications with the use of the non-policy modules of the integrated multiregion model. The activity involving the selection of these few proposals (which may be viewed as *setting the agenda*) via diverse compromises is depicted by the bold oval A where the political leaders, representatives of diverse interest groups and other behaving units, interact.

Step 2: Identification of Nature and Level of Instruments Required to Implement the Given Policy Mixes

In bold ovals B, C, D, and E, respectively

1. We determine the set of instruments μ_f ($f = 1, \ldots, F$) and levels of each required to effect each given policy mix (proposal) a_j ($j = 1, \ldots, J$);[2]
2. We consider whether the levels of instruments identified in (1) suggest the need to consider related policy issues not covered by these policy mixes.[3]
3. If there is such a need, we determine (identify) the constraints on the levels of the instruments required to ensure satisfactory outcomes regarding related policy issues;
4. We eliminate those policy mixes which cannot meet the constraints identified in (3). This yields a $J \times F$ matrix recording levels of the various instruments μ_f ($f = 1, \ldots, F$) required to implement the final small number of policy mixes a_j ($j = 1, \ldots, J$) to be considered in subsequent steps.

Step 3: Identification of Outcome Implications of Each Policy Mix

Each policy mix a_j can now be run through the non-policy modules of the integrated multiregion model indicated by the long rectangular box in Figure 15.1. This step is achieved by putting the set of relevant policy instruments at the corresponding levels identified in step 2. This yields a score for each policy mix (proposal) a_j ($j = 1, \ldots, J$) on a number of relevant outcome dimensions o_g ($g = 1, \ldots, G$).[4] See the shaded box 5 in Figure 15.1. We can then construct an outcome matrix (designated δ) of order $J \times G$, with typical element o_j recording the score of policy mix j on outcome dimension g. That is:[5]

$$
\begin{array}{ccc}
 & & \begin{array}{cccccc} o^1 & o^2 & \cdots & o^g & \cdots & o^G \end{array} \\
\begin{array}{c} a_1 \\ a_2 \\ \bullet \\ \bullet \\ \bullet \\ a_j \\ \bullet \\ \bullet \\ \bullet \\ a_J \end{array} &
\begin{array}{c} \rightarrow \\ \rightarrow \\ \\ \\ \\ \rightarrow \\ \\ \\ \\ \rightarrow \end{array} &
\begin{array}{c} o_1 \\ o_2 \\ \bullet \\ \bullet \\ \bullet \\ o_j \\ \bullet \\ \bullet \\ \bullet \\ o_J \end{array}
\left[\begin{array}{cccccc}
o_1^1 & o_1^2 & & o_1^g & & o_1^G \\
o_2^1 & o_2^2 & & o_2^g & & o_2^G \\
\bullet & \bullet & & \bullet & & \bullet \\
\bullet & \bullet & & \bullet & & \bullet \\
\bullet & \bullet & & \bullet & & \bullet \\
o_j^1 & o_j^2 & \cdots & o_j^g & \cdots & o_j^G \\
\bullet & \bullet & & \bullet & & \bullet \\
\bullet & \bullet & & \bullet & & \bullet \\
\bullet & \bullet & & \bullet & & \bullet \\
o_J^1 & o_J^2 & \cdots & o_J^g & & o_J^G
\end{array}\right] \\
 & & J \times G
\end{array}
$$

At this point we need to determine whether or not the outcome scores for any of these policy mixes suggest the need to consider implications for related policy issues not included in that policy mix.[6] See bold oval G. If not, we can proceed to step 5. Otherwise, we may need to change the levels of some of the instruments (bold oval D) associated with the given policy mixes. We may even need to impose other constraints[7] (bold oval F) which could result in the elimination of one or more policy mixes from further consideration. The integrated model must then be rerun for each non-eliminated policy mix under the new constraints, giving us a new δ matrix.

Step 4: Identification of the Relative Contribution of each Outcome Dimension to Achievement on Each Objective[8]

The evaluation of these contributions can be done via the use of the Saaty procedure for determining group priorities. (See shaded box 6.) For example, for a given objective, v_h, we might (after consulation with relevant experts) obtain a pairwise comparisons matrix ($G \times G$) in the following form:[9]

$$\begin{array}{c} \\ o_1 \\ o_2 \\ o_3 \\ \cdot \\ \cdot \\ \cdot \\ o_G \end{array}
\overset{\displaystyle\begin{array}{ccccc} o_1 & o_2 & o_3 & \ldots & o_G \end{array}}{\left[\begin{array}{ccc|c} 1.0 & 2.0 & 3.0 & \\ 0.5 & 1.0 & 2.0 & 0 \\ 0.33 & 0.5 & 1.0 & \\ \hline & 0 & & 0 \end{array}\right]}_{G \times G}
\ \text{which implies}\
\left[\begin{array}{c} 0.540 \\ 0.297 \\ 0.163 \\ \hline 0 \end{array}\right]_{G \times 1}
\ \text{when normalised}$$

As indicated above, this matrix yields a $G \times 1$ vector of 'weights' showing the relative contribution of outcome dimension o_g ($g = 1, \ldots, G$) to the achievement of objective v_h.[10] For each of the H objectives a comparable $G \times 1$ vector can be derived, and when these results are recorded in a single table this yields a matrix (designated α) of order $G \times H$.

Step 5: Identification of the Relative Contribution of Each Policy Mix to Achievements on Each Objective

To identify these contributions, premultiply the α matrix of order $G \times H$ derived in step 4 (shaded box 6) by the δ matrix of order $J \times G$ derived in step 3 (shaded box 5) to obtain, as noted in shaded box 7, a $J \times H$ matrix. The typical element of this latter matrix gives the relative contribution of policy mix j ($j = 1, \ldots, J$) to the achievement of objective h ($h = 1, \ldots, H$).

Step 6: Identification of the Relative Importance of Each Objective to Each Interest Group[11]

This identification process can be done via the use of the Saaty procedure for determining group priorities. (see shaded box 8.) For example, for a given interest group z_i we might (after consultation with relevant experts) obtain a pairwise comparisons matrix ($H \times H$) of the form:[12]

$$\begin{array}{c} \\ v_1 \\ v_2 \\ v_3 \\ \vdots \\ v_H \end{array}
\begin{array}{c} \begin{array}{cccccc} v_1 & v_2 & v_3 & \ldots & v_H \end{array} \\
\left[\begin{array}{ccc|c} 1.0 & 3.0 & 0.5 & \\ 0.33 & 1.0 & 0.2 & 0 \\ 2.0 & 5.0 & 1.0 & \\ \hline & 0 & & 0 \\ & & & \end{array}\right] \\ H \times H \end{array}
\text{ which implies }
\begin{array}{c} \\ \left[\begin{array}{c} 0.31 \\ 0.11 \\ 0.50 \\ \hline 0 \\ \end{array}\right] \\ H \times 1 \end{array}
\text{ when normalised}$$

As indicated above, this yields a $H \times 1$ vector of 'weights' showing the relative importance of objective v_h ($h = 1, \ldots, H$) to the interest group z_i.[13] For each of the I interest groups a comparable $H \times 1$ vector can be derived, and when these results are recorded in a single table this yields a matrix (designated β) of order $H \times$ I.

Step 7: Identification of the Relative Value of Each Policy Mix for Each Interest Group

We first premultiply the β matrix of order $H \times I$ derived in step 6 (shaded box 8) by the α matrix of order $G \times H$ derived in step 4 (shaded box 6) to obtain a $G \times I$ matrix which indicates the relative importance of each outcome dimension to each interest group. We can then premultiply this $G \times I$ matrix by the δ matrix of order $J \times G$ derived in step 3 (shaded box 5) to obtain, as noted in shaded box 9, a $J \times I$ matrix. The typical element of this latter matrix gives the relative value of policy mix j ($j = 1, \ldots, J$) to each interest group i ($i = 1, \ldots, I$).

Step 8: Identification of the Relative Importance (Influence) of each Interest Group in the Policy Formation Process[14]

This type of identification can be done via the use of the Saaty procedure for determining group priorities. See shaded box 10. For example, after consulting with experts on the policy formation process we might obtain a pairwise comparisons matrix of the form:[15]

$$
\begin{array}{c}
 \\ z_1 \\ z_2 \\ z_3 \\ \\ \bullet \\ \bullet \\ \bullet \\ z_I
\end{array}
\begin{array}{c}
\begin{array}{ccccc} z_1 & z_2 & z_3 & \ldots & z_I \end{array} \\
\left[\begin{array}{ccc|c}
1 & 2 & 7 & \\
0.5 & 1 & 6 & 0 \\
0.14 & 0.17 & 1 & \\
\hline
 & & & \\
 & 0 & & 0 \\
 & & & \\
\end{array}\right] \\
I \times I
\end{array}
\text{ which implies }
\begin{array}{c}
\left[\begin{array}{c}
0.58 \\ 0.25 \\ 0.07 \\
\hline
 \\ 0 \\ \\
\end{array}\right] \\
I \times 1
\end{array}
\text{ when normalised}
$$

As indicated above, this yields a $I \times 1$ vector of 'weights' showing the relative importance (influence) of each interest group z_i $(i = 1, \ldots, I)$ in selecting the policy mix to be put into effect by the given level of government. We designate this vector the γ vector.[16]

Step 9: Identification of the Relative Likelihood of Adoption of each Policy Mix

We next multiply the $J \times I$ matrix showing the relative value of each policy mix to interest groups derived in step 7 (shaded box 9) by the γ vector of order $I \times 1$ derived in the previous step (shaded box 10). As noted in the shaded box 11, this yields a $J \times 1$ vector (designated ε) whose typical element gives the relative likelihood of selection (importance) of policy mix a_j $(j = 1, \ldots, J)$ by the given level of government.[17] This is the desired outcome, and it may appear at first glance that we can now conclude that the policy mix (proposal) with the highest relative value be designated the 'most likely' compromise policy proposal. See the bold dotted oval H. However, we regard this conclusion as only temporary and suggest the adoption of one more step.

Step 10: Resetting of Agenda in the Light of the Results of Previous Steps

Next we may inquire (or the political leaders may consider) whether the nature of the most likely policy mix (identified in step 9) can be changed to yield an even more likely 'compromise' policy mix. See the bold dotted oval J. If not, they accept the results in shaded box 11 as final, thus moving to the bold dotted oval L. If yes, they, in effect, return (via box 16) to the agenda setting stage (bold ovals A through E), introduce an adjusted policy mix, and go through the entire process again. However, if the Saaty procedure is once again employed some new problems and inconsistencies may arise. We cannot discuss these further here due to limited space.

III CONCLUDING REMARKS

In this chapter we have briefly sketched how one conflict management procedure may be employed together with the non-policy modules of an integrated multiregion model to forecast the policy mix most likely to be selected (in the sense of being a best compromise). There are many other procedures that can be employed, as we have suggested in the footnotes. Each procedure has its strength and limitations and for further discussion the reader is referred to Isard and C. Smith [2] and references cited therein.

Notes

1. The policies selected for consideration are tariff policy, mineral development policy, immigration policy, foreign investment policy and the level and mix of federal government programmes (inclusive of taxation).
2. For example, the policy mix may relate to military expenditures, energy resource development, low income housing provision, transportation network extensions, etc. In this situation the relevant set of instruments would include military procurement by region, employment on military bases by region, subsidies on oil shale installations by region, dollar expenditures on federal housing by region, and expenditure on mass transit by region.
3. For example, should the level of tariffs be reduced across the board by 25 per cent in order to effect a trade policy, then we may need to consider the introduction of a manpower planning policy to avert the undesirable employment implications of this tariff change.
4. Typical outcome dimensions would be regional employment and/or income levels, unemployment rates, air pollution levels, etc.
5. Note that each row represents the outcome vector corresponding to a given policy mix, and that each column has been normalised (i.e., converted to a common scale).
6. For example, examining the results recorded in the δ matrix may reveal that a given policy mix (say, a_j') leads to an undesirably high level of unemployment

in a given region (or for a particular class of workers). This unemployment situation may then suggest the need for the simultaneous adoption of a region- (or class-) specific employment development programme.

7. For example, such constraints may relate to outcome dimensions such as oil imports, capital available for investment in housing and infrastructure, and disparities in regional per capital income. In some situations, political leaders (or their right hand men) may be unable to agree on a particular set of policy mixes to be considered, yet able to specify a minimum acceptable target score on each outcome dimension. The analyst may then be able to identify a set of policy mixes which are each capable of meeting these target scores.
8. If such relative contributions can not be identified using the Saaty procedure, then we must consider other alternatives. For example, (as indicated in boxes 12 and 13, respectively) should a ranking of policy mixes (or proposals) be available (or able to be constructed) for each interest group on the basis of the outcome scores identified in step 3, then the analyst may consider the use of rank-oriented conflict management procedures, which yield a temporary compromise policy proposal, as indicated in the light dotted oval I. Where such rankings are unavailable, the analyst may focus on the outcome scores or instrument dimensions directly and (via the use of one or more reasonable conflict management procedures) reach agreement on a compromise set of outcomes/instrument levels. These options are indicated in boxes 14 and 15, respectively. If successful, they too yield a temporary compromise policy proposal (light dotted oval I). On the other hand, should none of these approaches work, then the analyst will need to return via box 16 to the agenda-setting stage (bold oval A) and consider a revised set of policy mixes.
9. Note that the 3.0 in the cell o_{13} implies that o_1 makes a 'weakly more important' contribution to objective v_h than o_3, and in particular is three times as important as o_3 when using Saaty's scale and method. For a full discussion of the scale used in making such pairwise comparisons, see Saaty and Khouja [6] and Isard and C. Smith ([2]: 148–50).
10. For example, the 0.540 and 0.297 in the first and second rows of this vector indicate that compared with a unit of o_2, a unit of o_1 makes a (0.540/0.297) times greater contribution to objective v_h.
11. If such relative importance cannot be identified using the Saaty procedure, then we must consider other alternatives. For example, (as indicated in box 17) should a ranking of policy mixes (or proposals) be available (or able to be constructed) for each interest group on the basis of the objective achievement scores identified in step 5, then the analyst may consider the use of rank-oriented conflict management procedures. (This yields a temporary compromise policy proposal, as indicated in the light dotted oval I). Where such rankings are unavailable, then the analyst may focus directly on the objective achievement scores and (via the use of one or more reasonable conflict management procedures) reach a compromise set of objective achievements. This option is indicated in box 18 and, if successful, it too yields a temporary compromise policy proposal. On the other hand, should neither of the above approaches work, then the analyst will need to consider one or more of the options discussed in note 8.
12. Note that the 5.0 in the cell v_{23} implies that interest group j considers that v_2 is of 'strong' importance relative to v_3, and in particular is five times as important as v_3 when using Saaty's scale and method. For a full discussion of the scale used in making such pairwise comparisons, see the references cited in note 9.
13. For example, the 0.31 and 0.11 in the first and second rows of this vector

indicate that for interest group z_i a unit of objective v_1 is regarded as (0.31/0.11) times more important than a unit of objective v_2.

14. If such relative influence (importance) cannot be identified using the Saaty procedure, then we must consider other alternatives. For example (as indicated in box 19 outlined by dashes), the analyst may consider the use of outcome-oriented conflict management procedures based on the relative preference values derived for each interest group in step 7. These procedures yield a temporary compromise policy proposal (light dotted oval I). If no acceptable outcome-oriented CMPs can be found, then one or more of the approches discussed in notes 9 and 12 need to be employed.
15. Note that the 7.0 in the cell z_{13} implies that, for the given government unit, interest group z_3 is considered to have 'demonstrated' influence relative to z_1 and, in particular, is seven times as important as z_1 when using Saaty's scale and method. For a full discussion of the scale used in making such pairwise comparisons, see references cited in note 9.
16. In later manuscript, where we generalise the analyses to cover many government units $_d$ (d = 1, . . . , D), we obtain a η matrix of order $I \times D$ where each column gives the relative influence of interest groups within any given government unit. We then need to take the additional step of identifying a $D \times 1$ vector showing the relative importance of each government unit in determining the nature of the policy mix adopted.
17. Mathematically, we may summarise the operations performed up to this point as $\underset{J \times 1}{[\varepsilon]} = \underset{J \times G}{[\delta]} \underset{G \times H}{[\alpha]} \underset{H \times I}{[\beta]} \underset{I \times 1}{[\gamma]}$.

References

[1] Isard, W. and L. Anselin, 1982. 'Integration of Multiregional Models for Policy Analysis', *Environment and Planning* A, vol. 14, pp. 359–76.

[2] Isard, W. and C. Smith, 1982. *Conflict Analysis and Practical Conflict Management Procedures: An Introduction to Peace Science* (Ballinger, Cambridge, MA).

[3] Isard, W. and C. Smith, 1982a. 'Linked Integrated Multiregion Models at the International Level', *Papers, Regional Science Association*, vol. 50.

[4] Isard, W. and C. Smith, 1982b. 'Explorations on an Empirical Framework of the Integrated Multiregion Model for USA,' Paper presented at North American Meetings of Regional Science Association, Pittsburgh, November.

[5] Isard, W. and C. Smith, 1983. 'Incorporation of Conflict and Policy Analysis in an Integrated Multiregion Model'. To appear in A. A. Andersson, W. Isard and T. Puu (eds), *Regional and Industrial Development Theories, Models and Empirical Evidence*, (North-Holland, Amsterdam).

[6] Saaty, T. L. and M. W. Khouja, 1976. 'A Measure of World Influence,' *Journal of Peace Science*, vol. 2, no. 1, pp. 31–48.

[7] Smith, C., 1983. *Integration of Multiregional Models for Policy Analysis: An Australian Perspective* (North-Holland, Amsterdam).

[8] Smith, C., 1982a. 'An Empirical Framework for an Integrated Multiregion Model for Australia', Paper presented at North American Meeting of Regional Science Association, Pittsburgh.

Author Index

Subject Index